Springer Proceedings in Physics

Volume 396

Indexed by Scopus

The series Springer Proceedings in Physics, founded in 1984, is devoted to timely reports of state-of-the-art developments in physics and related sciences. Typically based on material presented at conferences, workshops and similar scientific meetings, volumes published in this series will constitute a comprehensive up to date source of reference on a field or subfield of relevance in contemporary physics. Proposals must include the following:

- Name, place and date of the scientific meeting
- A link to the committees (local organization, international advisors etc.)
- Scientific description of the meeting
- List of invited/plenary speakers
- An estimate of the planned proceedings book parameters (number of pages/articles, requested number of bulk copies, submission deadline).

Please contact:

For Americas and Europe: Dr. Zachary Evenson; zachary.evenson@springer.com
For Asia, Australia and New Zealand: Dr. Loyola DSilva; loyola.dsilva@springer.com

Abhilash Suryan · Minoru Yaga · Han Seo Ko ·
Zhang Guang
Editors

Proceedings of the 9th Asian Joint Workshop on Thermophysics and Fluid Science, 27–30 November 2022, Utsunomiya, Japan

Asian Research in Thermal and Fluid Sciences

Springer

Editors
Abhilash Suryan
Department of Mechanical Engineering
College of Engineering Trivandrum
Thiruvananthapuram, Kerala, India

Minoru Yaga
Faculty of Engineering
University of the Ryukyus
Nishihara, Okinawa, Japan

Han Seo Ko
Department of Mechanical Engineering
Sungkyunkwan University
Suwon, Korea (Republic of)

Zhang Guang
Department of Mechanical Engineering
Zhejiang Sci-Tech University
Hangzhou, China

ISSN 0930-8989 ISSN 1867-4941 (electronic)
Springer Proceedings in Physics
ISBN 978-981-99-9469-4 ISBN 978-981-99-9470-0 (eBook)
https://doi.org/10.1007/978-981-99-9470-0

This Springer imprint is published by the registered company Springer Nature Singapore Pte Ltd.
The registered company address is: 152 Beach Road, #21-01/04 Gateway East, Singapore 189721, Singapore

Isn't it a pleasure to study and practice what you have learned?
Confucius

To

All the Teachers

from whom we have learned

Foreword

The 9th Asian Joint Workshop on Thermophysics and Fluid Science (AJWTF) was held in a hybrid format in Utsunomiya, Tochigi Prefecture, Japan, at the end of November last year. It was a challenging time due to the COVID-19 pandemic, but thanks to the outstanding leadership of the Workshop Chairperson, Prof. Makoto Yamamoto, and the dedicated efforts of the Local Organizing Committee, more than 200 papers were successfully presented with great enthusiasm.

The 9th AJWTF covered various topics, including viscous flows, steady/unsteady flows, heat transfer, flow-induced noise/vibration, aerodynamics, turbomachinery, compressible flows, microfluidics, bio-fluids, reacting flows, particle-laden flows, etc. It received significant attention from many researchers who participated in person and virtually.

We are delighted to announce that, on behalf of the National Representatives of AJWTF, we will be publishing the excellent papers presented at the 9th AJWTF as a proceeding book with the prestigious publisher, Springer. We would like to express our sincere gratitude and appreciation to the National Representatives, Prof. Minoru Yaga (Japan), Prof. Abhilash Suryan (India), Prof. Guang Zhang (China) and Prof. Han-Seo Ko (Korea) for their meticulous efforts for editing the book.

Prof. Heuy Dong Kim
Department of Mechanical Engineering
Andong National University
Andong-si, South Korea

Prof. Toshiaki Setoguchi
Department of Mechanical Engineering
Saga University
Saga, Japan

Preface

This book is a collection of the selected papers from the proceedings of the ninth Asian Joint Workshop on Thermophysics and Fluid Science (AJWTF9) organized in Utsunomiya, Japan, during November 27–30, 2022. The papers include the research outputs from the Asian countries, India, China, Japan and Korea. The book will benefit the academicians, researchers and students working on research problems on thermal and fluid sciences.

The Asian Joint Workshop on Thermophysics and Fluid Science (AJWTF) is an international conference organized biennially. The conference was established by the collaborative efforts of three distinguished professors—Prof. Heuy Dong Kim from the Republic of Korea, Prof. Toshiaki Setoguchi from Japan and Prof. Shen Yu from the People's Republic of China. AJWTF9 was the ninth in the series of highly successful biennial colloquia that started at Qufu, China (2006), followed by Luoyang, China (2008), Matsue, Japan (2010), Busan, South Korea (2012), Nagasaki, Japan (2014), Guilin, China (2016), Trivandrum, India (2018), and Daegu, Korea (2020). The conference has over the years facilitated close cooperation and intellectual exchange among a large number of experts from Asian countries. The distinguished speakers at the conference hailed from academia, leading R&D institutions, government agencies and the industry.

Topics covered in the current edition of the conference included Aerospace engineering, Aerodynamics, Propulsion, Transonic and Supersonic flows, Aeroacoustics, Fluid dynamics, Thermodynamics, Combustion, Heat and Mass transfer and Turbomachinery. Different aspects of theoretical, computational and experimental methods in fluids and thermal sciences were presented and discussed. Careful consideration was given in selecting the papers for the book. A peer-review process was conducted initially, followed by an assessment by the session chairs and the editorial board.

I wish to extend sincere gratitude to the founders of the conference and the international scientific committee for their guidance in bringing out this book. I am also grateful to my fellow editors Prof. Minoru Yaga, Prof. Han Seo Ko and Dr. Guang Zhang and all the researchers who helped with the review process for their invaluable help in selecting the papers. Special thanks to my colleague Suresh Lal S. R. and my students Jagadeesan A. K. and Abhinav L. R. for their assistance in preparing the proceedings.

Thiruvananthapuram, India

Abhilash Suryan
Chief Editor

Contents

Editors and Contributors

About the Editors

Dr. Abhilash Suryan received his B.Tech. in Mechanical Engineering from College of Engineering Trivandrum, University of Kerala, in 1995 and M.Tech. in Propulsion Engineering from the same institution, in 1997. He received his Ph.D. degree in Gas Dynamics from Andong National University, Republic of Korea, in 2012. Currently, he is an Associate Professor at his alma mater, College of Engineering Trivandrum. His research interests include hypersonic and supersonic flows, nozzle flows, propulsion engineering, renewable energy and hydrogen energy. He has published more than 50 papers in reputed international journals.

Prof. Minoru Yaga is a Professor at the Department of Mechanical Systems Engineering, University of the Ryukyus, Okinawa, Japan. He had received his B.Eng. from the Department of Mechanical Engineering, Faculty of Engineering, University of the Ryukyus, Japan, in 1984. He was awarded M.Eng. and Ph.D. by the Interdisciplinary Graduate School of Engineering sciences, Kyushu University, Japan, in 1986 and 1989, respectively. He had been a visiting associate at the California Institute of technology during 1994–95. His major research topics include development of high performance of submerged water jet, active control of oscillating shockwave in a diffuser, behavior of propagating shock wave in an elliptical cell and mixing enhancement of jet with main supersonic flow.

Prof. Han Seo Ko received his B.S. degree from Yonsei University, Korea, in 1990, and M.S. and Ph.D. degrees in Mechanical Engineering from Texas A&M University, USA, in 1994 and 1998, respectively. In 2001, he joined the faculty of the School of Mechanical Engineering, Sungkyunkwan University, Korea, following research fellow work at the University of Tokyo, Japan. He is interested in flow control, microfluidics, optical tomography, air/water plasma control, micro-droplet ejection, and heat and mass control. He is hosting more than 5 research projects on multiphase flow by electromagnetics, water purification system, electric propulsion system, and heat transfer for wafer batch system by National Research Foundation and Ministry of Trade, Industry and Energy, Korea. He has published more than 90 international journal papers.

Dr. Zhang Guang received his B.S. degree in Mechanical Engineering from Three Gorges University, China, in 2012 and dual M.S degrees in Mechanical Engineering from Zhejiang Sci-Tech University, China, and Andong National University, Republic of Korea, in 2015. He then received his Ph.D. degree in Mechanical Engineering from Andong National University, Republic of Korea, in 2018. Currently, Dr. Zhang Guang is an Associate Professor in Mechanical Engineering in Zhejiang Sci-Tech University, China. He had rich experience in experimentally and numerically studying compressible flows in gas dynamics and cavitation flows in fluid machinery. His research interests include Shock Wave Dynamics, Flow Measurement, Micro Pressure Wave induced by High-Speed Trains, Particle Erosion and Cryogenic Cavitation. He is hosting more than 5 research projects on particle erosion and cryogenic cavitation provided by National Natural Science Foundation of China (NSFC) and Natural Science Foundation of Zhejiang (NSFZ), respectively. He has published more than 20 international journal papers.

Contributors

M. M. A. Alam Department of Mechanical Engineering for Transportation, Faculty of Engineering, Osaka Sangyo University, Daito, Osaka, Japan

J. P. Ananthu Mechanical Engineering Discipline, Indian Institute of Technology Gandhinagar, Palaj, Gujarat, India

A. Anuroopa Department of Electrical and Electronics Engineering, ACE College of Engineering, Thiruvananthapuram, India;
Department of Electrical and Electronics Engineering, Noorul Islam Centre for Higher Education, Kanyakumari, India

Pradeep Arjunan Noorul Islam Center for Higher Education, Kanyakumari, Tamil Nadu, India;
College of Engineering Trivandrum, Kerala, India

Praveen Arjunan School of Engineering, Cochin University of Science and Technology, Kochi, Kerala, India;
Micro/Nanofluidics Research Laboratory, Department of Mechanical Engineering, College of Engineering Trivandrum, APJ Abdul Kalam Technological University Kerala, Thiruvananthapuram, India

Suda Atsushi National Institute of Technology, Nara College, Nara, Japan

Ishijima Ayumu University of Tokyo, Tokyo, Japan

Dijo George Biju Micro/Nano Fluidics Research Laboratory, Department of Mechanical Engineering, College of Engineering Trivandrum (Government of Kerala), Thiruvananthapuram, Kerala, India

K. S. Biju Department of Electronics and Communication Engineering, College of Engineering Trivandrum, Thiruvananthapuram, Kerala, India

K. Nandakumar Chandran Micro/Nano Fluidics Research Laboratory, Department of Mechanical Engineering, College of Engineering Trivandrum (Government of Kerala), Thiruvananthapuram, Kerala, India

Longrui Chang School of Power and Energy, Northwestern Polytechnical University, Xi'an, China;
National Key Laboratory of Aerodynamic Design and Research, Northwestern Polytechnical University, Xi'an, China

Keitaro Deyama Faculty of Advanced Engineering, National Institute of Technology, Nara College, Nara, Japan

Masaki Endo Division of Mechanical Engineering, School of Science and Engineering, Tokyo Denki University, Ishizaka, Hatoyama, Hiki, Saitama, Japan

Yuan Feng Tsinghua University, Beijing, China

Takamitsu Fujino Tokyo Denki University, Ishizaka, Hatoyama, Hiki, Saitama, Japan

Koji Fukudome Department of Mechanical Engineering, Tokyo University of Science, Tokyo, Japan

Hiroshi Fukuoka National Institute of Technology, Nara College, Nara, Japan

Naoya Fukushima Tokai University, Hiratsuka Kanagawa, Japan

Shomaru Fukuzato Kumamoto University, Kumamoto, Japan

Limin Gao School of Power and Energy, Northwestern Polytechnical University, Xi'an, China;
National Key Laboratory of Aerodynamic Design and Research, Northwestern Polytechnical University, Xi'an, China

Chunwei Gu Tsinghua University, Beijing, China

Hiromitsu Hamakawa Oita University, Oita, Japan

Yutaka Hara Tottori University, Tottori, Japan

Hiroki Harada Oita University, Oita, Japan

B. Harish Kumar Department of Mechanical Engineering, College of Engineering Trivandrum, Thiruvananthapuram, India;
APJ Abdul Kalam Technological University, Thiruvananthapuram, India

G. M. Joselin Herbert Department of Mechanical Engineering, Noorul Islam Center for Higher Education, Kanyakumari, Tamil Nadu, India

Yoshihito Higa Graduate School of Engineering, Kyushu University, Fukuka, Japan

Fukuoka Hiroshi National Institute of Technology, Nara College, Nara, Japan

Ryo Iida Graduate School of Engineering, Kyushu University, Fukuka, Japan

Natsuki Ishida Oita University, Oita, Japan

Abhinanth Jayakumar Micro/Nanofluidics Research Laboratory, Department of Mechanical Engineering, College of Engineering Trivandrum, APJ Abdul Kalam Technological University Kerala, Thiruvananthapuram, India

Agnes Jinu Department of Electronics and Communication Engineering, College of Engineering Trivandrum, Thiruvananthapuram, Kerala, India

Yoshifumi Jodai National Institute of Technology (KOSEN), Kagawa College, Takamatsu, Japan

Masanao Kaneko Tokyo Denki University, Ishizaka, Hatoyama-Machi, Hiki-Gun, Saitama, Japan

Masataka Kaneshi Graduate School of Mechanical Engineering, Tokyo University of Science, Tokyo, Japan

Akimaro Kawahara Kumamoto University, Kumamoto, Japan

Nakagawa Keiichi University of Tokyo, Tokyo, Japan

Heuy Dong Kim Department of Mechanical Engineering, Andong National University, Andong, Korea

Toshiki Kimura Graduate School of Engineering and Science, Nishihara, Okinawa, Japan

Yoichi Kinoue Saga University, Saga, Japan

P. Pradeep Kumar Department of Aerospace Engineering, Indian Institute of Space Science and Technology, Trivandrum, India

Ranjith S. Kumar Micro/Nanofluidics Research Laboratory, Department of Mechanical Engineering, College of Engineering Trivandrum (Government of Kerala), APJ Abdul Kalam Technological University Kerala, Thiruvananthapuram, India

Eru Kurihara Oita University, Oita, Japan

Xiangfu Lei School of Power and Energy, Northwestern Polytechnical University, Xi'an, China;
National Key Laboratory of Aerodynamic Design and Research, Northwestern Polytechnical University, Xi'an, China

Shanshan Li China United Gas Turbine Technology, Beijing, China

Xuesong Li Tsinghua University, Beijing, China

Antonius Luy Intern, College of Engineering Trivandrum, Thiruvananthapuram, Kerala, India;
Karlsruhe Institute of Technology, Karlsruhe, Germany

Xuan Lv China United Gas Turbine Technology, Beijing, China

Sanjeev Kumar Mall Mayekawa Mfg. Co. Ltd, Ibaraki, Japan

Xiaochen Mao School of Power and Energy, Northwestern Polytechnical University, Xi'an, China;
National Key Laboratory of Aerodynamic Design and Research, Northwestern Polytechnical University, Xi'an, China

Junpei Maruo OSU Company, Daito, Osaka, Japan

Mizue Munekata Kumamoto University, Kumamoto, Japan

Tengen Murakami Saga University, Saga, Japan

Adarsh R. Nair Micro/Nano Fluidics Research Laboratory, Department of Mechanical Engineering, College of Engineering Trivandrum (Government of Kerala), Thiruvananthapuram, Kerala, India

Prasanth P. Nair Mechanical Engineering Discipline, Indian Institute of Technology Gandhinagar, Palaj, Gujarat, India

Shigeto Nakamura National Institute of Technology, Nara College, Nara, Japan

Vinod Narayanan Mechanical Engineering Discipline, Indian Institute of Technology Gandhinagar, Palaj, Gujarat, India

Kosuke Nasu Kumamoto University, Kumamoto, Japan

K. Niju Mohammed Micro/Nanofluidics Research Laboratory, Department of Mechanical Engineering, College of Engineering Trivandrum, APJ Abdul Kalam Technological University Kerala, Thiruvananthapuram, India

Tatsuya Ohkawa Tokyo Denki University, Hatoyama, Japan

Kenji Ono Kyushu University, Fukuoka, Japan

Jiang Qin School of Energy Science and Engineering, Harbin Institute of Technology, Harbin, China

Xiaodong Ren Tsinghua University, Beijing, China

Masaki Sakaguchi Graduate School of Saga University, Saga, Japan

Yoko Sakakibara Division of Mechanical Engineering, School of Science and Engineering, Tokyo Denki University, Ishizaka, Hatoyama, Hiki, Saitama, Japan

Yosuke Shimoda Tokai University, Hiratsuka Kanagawa, Japan

Norimasa Shiomi Saga University, Saga, Japan

Yosuke Shiromoto Graduate School of Engineering, Osaka Sangyo University, Daito, Osaka, Japan

Sandeep Soman College of Engineering Trivandrum, Thiruvananthapuram, Kerala, India

Subramanya U R Rao Satellite Centre, Bangalore, India

Atsushi Suda National Institute of Technology, Nara College, Nara, Japan

S. R. Suresh Lal Department of Mechanical Engineering, Noorul Islam Center for Higher Education, Kanyakumari, India;
Department of Mechanical Engineering, College of Engineering Trivandrum, Thiruvananthapuram, India;
APJ Abdul Kalam Technological University, Thiruvananthapuram, India

Abhilash Suryan College of Engineering Trivandrum, Trivandrum, Kerala, India

Hiromasa Suzuki Tokyo Metropolitan College of Industrial Technology, Shinagawa, Tokyo, Japan

Manabu Takao National Institute of Technology, Matsue College, Matsue, Shimane, Japan

Yuki Tanaka Kumamoto University, Kumamoto, Japan

Kono Tatsuya Tokyo Denki University, Ishizaka, Hatoyama, Hiki, Saitama, Japan

Ikeyama Tetsuro National Institute of Technology, Nara College, Nara, Japan

Nitta Tetsuro National Institute of Technology, Nara College, Nara, Japan

B. S. Gireeshkumaran Thampi School of Engineering, Cochin University of Science and Technology, Kochi, Kerala, India

Sujatha Therese Department of Electrical and Electronics Engineering, Noorul Islam Centre for Higher Education, Kanyakumari, India

P. S. Tide School of Engineering, Cochin University of Science and Technology, Kochi, Kerala, India

Shin-ichi Tsuda Department of Mechanical Engineering, Kyushu University, Fukuka, Japan

Hoshio Tsujita Hosei University, Tokyo, Japan

Wakana Tsuru Saga University, Saga, Japan

Sankalp Vishnoi Vikram Sarabhai Space Centre, Trivandrum, India

Kai Wang National-Provincial Joint Engineering Laboratory for Fluid Transmission System, Zhejiang Sci-Tech University, Hangzhou, China

Sibo Wang School of Energy Science and Engineering, Harbin Institute of Technology, Harbin, China

Ziye Wang China United Gas Turbine Technology, Beijing, China

Satoshi Watanabe Department of Mechanical Engineering, Kyushu University, Fukuka, Japan

Takashi Watanabe Shizuoka Plant, Shizuoka, Japan

Zeyong Wu National-Provincial Joint Engineering Laboratory for Fluid Transmission System, Zhejiang Sci-Tech University, Hangzhou, China

Minoru Yaga University of the Ryukyus, Nishihara, Okinawa, Japan

Kazuyoshi Yamada Kumamoto University, Kumamoto, Japan

Osamu Yamada OSU Company, Daito, Osaka, Japan

Makoto Yamamoto Department of Mechanical Engineering, Tokyo University of Science, Tokyo, Japan

Satoshi Yamazaki Mechanical Engineering, Graduate School of Science and Engineering, Tokyo Denki University, Ishizaka, Hatoyama, Hiki, Saitama, Japan

Guanhua Yang School of Power and Energy, Northwestern Polytechnical University, Xi'an, China;
National Key Laboratory of Aerodynamic Design and Research, Northwestern Polytechnical University, Xi'an, China

Yukihiro Yonemoto Kumamoto University, Kumamoto, Japan

Hiroyuki Yoshikawa Kumamoto University, Kumamoto, Japan

Guang Zhang National-Provincial Joint Engineering Laboratory for Fluid Transmission System, Zhejiang Sci-Tech University, Hangzhou, China

Haodian Zhang National-Provincial Joint Engineering Laboratory for Fluid Transmission System, Zhejiang Sci-Tech University, Hangzhou, China

Chapter 1
Effect of Multi-Layered Piezo Ceramic Actuator on Transonic Flow

Toshiki Kimura, Sanjeev Kumar Mall, and Minoru Yaga

Abstract An experimental study is done to check the influence of multi-layered piezoceramic actuator at the throat and downstream of transonic diffuser. The diffuser has circular arc upper wall and flat bottom wall where the piezo ceramic actuator is installed. The diffuser throat height h* is 3 mm and multi-layered piezoceramic actuator is driven at the frequency of 100–600 Hz at 90 V. Pressure ratio across the diffuser is set up from 1.1 to 1.6. Schlieren technique is used for flow visualization. The displacement of the piezoceramic actuator is 45micro meter for the supplied voltage of 120 V, which is quite small for the usual flow activation. As a result, even though the displacement of the piezoceramic actuator is invisible, the effect of the piezoceramic actuator on the throat pressure fluctuation is found to be clear. The rectangular input voltage signal causes the similar pressure fluctuation at the throat. And the dominant frequency of the throat pressure fluctuation corresponds to the input voltage signal. Moreover, its effect on the pressure fluctuation of the diffuser throat depends on the pressure ratio.

1.1 Introduction

It is well known that shock waves are generated in jet engines inside high-pressure pipes and there is a high-pressure difference before and after the shock wave. This shock wave vibrates irregularly causing disturbance to the flow field which may cause noise and fatigue to the equipment. There are a lot of reports about the unsteady

T. Kimura
Graduate School of Engineering and Science, 1 Senbaru, Nishihara, Okinawa 903-0213, Japan
e-mail: k228430@eve.u-ryukyu.ac.jp

S. K. Mall
Mayekawa Mfg. Co. Ltd, Tatsuzawa 2000, Ibaraki 302-0118, Japan
e-mail: sanjeev-kumar@mayekawa.co.jp

M. Yaga (✉)
University of the Ryukyus, 1 Senbaru, Nishihara, Okinawa 903-0213, Japan
e-mail: yaga@tec.u-ryukyu.ac.jp

A. Suryan et al. (eds.), *Proceedings of the 9th Asian Joint Workshop on Thermophysics and Fluid Science, 27–30 November 2022, Utsunomiya, Japan*, Springer Proceedings in Physics 396, https://doi.org/10.1007/978-981-99-9470-0_1

shock wave behaviours and pressure fluctuations. Also, the technique to control the flow fields are proposed such as passive control, synthetic jet, plasma actuator, and flexible plate [1–5]. On the other hands, the effect of the bimorph type piezo ceramic actuator on transonic diffuser was already confirmed as reported by Yaga et al. [6], however, it was found that the bimorph type piezo ceramic actuator is easily driven by the pressure difference due to large area around the throat. In the present study in order to use the high frequency signals, the bimorph type piezo ceramic actuator has been replaced with the multi-layered piezo ceramic actuator which has high resonance frequency. Despite of the high resonance frequency, its displacement is quite small which makes it important to confirm the effect of the multi layered piezo ceramic on the transonic diffuser so that its future application to high-speed devices can be recommended as shown in the report [7]. A fundamental investigation of the response of multi-layered piezo ceramic actuator at the throat of the transonic diffuser is done. Experiments have been conducted under the conditions of various range of frequencies and pressure ratio (ratio of stagnation pressure and atmospheric pressure) to produce a series of test cases. The experimental technique of Schlieren photography and unsteady measurements have been used to characterize the shock wave and downstream flow.

1.2 Experimental Apparatus and Method

1.2.1 Wind Tunnel Facility

Experiments have been performed in the blowdown-type supersonic wind tunnel [6]. Figures 1.1 and 1.2 show the wind tunnel, the details of the test section, respectively. The experimental apparatus consists of a 0.7 MPa compressor, a settling chamber, a regulator valve, data acquisition system, function generator, DC amplifiers as show in Fig. 1.1. A circular arc transonic half diffuser which has three section and multi-layered piezo ceramic actuator is fitted in the middle section. In Fig. 1.2 it can be seen that the multilayered piezo ceramic actuator is fitted at the throat which is opposite to the sensor tap. Two multilayered piezo ceramic actuators are fitted and are driven by Piezo DC driver amplifier at various frequencies. The throat height of the diffuser $h^* = 3$ mm. The test section has a constant width of 27 mm. The circular arc diffuser has different holes to measure pressure fluctuations at different distances from the throat i.e., at 20 mm, 30 mm, 40 mm, 50 mm, 60 mm along the bottom straight wall as show in Fig. 1.2. Schlieren method was used to capture time-resolved images of the diffuser with Nikon D7200 digital camera. Unsteady wall static pressure fluctuations are measured at settling chamber at positions $x = 20$mm and throat. A LabVIEW device controls the entire process of reading and storing data from the pressure transducers. The signals from the pressure transducer are digitized at a sampling frequency of 10 kHz with 16-bit accuracy and are analyzed to evaluate the effect of the multilayered piezo ceramic actuator on flow fields. The total pressure at the

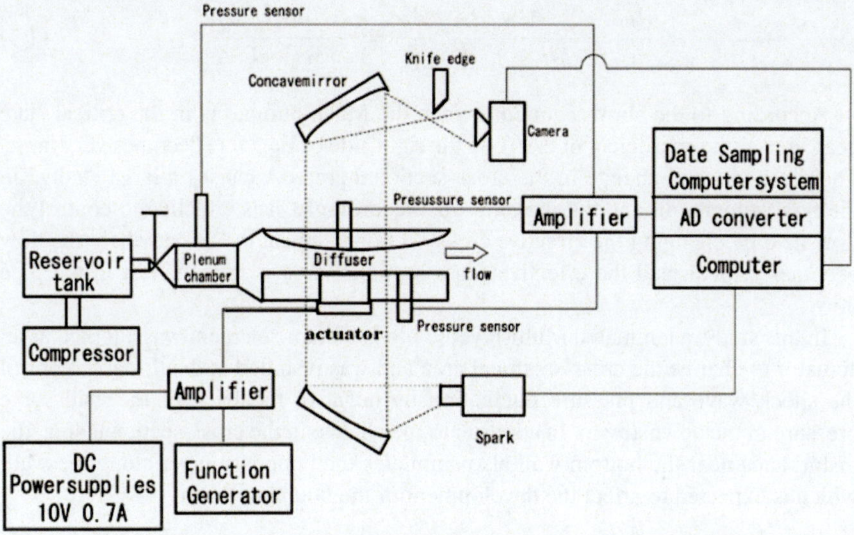

Fig. 1.1 Schematic diagram of experimental setup

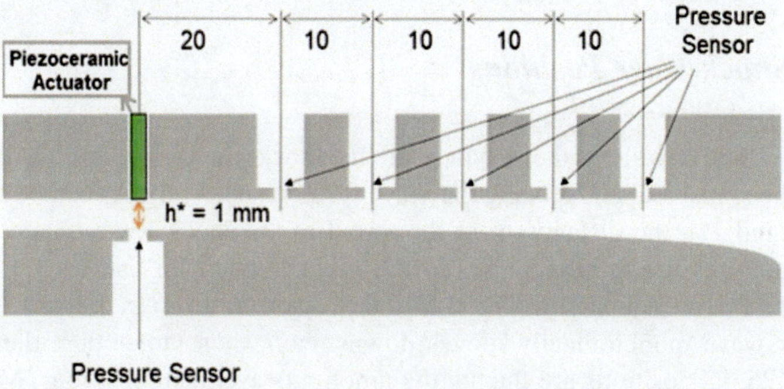

Fig. 1.2 Arrangement of multi-layered piezo ceramic actuator and pressure sensors

settling chamber is monitored during the experiments by LabVIEW program (Data sampling computer system), which controls the timing of the pressure measurements and flow visualizations.

1.3 Equations

Assuming that the flow field is isentropic in supersonic flow, the relationship between the change in cross-sectional area and the velocity and pressure is as shown in Eq. (1.1).

$$\frac{du}{u} = \frac{1}{M^2 - 1}\frac{dA}{A};\ \frac{dp}{p} = -\frac{\gamma M^2}{M^2 - 1}\frac{dA}{A} \tag{1.1}$$

According to the above equation, when the Mach number is in the critical state near $M = 1$, the coefficient of dA/A on the right side of Eq. (1.1) becomes very large. Therefore, a small change in the cross-sectional area dA causes a large change in the flow velocity du and the pressure dp. Accordingly, it is effective to control the flow field by changing the effective cross-sectional area at the throat where the flow becomes critical, and the effective cross-sectional area is the smallest in the pipe flow.

In this study, a laminated Multi-layered piezo ceramic element was adopted as an actuator to change the cross-sectional area and was installed at the throat to control the shock wave and pressure fluctuation by negative feedback of the wall static pressure in the downstream. In addition to the change in the cross-sectional area, the disturbance near the bottom wall also simulates the boundary layer along the wall, which is expected to affect the development of the boundary layer.

1.4 Results and Discussion

1.4.1 Shock Wave Positions

Figures 1.3 and 1.4 show the distance of the shock wave from the throat for the cases of the sinusoidal and rectangular wave form, respectively. As can be seen from Figs. 1.3 and 1.4, the distance from the throat to the shock wave increases as the wind tunnel pressure ratio increases for all input frequencies and wave forms. For the case of 0 Hz frequency in Fig. 1.3, the pressure ratio from 1.1 to about 1.25, the shock wave monotonically moves downstream, then larger than the pressure ratio of 1.25, its positions are fluctuating among its averaged positions. Monotonic displacements of the shock wave for cases of two frequencies seem to be the same up to the pressure ratio about 1.38 then it starts to fluctuate at a large amplitude. According to the reports about the oscillation of the shock wave, its oscillations are caused by interactions between the shock wave and the boundary layer, therefore the interactions are considered to become severe because of the increase in the shock wave strength. The difference in the pressure ratio at which the shock wave starts to fluctuate is caused by an unsteady stimulation of the multi-layered piezo ceramic at the throat. In other words, the shock waves are stabilized by the multi-layered piezo ceramic actuator at the throat up to the pressure ratio of 1.35. For different wave form shown in Fig. 1.4, the displacements of the shock waves are the same trend as the case of rectangular one. That is, the large amplitude fluctuations of the shock wave are shown in the pressure ratio above 1.35. Also, it is found that at the same input frequency of each waveform, there is no clear characteristic difference in the distance of the shock wave. Therefore, it is considered that the position of the

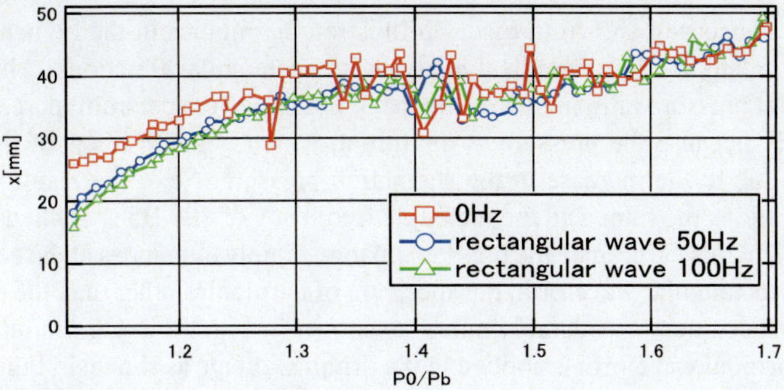

Fig. 1.3 Comparison of shock wave distance from throat between rectangular Wave 0 Hz, 50 Hz and 100 Hz

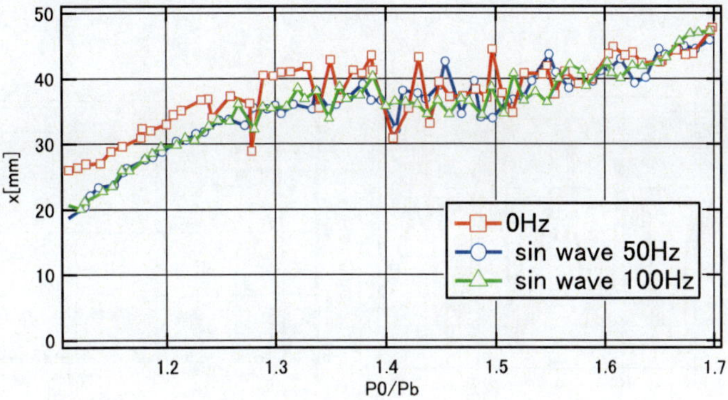

Fig. 1.4 Comparison of shock wave distance from throat between sinusoidal wave 0 Hz, 50 Hz and 100 Hz

shock wave is not affected by the change in the waveform of the driving signals of the Multi-layered piezo ceramic actuator.

1.4.2 Pressure Fluctuation and Frequency Analysis

Figures 1.5a and b, 1.6a and b, 1.7a and b, and 1.8a and b show the pressure fluctuations at the throat and its power spectrum of the frequencies, respectively. The pressure fluctuations are the ratio of the difference from the time-averaged pressure divided by atmospheric pressure. In the figures, the black line and the red line show the cases of pressure ratio of 1.11 and 1.7, respectively. From Fig. 1.5a, in the case of a rectangular wave of 50 Hz, the pressure fluctuation is fairly close to the input signal of the rectangular waveform. The power spectrum is close to the theoretical power spectrum of the so-called rectangle with overshoot and undershoot noises.

The power spectrum shown in Fig. 1.5b illustrate harmonics in the higher frequencies as is calculated by theoretical analysis for rectangular functions. Also, as the wind tunnel pressure ratio increases, the peak of the power spectrum increases. This is probably because the pressure at the throat, which is 0.528 times of that value, increases due to the increase in the stagnation pressure. Next, the result when the pressure ratio is the same and the rectangle frequency of 100 Hz is applied is shown in Fig. 1.6. In this case, since the pressure change rapidly attenuates at the edge peculiar to the rectangular waveform, the spectrum of harmonics other than the dominant frequency becomes considerably small as shown in Fig. 1.6b. On the other hand, when sinusoidal waveform is applied to the driving voltage as shown in Figs. 1.7 and 1.8, although there are small fluctuations due to noises, the sinusoidal waveform is considerably kept at 50 Hz and 100 Hz. This is relatively clear from the result of the power spectrum in each figure (b) and is also clear from the fact that the harmonics other than the dominant frequency are hardly seen compared with the case of the rectangular wave.

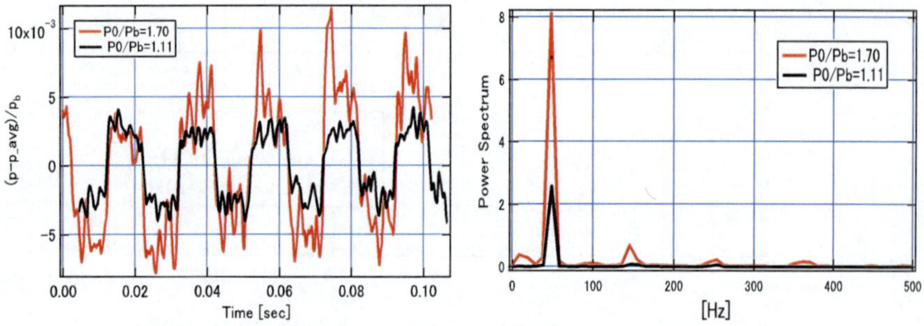

(a) Pressure fluctuations (b) FFT analysis

Fig. 1.5 Pressure fluctuation and FFT analysis for rectangular wave form of input frequency 50Hz

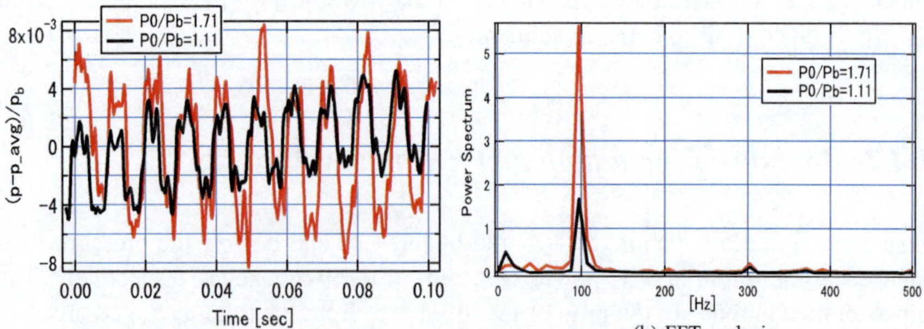

(a) Pressure fluctuations (b) FFT analysis

Fig. 1.6 Pressure fluctuation and FFT analysis for rectangular wave form of input frequency 100 Hz

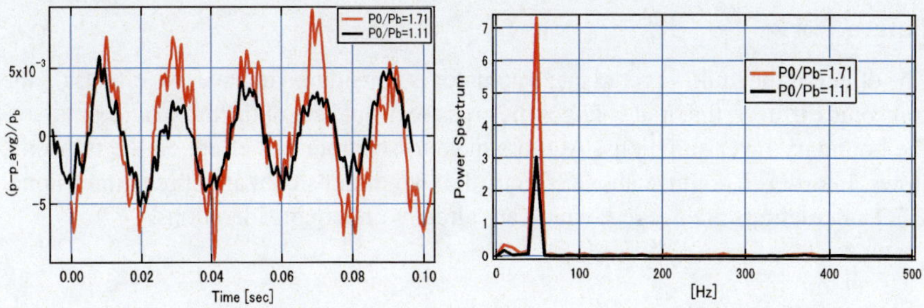

(a)Pressure fluctuations (b) FFT analysis

Fig. 1.7 Pressure fluctuation and FFT analysis for sinusoidal wave form of input frequency 50 Hz

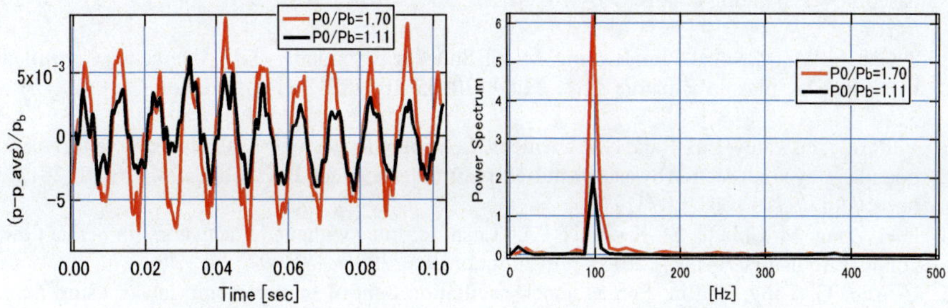

(a)Pressure fluctuations (b) FFT analysis

Fig. 1.8 Pressure fluctuation and FFT analysis for sinusoidal wave form of input frequency 100 Hz

1.5 Conclusion and Future Work

The pressure fluctuation and the shock wave distance were compared when the multi-layered piezo ceramic actuator was driven at a driving voltage of 90 V at the ranges of 0 Hz, 50 Hz, and 100 Hz by rectangular waveform and sinusoidal waveform. The results obtained are shown below.

1. The starting shock wave move downstream as the wind tunnel pressure ratio increases.
2. The positions of the starting shock wave are affected by the existence of the multi-layered piezo ceramic acutator where as the frequencies of the input signal seems to have no effect on the averaged shock positions.
3. The pressure fluctuations at the throat are confirmed to have well response to the multi-layered piezo ceramic actuator especially at the frequency of 50 Hz. However, for the 100 Hz input frequency the rectangular waveform, the pressure fluctuation has some attenuation according to the high frequency filtering effect during the transmission from the actuator to the pressure sensor.

Future Work

The effect of the multi-layered piezo actuator is considered to have two effects on the transonic diffuser; that it, it changes the cross sectional are of the throat and simulates the boundary layer situations. Authors have to separate this effect on the transonic flows. In order to confirm above effects, the two dimensional and three dimensional CFD calculations are needed, which are already under considerations.

References

1. B.M. Reese, E.G. Collins, E. Fernandez, F.S. Alvi, Nonlinear adaptive approach to microjet-based flow separation control. AIAA J. **54**(10), 3002–3014 (2016). https://doi.org/10.2514/1J054307
2. T. Gan, Y. Wu., Z. Sun, D. Jin, H. Song, M. Jia, Shock wave boundary layer interaction controlled by surface arc plasma actuators. Phys. Fluids **30**, 055107 (2018). https://doi.org/10.1063/1.5013166
3. V. Narayanaswamy, L.L. Raja, N.T. Clemens, Control of unsteadiness of a shock wave/turbulent boundary layer interaction by using a pulsed-plasma-jet actuator. Phys. Fluids **24**, 076101 (2012). https://doi.org/10.1063/1.4731292
4. H.H. Zong, M. Chiatto, M. Kotsonis, L.D. Luca, Plasma synthetic jet actuators for active flow control. Actuators **7**(4), 77 (2018). https://doi.org/10.3390/act7040077
5. C. Yao, G. Zang, Z. Liu, Forced shock oscillation control in supersonic intake using fluid -structure interaction. AIAA J. **55**(8), 2580–2956 (2017). https://doi.org/10.2514/1.J055627
6. M. Yaga, Y. Uechi, H. Ozono, M. Ishikawa, I. Teruya, Experimental study of the response of transonic diffuser flow to a piezoceramic actuator at diffuser throat. Open J. Fluid Dyn. **3**(2A), 14–21 (2013). https://doi.org/10.4236/ojfd.2013.32A003
7. S.K. Mall, M. Yaga, Effect of piezoceramic actuator at throat on transonic flow, in *Proceedings of the 34th Computational Fluid Dynamic Symposium*, Paper No. A10–1 (Okinawa, 2020), pp. 1–6

Chapter 2
Numerical Simulation for Prediction of Secondary Droplets by Water Droplet Impingement on Thin Water Film Using E-MPS Method

Masataka Kaneshi, Koji Fukudome, and Makoto Yamamoto

Abstract Ice accretion is a phenomenon in which super-cooled water droplets or ice particles in the atmosphere impinge on a solid surface, and an ice layer is formed there. The ice accretion on an aircraft threatens navigation safety because it causes deformation of the wing shapes and reduces the aerodynamic performance. Hence, it is a significant technical issue to predict icing during the design phase of an aircraft. When the droplet diameter is over 40 μm, it is called a super-cooled large droplet (i.e. SLD). It is well known that splash and rebound occur when a SLD impinges on a wall. Especially, as splash generates secondary droplets, it forms very complex ice shapes due to the secondary droplets. Furthermore, liquid film on a wall significantly affects the impact behavior. Therefore, it is necessary to consider the secondary droplet generation and the existence of liquid film on the impact behavior of an SLD in aircraft icing. Considering these backgrounds, we simulated the droplet impact behavior on a thin water film by using the Explicit-Moving Particle Simulation (E-MPS) method. Through the present study, the following insights are obtained; (1) In the case of vertical impact, each dimension of the crown shows a similar tendency to change with time as in the experiment. (2) In the case of oblique impact, secondary droplets are generated at the impact angle between 75 to 31 degrees, and the total mass of secondary droplets reaches a maximum of around 39 degrees.

M. Kaneshi
Graduate School of Mechanical Engineering, Tokyo University of Science, Niijuku 6-3-1, Katsushika-Ku, Tokyo 125-8585, Japan

K. Fukudome · M. Yamamoto (✉)
Department of Mechanical Engineering, Tokyo University of Science, Niijuku 6-3-1, Katsushika-Ku, Tokyo 125-8585, Japan
e-mail: yamamoto@rs.tus.ac.jp

A. Suryan et al. (eds.), *Proceedings of the 9th Asian Joint Workshop on Thermophysics and Fluid Science, 27–30 November 2022, Utsunomiya, Japan*, Springer Proceedings in Physics 396, https://doi.org/10.1007/978-981-99-9470-0_2

2.1 Introduction

Ice accretion is a phenomenon in which super-cooled water droplets or ice particles in the atmosphere impinge on a solid surface, and an ice layer is formed there. The ice accretion on an aircraft threatens navigation safety because it causes deformation of the wing shapes and reduces the aerodynamic performance. Hence, it is a significant technical issue to predict icing during the design phase of an aircraft.

The glaze ice is formed where the ambient temperature is about − 10 to −3 °C which is a relatively mild temperature in aircraft icing. In the glaze ice conditions, the impinged super-cooled droplets do not freeze instantly and runback along the wing surface as a liquid film. As a result, they form a characteristic icing shape such as a horn. In addition, splashing and rebounding occur, which generate secondary droplets if super-cooled large droplets (SLD) whose diameters are more than 40 μm impinge on an aircraft. Therefore, it forms very complex ice shapes due to secondary droplets in SLD icing conditions [1]. Furthermore, liquid film on a solid surface significantly affects the impact behavior, such as splashing and rebounding [2]. Therefore, it is necessary to consider the existence of liquid film on the impact behavior of the SLD in aircraft icing. In addition, it is also important to predict the behavior of secondary droplets by considering these environmental conditions.

To precisely simulate the SLD icing under glaze ice conditions, it is necessary to clarify the impact behavior of SLD on a solid wall accompanied by a liquid film. Many experimental and numerical studies have been conducted on droplet impingement on the wall with a liquid film. For example, Zhu et al. [3] experimented with droplet impingement on thin water film and clarified the effects of water film thickness and impact velocity for vertical impact. However, few studies focused on secondary droplets due to droplet impingement on liquid film. Therefore, this study aims to characterize secondary droplets generated by the droplet impingement on thin water film by using the Explicit-Moving Particle Simulation (E-MPS) method.

2.2 Numerical Method

The Explicit-Moving Particle Simulation (E-MPS) method [4] is based on a Lagrangian framework with governing equations of the motion for each computational particle. It is a meshless method and suitable for analyzing phenomena with large deformation of the interface between multi-phase fluids. In addition, the E-MPS method strictly satisfies the conservation of mass. Furthermore, the computational cost of the E-MPS method is lower than the original Moving Particle Semi-implicit (MPS) method [5] because of adapting the explicit pressure solver. In this study, we used the three-dimensional E-MPS method with the incompressible Navier–Stokes equation and the mass conservation equation (i.e., continuity equation) as the governing equations. The effective radius, which determines the domain of particles interacting with each other, is set up with 3.1 times as large as the computational particle diameter [6]. Furthermore, we used the potential-based model proposed by Kondo et al. [7] to estimate the surface tension and the polygon wall boundary condition [6]. The polygon wall boundary condition has many advantages over the wall particles method, such as smooth and easy handling of wall surfaces, and reduced computational cost.

2.3 Numerical Condition

Figure 2.1 shows the schematic of the computational setup, a droplet with a diameter of d_0 and water film with a thickness *h*. Computational conditions are shown in Table 2.1. The impact angle in Table 2.1 defines the angle between the initial velocity vector of the droplet and the surface of the water film (see Fig. 2.2). A droplet moves in the positive direction of the x-axis. In this study, we used about 1,640,000 particles, composing 1.6 million particles for the water film and about 36,000 particles for the droplet. In addition, the effect of airflow is ignored in this study under the assumption that airflow has a negligible effect on impact behavior.

Figure 2.3 shows the schematic of the crown formed by droplet impingement on water film and defines the names of the main components. The free rim indicates the rim formed at the top of the crown, and the finger indicates the jet generated from the free rim. The crown shape is characterized by the crown diameter in the horizontal direction D_c, and the crown height H_c. These dimensions show the results for the crown as a whole, including the free rim and finger. A non-dimensionalization of

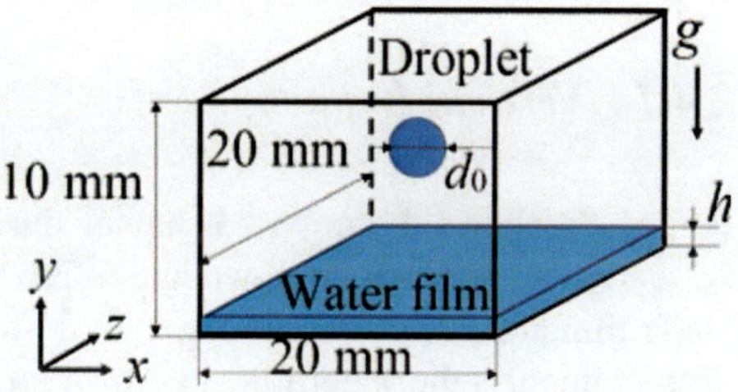

Fig. 2.1 Schematic of the computational setup

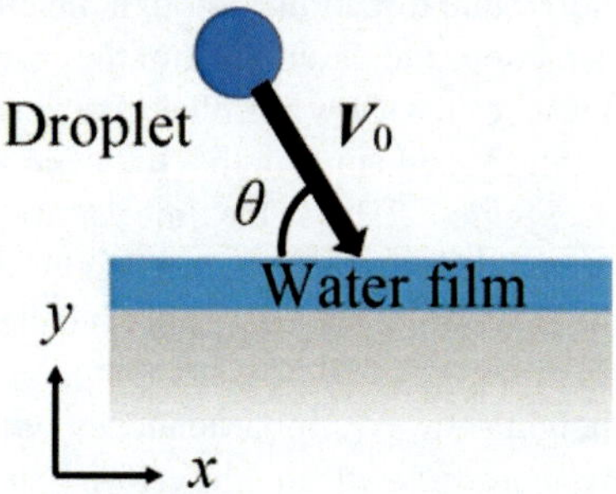

Fig. 2.2 Droplet impact angle θ

Table 2.1 Computational conditions

Initial droplet diameter d_0 [mm]	1.95	Water temperature [°C]	20, 0.01
Initial water film thickness h [mm]	0.449	Contact angle [deg.]	90
Impact velocity V_0 [m/s]	3.75	Impact angle θ [deg.]	15–90
Acceleration of gravity g [m/s^2]	9.81	Comp. particle size [μm]	50.0

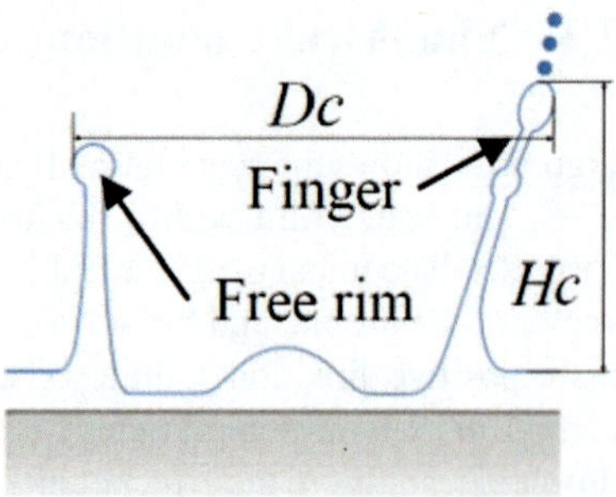

Fig. 2.3 Schematic of the crawn

the crown diameter, crown height, and thickness of the water film is applied by the initial droplet diameter d_0. An asterisk (*) denotes a non-dimensional quantity. In the present study, the non-dimensional thickness of water film h^* is 0.23.

2.4 Numerical Results and Discussion

2.4.1 Vertical Impact

Figure 2.4 shows the impact behavior during a vertical impact for the case of 20 °C water temperature. A crown and a free rim are formed due to droplet impact on water film at $t = 1.5$ ms (see Fig. 2.4a). The time t is based on the moment when a droplet impacts the water film. A crown grows vertically and horizontally with time progression (see Fig. 2.4b). For $t = 6.0$ ms (Fig. 2.4c), the free rim expands in the horizontal direction, although the crown except the free rim decreases in the vertical direction, and the collapse of the crown progresses. Although on a different timescale, these results show a similar time development with the experiment conducted by Zhu et al. [3] and confirm that the generation of the crown is reproduced.

Figure 2.5 shows the time variation of crown dimensions (crown diameter for horizontal direction and crown height) at each water temperature. The solid and dashed lines show the present results and the closed circles show the experimental results [3]. For the case of 20 °C water temperature, the time development of crown diameter and height shows a similar tendency with the experiment. However, the crown diameter is overestimated, and the crown height decreases faster than in the experiment. It has

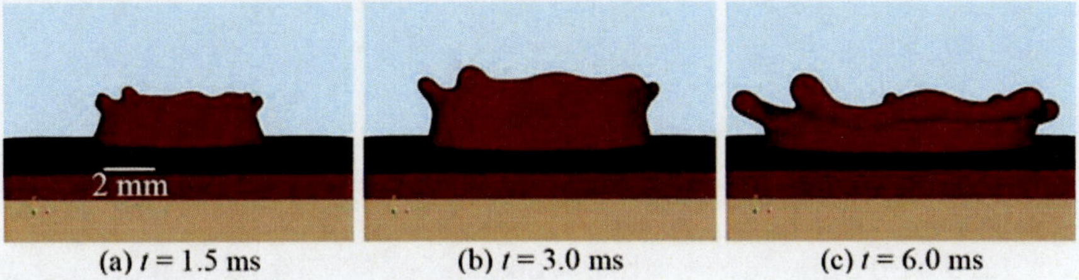

Fig. 2.4 Results of vertical impact on a wall with a thin water film

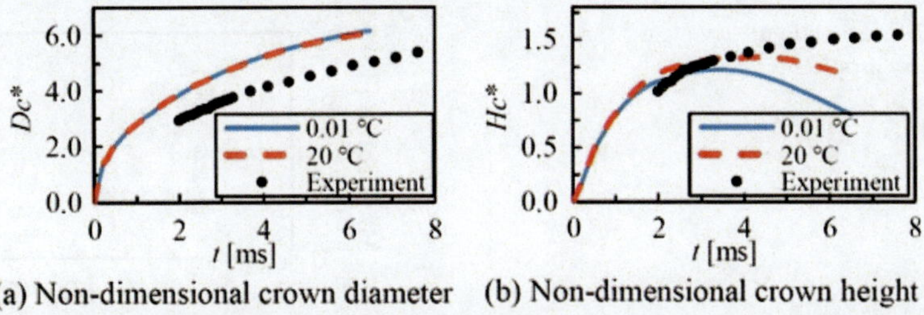

Fig. 2.5 Development of crown dimensions. Lines and symbols represent present and experimental results [3], respectively

been shown by the experiments [3, 8] that the crown height increases with the Weber number of the droplet. It has also been pointed out that the potential model used to evaluate surface tension in this study overestimates the effect of surface tension [9]. The overestimation of surface tension is also confirmed in this study, resulting in a faster decrease in crown height. The crown height does not change significantly with water temperature, although the crown height at the case of 0.01 °C water temperature is smaller than the case of 20 °C water temperature. This is due to the decrease in the Weber number of the droplet as the water temperature decreases.

2.4.2 *Oblique Impact*

This section focuses on the case of 0.01 °C water temperature. Figure 2.6 shows the calculation results with the impact angles of 30, 60, and 90 degrees at t = 3.0 ms. In Fig. 2.6a, the crown and free rim formation can be seen in the middle of the figure, but the crown does not rise downstream (right side of the figure). The crown at the downstream and upstream (left side of the figure) rise as the impact angle increase (see Fig. 2.6b and c). The finger is seen in the case of a 60-degree impact angle.

Figure 2.7 shows the ratio of the total mass of secondary droplets to the mass of the initial droplet at the impact angle of 15 to 90 degrees. Here, simulations for the impact angle from 44 to 31 degrees are conducted three times to remove the initial droplet position dependency. The secondary droplet does not occur in vertical impact.

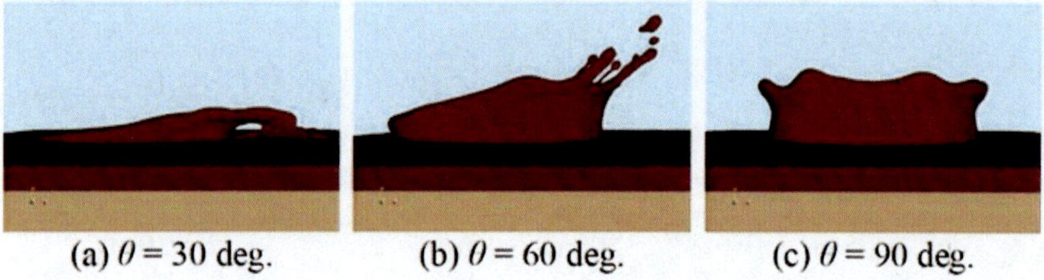

Fig. 2.6 Effect of the impact angle on the crown dynamics at t = 3.0 ms

Fig. 2.7 Mass ratio of secondary droplets

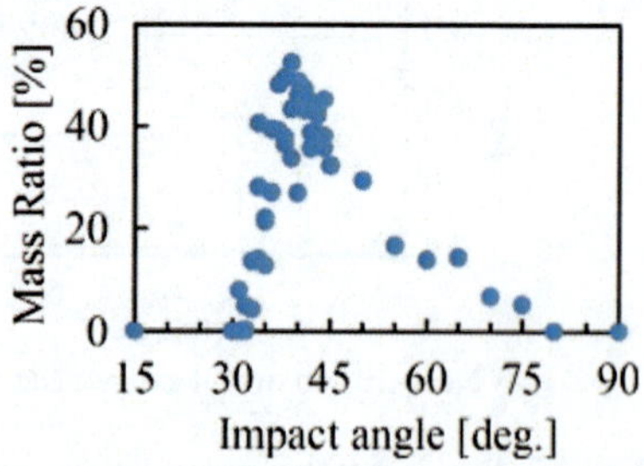

In oblique impact, the secondary droplets occur at the impact angle between 75 to 31 degrees. In the case that the impact angle is lower than 30 degrees, the secondary droplets were not produced. For the cases where secondary droplets formed, the total mass of secondary droplets increases as the impact angle decreases and attains a maximum of about 39 degrees. According to the experiment conducted by Samenfink et al. [10], the total mass of the secondary droplets reached a maximum at an impact angle of around 30 degrees. These results may be attributed to two effects. One is the effect of an increase in the total mass of the secondary droplets due to an increase in the velocity component in the horizontal direction of the initial droplet as the impact angle decreases. The other is the effect of the decrease due to the increase in the difference of horizontal velocity between the initial droplet and water film, which promotes vortex formation due to the Kelvin–Helmholtz instability. The Kelvin–Helmholtz instability between the droplet and water film has also been pointed out by Guo et al. [11]. In this study, vortices that may be caused by Kelvin–Helmholtz instability are confirmed when the impact angle is less than 41 degrees, and these vortices may promote the mixing of momentum, and thus prevent the formation of secondary droplets.

Figure 2.8 shows the relationship between the formation time and scattering velocity of the secondary droplets. The non-dimensional secondary droplet formation time T^* [–] and non-dimensional secondary droplet scattering velocity v^* [–] are calculated using the following formulas: secondary droplet formation time T [s] and secondary droplet scattering velocity v [m/s] (see Eq. (2.1) and (2.2)). Secondary droplet formation time is the time from the droplet impingement on water film to the moment the finger splits completely and a secondary droplet is formed. Figure 2.9

shows the relationship between the formation time and the angle between the scattering velocity vector of a secondary droplet and water film. The definition of θ (initial droplet impact angle) and θ' (secondary droplet scattering angle) is shown in Fig. 2.10. Secondary droplets with θ' smaller than 0 degrees indicate that droplets are scattered in the direction of the water film when they break up from finger. Figures 2.8 and 2.9 show that the scattering velocity of the secondary droplet is faster, and the scattering angle is larger as the secondary droplet formation time is faster.

$$T^* = T\frac{V_0 sin\theta}{d_0} \tag{2.1}$$

$$v^* = \frac{v}{V_0} \tag{2.2}$$

Fig. 2.8 Relationship between v*–T*

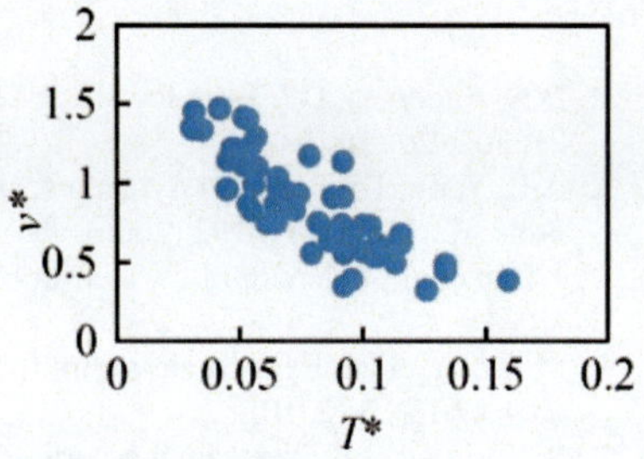

Fig. 2.9 Relationship between θ'–T*

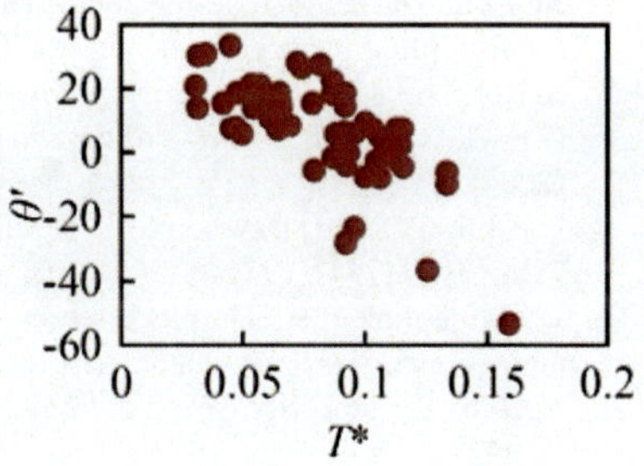

Fig. 2.10 Definition of θ and θ'

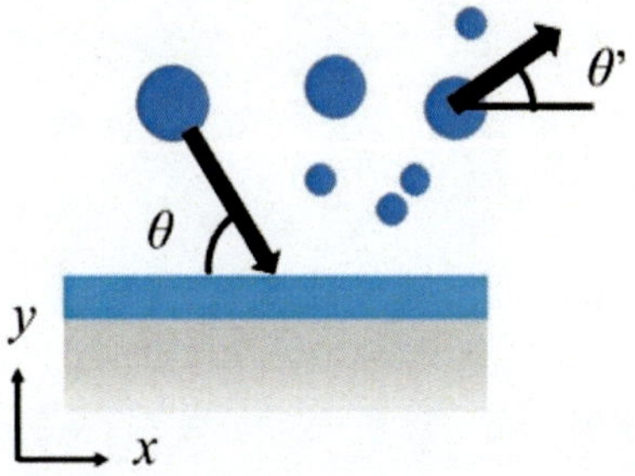

2.5 Concluding Remarks

In this study, we performed simulations of droplet impingement on water film using the E-MPS method and evaluated the secondary droplet depending on the impact angle. In the case of vertical impact, each dimension of the crown shows a similar tendency to change with time as in the experiment. However, these values need to be improved in the future. For oblique impact, secondary droplets are generated at the impact angle between 75 to 31 degrees, and the total mass of secondary droplets reaches a maximum around 39 degrees. The scattering velocity of the secondary droplets becomes faster and the scattering angle becomes larger as the secondary droplet formation becomes faster.

References

1. D.N. Anderson, J.C. Tsao, Ice Shape Scaling for Aircraft in SLD Conditions, NASA/CR-2008–215302 (2008)
2. A.L. Yarin, Drop impact dynamics: splashing, spreading, receding, bouncing…. Annu. Rev. Fluid Mech. **38**, 159–192 (2006)
3. J. Zhu et al., Behavior of a water droplet impacting a thin water film. Exp. Fluids **62**(143), 13 (2021)
4. M. Oochi et al., Explicit algorithm for free surface flow analysis (in Japanese). Trans. JSCES (20100013), 8 (2010)
5. S. Koshizuka, Y. Oka, Moving-particle semi-implicit method for fragmentation of incompressible fluid. Nucl. Sci. Eng. **123**, 421–434 (1996)
6. M. Kondo et al., Surface Tension Model using Inter-particle Potential Force in Moving Particle Semi-implicit method (in Japanese). Trans. JSCES (20070021), 6 (2007)
7. S. Koshizuka, et al., *Introduction for Particle Method (in Japanese).* Maruzen shuppan (2014)
8. G.E. Cossali et al., The role of time in single drop splash on thin film. Exp. Fluids **36**, 888–900 (2004)
9. E. Ishii, T. Sugii, Development of surface tension for particle method (in Japanese). Trans. JSME.B **78**(794), 1710–1725 (2012)
10. W. Samenfink et al., Droplet interaction with shear-driven liquid films: analysis of deposition and secondary droplet characteristics. Int. J. Heat Fluid Flow **20**, 462–469 (1999)
11. Y. Guo, Y. Lian, High-speed oblique impact on thin liquid films. Phys. Fluids **29**(082108), 11 (2017)

Chapter 3
Experimental Observation of Collision Process Between Converging Shock Wave and Supersonic Jet

Keitaro Deyama, Hiroshi Fukuoka, Atsushi Suda, and Shigeto Nakamura

Abstract Shock wave has the properties of waves that reflect when they collide with objects, walls, and supersonic jet. Focusing on this property, a technique for the combustion of methane-based mixtures using the temperature increase caused by a single reflection of shock waves has been reported. In this paper, we visualize and observe the process of reflection of shock waves emitted from the shock tube between the reflector and the supersonic jet. The Schlieren method is used to visualize the flow field. The lengths of the shock tube's high-pressure and low-pressure chambers are 10 mm and 20 mm, respectively. The diameter of the shock tube is 10 mm. The experimental parameters were the distance between the reflector and open end of the shock tube 80, 100, 120 mm, the pressure ratio between the high and low-pressure chambers of the shock tube 10.7, 22.6, 35.0, 46.0 and the radius of the reflector ∞, 40 mm. From the experiment, it was confirmed that shock waves reflected by the concave reflector converge and impinge on the jet. For other pressure ratios, it was confirmed that the converging shock waves converge at the same location regardless of the distance from open end of the shock tube to the reflector.

3.1 Introduction

A shock wave induced by supersonic events is high-energy pressure wave with discontinuous gradients of pressure, temperature, and density [1, 2]. And thereby, the multiple reflections of the shock wave are expected to make the high-pressure and high-temperature area. Miyaoku et al. reported the multiple reflections of the shock wave between a jet of head generated by a shock tube and a reflector [3]. However,

K. Deyama
Faculty of Advanced Engineering, National Institute of Technology, Nara College, 22 Yata, Yamatokoriyama, Nara 639-1080, Japan

H. Fukuoka (✉) · A. Suda · S. Nakamura
National Institute of Technology, Nara College, 22 Yata, Yamatokoriyama, Nara 639-1080, Japan
e-mail: fukuoka@mech.nara-k.ac.jp

A. Suryan et al. (eds.), *Proceedings of the 9th Asian Joint Workshop on Thermophysics and Fluid Science, 27–30 November 2022, Utsunomiya, Japan*, Springer Proceedings in Physics 396, https://doi.org/10.1007/978-981-99-9470-0_3

the factor for the reflection shock wave mechanization is not clear. This study aims to clarify the reflection mechanism of the shock wave by focusing on the converging phenomena that thought to generate wave of various strengths. This paper examines the position relationship of the supersonic jet and the converging shock wave. We generated the supersonic jet and the shock wave by using the small chamber shock tube. We formed the converging shock wave by using the curvature reflector.

3.2 Experimental Method

Figure 3.1 shows the experimental apparatus. The aluminum reflector was installed face to face the shock tube to reflect the shock wave generated by the shock tube. Driver gas was regulated by a pressure regulator and filled in the high-pressure chamber with a pressure sensor. In this study, the Schlieren method was utilized for visualizing. Figure 3.2 shows the details of the shock tube. In Fig. 3.2, the shock tube diameter D = 10 mm, the high-pressure chamber length $L_h = 10$ mm, the low-pressure chamber length $L_b = 20$ mm, the radius of the reflector $R = 40$ mm, and the distance between the end of the shock tube to the reflector $L = 100$ mm. Also, P_h and P_b are the pressure of the high-pressure chamber and the low-pressure chamber, respectively. The diaphragm was utilized the 25, 50, 75, and 100 μm PET film. The diaphragm of 25, 50, 75 and 100 μm corresponds to $P_h/P_b = 10.7$, 22.6, 35.0 and 46.0, respectively.

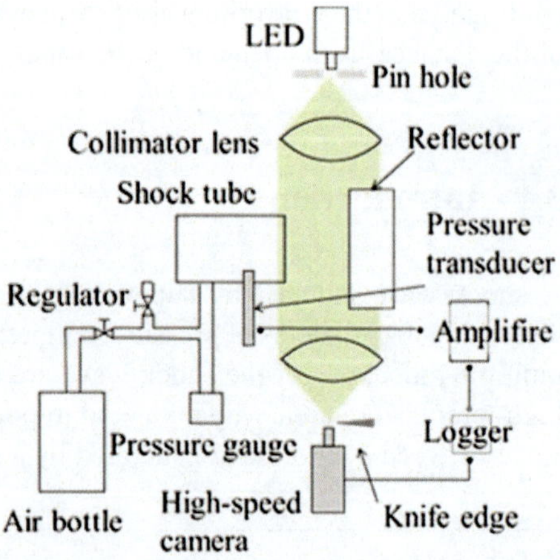

Fig. 3.1 Experimental apparatus

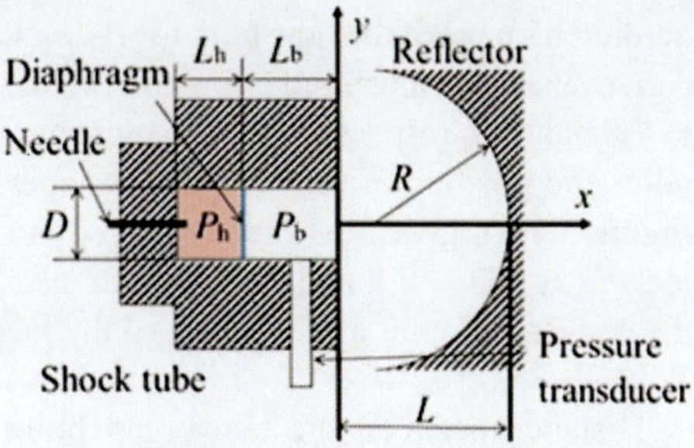

Fig. 3.2 Shock tube and reflector

3.3 Results and Discussion

Figure 3.3 shows the Schlieren image of the flow field for $L = 100$ mm and $P_h/P_b = 35.0$. In Fig. 3.3, where t denotes the time elapsed from the moment of changes appear in the visualization range. Figure 3.3a shows that the shock wave ejected by the shock tube is propagating downstream. Figures 3.3b and c show that the shock wave reflects and converges. Figure 3.3c gives that the converged point and the head of the supersonic jet are located at $x/D = 7.8$ and 6.6, respectively. And then Fig. 3.3d shows the converged shock wave collided with the supersonic jet.

The shock wave is converged and collided with the supersonic jet for $L/D = 10$ and $P_h/P_b = 35.0$. Following this result, confirm the behavior of the supersonic jet and the converging shock wave by varying P_h/P_b. Figure 3.4a–d and (e)-(h) show the

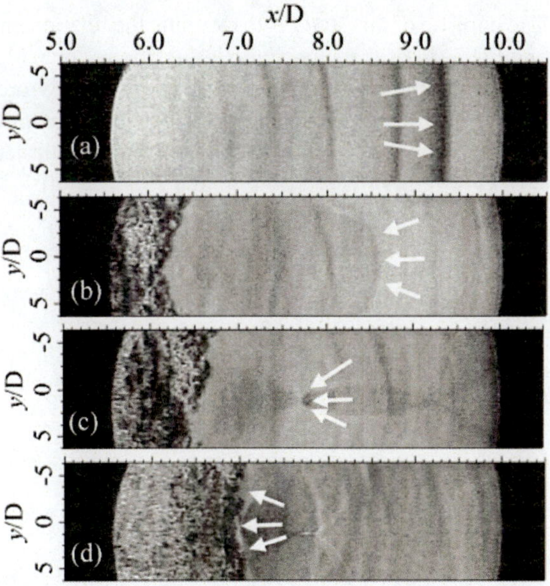

Fig. 3.3 Schlieren Image for $L = 100$ mm, $P_h/P_b = 35.0$ and $R = 40$ mm **a** $t = 100$ μs, **b** $t = 160$ μs, **c** $t = 180$ μs and **d** $t = 210$ μs

Schlieren image of the flow field for $P_h/P_b = 10.7$ and 46.0, respectively. Figure 3.4a–d gives that the converged point and the head of the supersonic jet are located at x/D = 7.8 and 5.9, respectively. This indicates that the distance between the converged point and position of the head of the supersonic jet is increaser than $P_h/P_b = 35.0$. Figure 3.4e–h gives that the converged point and the head of the supersonic jet are located at x/D = 7.8 and 7.2, respectively. This indicates that the distance between the converged point and position of the head of the supersonic jet is decreaser than $P_h/P_b = 35.0$.

Distance between the converged point and the head of the supersonic jet is confirmed changing by P_h/P_b. Following this result, attention will now be given to the converged point and position of the head of the supersonic jet. Figure 3.5 shows the position of the converged point and the head of the supersonic jet for $P_h/P_b = 10.7, 22.6, 35.0$ and 46.0. The vertical axis shows the distance from the reflector L_{ref}/D. The red and blue symbol in Fig. 3.5 show the position of the head of the supersonic jet and the converged point, respectively. From Fig. 3.5, the converged point is nearly identical. On the other hand, the distance between the converged point and position of the head of the supersonic jet decreased with increasing P_h/P_b. Because of while the shock wave not attenuate, the supersonic jet attenuate. It has been felt that the relationship of the collision position between the converging shock wave and the supersonic jet enable to control by P_h/P_b.

3.4 Conclusion

The purpose of this study is to examine the relationship between the converged point and the head of the supersonic jet position. The parameter of the experiment is the pressure of the high-pressure chamber to the low-pressure chamber ratio of P_h/P_b. It was confirmed that the shock wave ejected from the small chamber shock tube converged and then collided with the supersonic jet for L = 100 mm and P_h/P_b = 10.7–46.0. For L = 100 mm and P_h/P_b = 10.7–46.0 while the shock wave not attenuate, the supersonic jet attenuate.

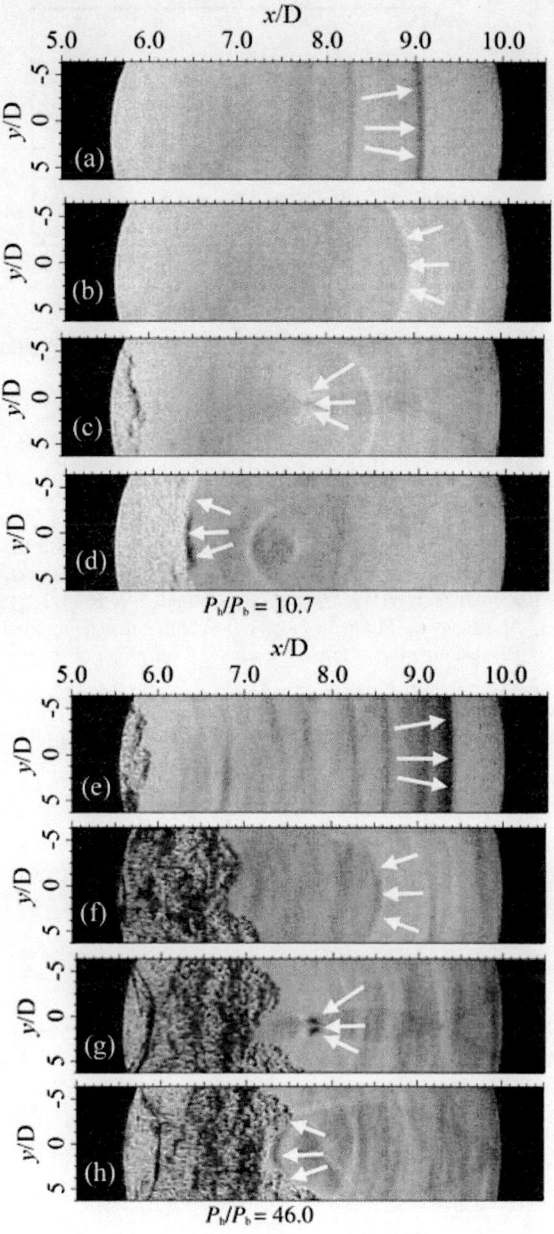

Fig. 3.4 Schlieren image for $L = 100$ mm and $R = 40$ mm. (**a** $t = 100$ μs, **b** $t = 160$ μs, **c** $t = 190$ μs, **d** $t = 230$ μs, **e** $t = 100$ μs, **f** $t = 160$ μs, **g** $t = 180$ μs and **h** $t = 190$ μs)

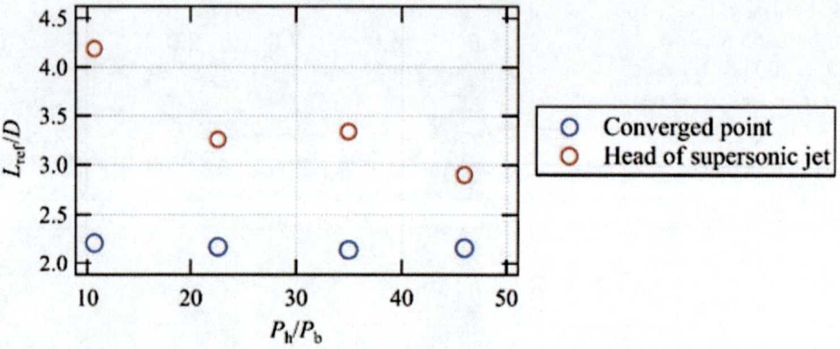

Fig. 3.5 Converged Point and Head of Supersonic Jet Position for $P_h/P_b = 10.7, 22.6, 35.0$ and 46.0 and $L = 100$ mm

References

1. B. Zhang et al., Analysis of the ignition induced by shock wave focusing equipped with conical and hemispherical reflectors. Combust. Flame **236**, 111763 (2022)
2. H. Wang et al., Effect of laser shock peening without protective coating on the surface mechanical properties of NiTi alloy. J. Alloy. Compd. **896**, 163011 (2021)
3. M. Koki et al., Study on shock wave confinement using small high-pressure chamber shock tube, *Proceeding of the 31st International Symposium on Transport Phenomena, A5–131* (2020)

Chapter 4
Reduction of Aerodynamic Noise Radiated from Multicopter

Natsuki Ishida, Hiroki Harada, Hiromitsu Hamakawa, and Eru Kurihara

Abstract The present paper is focused on reducing the aerodynamic noise emitted by multicopter. The effect of the serrations and the dimples of the propeller blades on the aerodynamic noise has been studied experimentally. In addition, we discussed the interaction noise between the propeller and the arm of the multicopter. Multiple peaks were formed in the spectra of SPL. As the distance or height decreased, the peak SPLs and broadband noise increased. The aerodynamic noise was increased when the arm was placed in front of the propeller. We considered the effective countermeasures to reduce the aerodynamic noise emitted by multicopters. The arm located on the downstream side of the propeller is effective in reducing the aerodynamic noise.

4.1 Introduction

Low noise level is an important selling point of the various types of machines, as well as high performance and miniaturization. This also applies to propellers used in multicopters, for example.

Recently, UAV drone has attracted a lot of attention. Drone technology is expected to be used for land delivery, pesticide spraying, aerial photography, and flying cars in the near future. However, the weak point is the instability in the strong wind, the aerodynamic noise, the obstacle avoidance and so on. There have been many studies to reduce the aerodynamic noise emitted by a propeller [1–4]. However, the characteristics of the aerodynamic noise emitted by the multicopter are not clear.

The purpose of the present study is to clarify the characteristics of the aerodynamic noise emitted by the multicopter. We also discussed techniques for reducing the aerodynamic noise emitted by the multicopter.

N. Ishida · H. Harada · H. Hamakawa (✉) · E. Kurihara
Oita University, 700 Dannoharu, Oita 870-1192, Japan
e-mail: hamakawa@oita-u.ac.jp

A. Suryan et al. (eds.), *Proceedings of the 9th Asian Joint Workshop on Thermophysics and Fluid Science, 27–30 November 2022, Utsunomiya, Japan*, Springer Proceedings in Physics 396, https://doi.org/10.1007/978-981-99-9470-0_4

4.2 Experimental Apparatus and Procedure

4.2.1 Field Experiment

Figure 4.1a shows an overview of the measurement method of the aerodynamic noise emitted by the flying multicopter in the airspace of the real baseball field. The distance between the test multicopter and the microphone, L, was 2, 4, 6, 8, 10, 20 and 30 m. The height of the multicopter, h, was 2, 4, 6, 8, 10 and 12 m. The multicopter used in the test was a commercial product of a quadcopter. It has four propellers without propeller guards. The two propellers rotate clockwise and the others counterclockwise. The effect of serrations, dimples on the aerodynamic noise radiated by the blade of a test multicopter was studied.

Fig. 4.1 Measuring method of aerodynamic noise

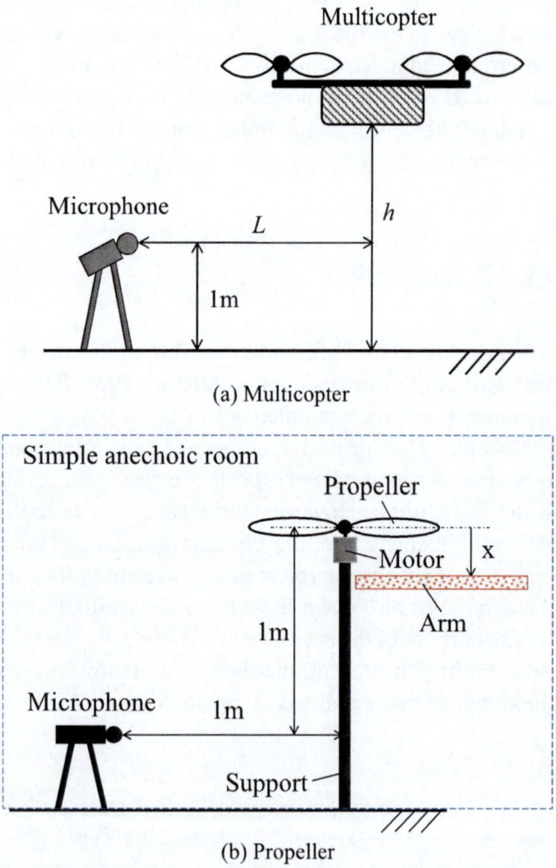

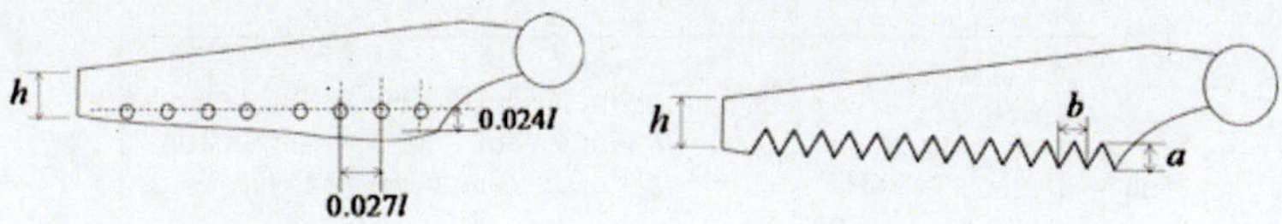

(a) Dimples (b) Trailing edge serrations

Fig. 4.2 Shape of test blade

4.2.2 *Experimental Apparatus*

Figure 4.1b shows an overview of the measurement method of the aerodynamic noise emitted by the test propeller. The experimental apparatus was located in the simple anechoic room. The distance between the test propeller and the microphone was 1.4 m. The test propeller consisted of two blades. The speed has been changed from 0 to 5000 rpm. The distance between the test propeller and the arm of multicopter, x, was -10 mm, -5 mm, 5 mm and 10 mm. The effect of distance, x, cross section shape and material of the arm on the aerodynamic noise emitted by the propeller was studied.

Figure 4.2 is the shape of test airfoil. The blade length, l, is 100 mm. The effect of serrations or dimples on aerodynamic noise has also been studied. Figure 4.2a shows test airfoil with dimples. The dimples were installed near the trailing edge on the surface of the suction or pressure side of the test airfoil. The pitch of dimples is 2.7 mm, the depth is 2.4 mm from trailing edge. Figure 4.2b shows the test airfoil with trailing edge serration. The serration was mounted on the trailing edge of the test airfoil. The pitch of serration, a, was 2.7 mm, the depth, b, was 2.4 mm.

4.3 Results and Discussion

Figure 4.3 shows the spectra of the SPL emitted by the flying multicopter. Multiple peaks were formed in the spectra of SPL. The blade passing frequency (BPF) was 184.6 Hz. The peak frequency of 552.2 Hz was three times the BPF. As the distance, L, or height, h, decreased, the peak SPLs and broadband noise increased.

Figure 4.4 is the comparison of the spectra of the SPL. The brown line in Fig. 4.4 is the result of the trailing edge serrations, and the blue line is the normal type of propeller. The peak SPL was 4.7 dB lower than the normal blade at BPF. However, the maximum peak level of the serrations agreed well with the normal blade.

Figure 4.5 shows the effect of dimples on SPL spectrum. The peak SPL of the dimples was consistent with the normal blade.

Figure 4.6 shows the effect of the distance between the arm and the propeller on the SPL spectra. The blade passing frequency (BPF) was 150.9 Hz. Multiple peaks of the harmonic overtone of the BPF were formed in the spectra of the SPL. The maximum peak SPL of $x = -5$ mm was 64.2 dB at 1211.4 Hz, which was 8 times

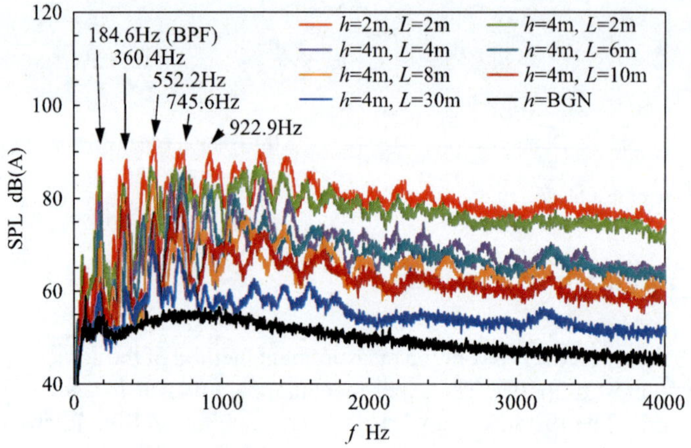

Fig. 4.3 Spectra of SPL of multicopter [3]

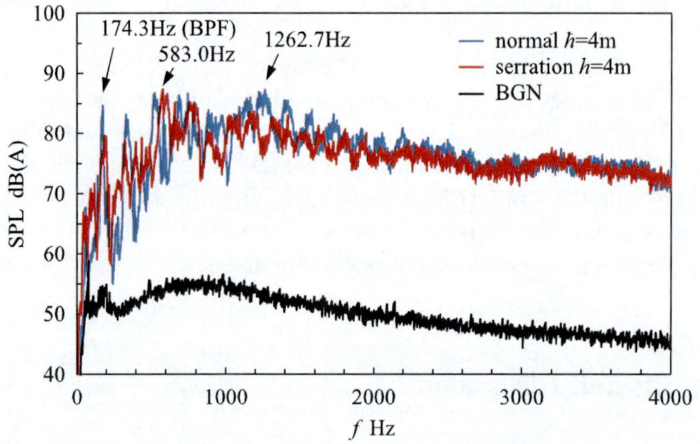

Fig. 4.4 Effect of trailing edge serrations on SPL spectrum [3]

that of the BPF. This peak SPL was 22.4 dB, 19.2 dB greater than that of $x = 5$ mm and 10 mm, respectively. The same tendencies were obtained for the harmonic overtone of the BPF. This means that the aerodynamic noise increases when the arm is in front of the propeller.

Figure 4.7 shows the effect of the arm size on the SPL spectrum. As the size of the arm increased, the peak level of the SPL also increased. This is caused by increased turbulence in the main flow entering the propeller. The arm downstream of the propeller is effective to reduce aerodynamic noise.

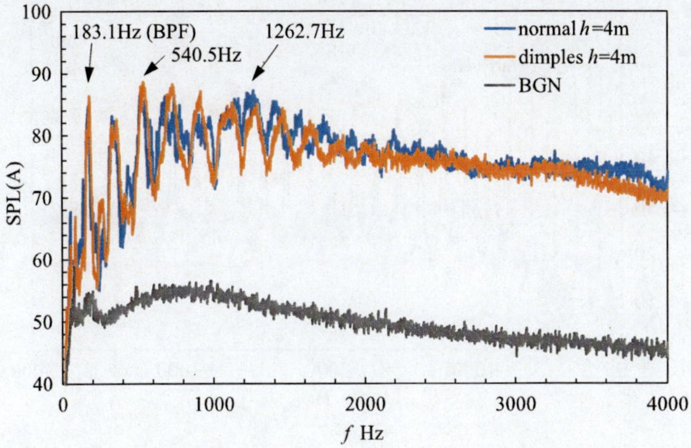

Fig. 4.5 Effect of dimples on SPL spectrum

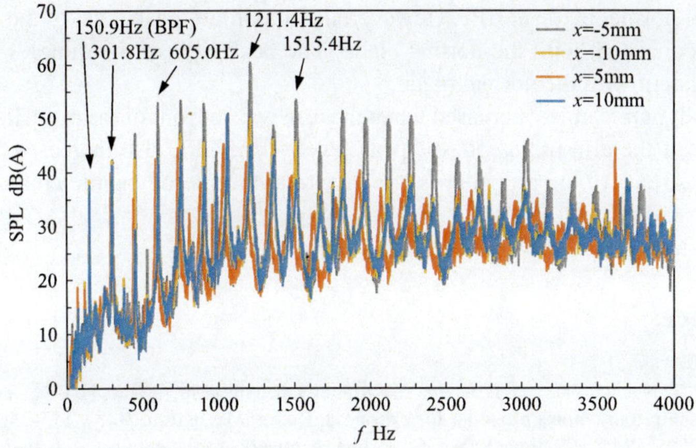

Fig. 4.6 Effect of distance between arm and propeller on SPL spectrum

4.4 Conclusions

The characteristics of the aerodynamic noise emitted by the multicopter were studied experimentally. We discussed the effective countermeasures to reduce the aerodynamic noise emitted by multicopters.

(1) In the normal multicopter, the peak SPLs and broadband components increased as the distance or altitude decreased. In the SPL spectra, multiple peaks were formed at the harmonic overtone of the BPF.

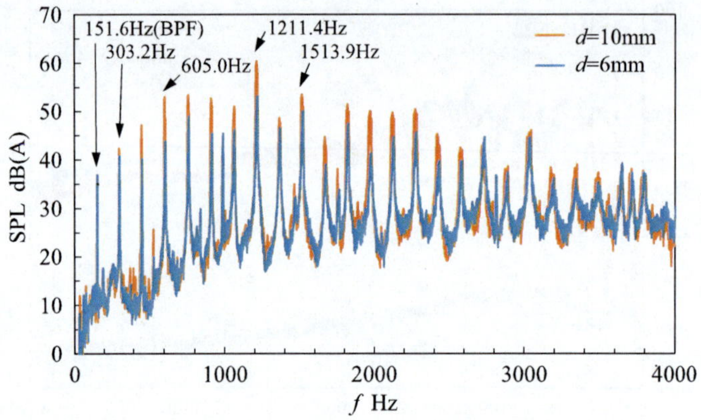

Fig. 4.7 Effect of arm size on SPL spectrum

(2) In the multicopter, the peak SPL of the serrations was 4.7 dB lower than that of the normal blade at BPF. However, the maximum peak level of the serration was consistent with the normal blade. The peak SPL of the dimples was also consistent with the normal blade.
(3) Aerodynamic noise increased when the arm was in front of the propeller. As the scale of the arm increased, the peak level of the SPL also increased. The arm downstream of the propeller is effective to reduce aerodynamic noise.

References

1. R. Noda, T. Nakata, T. Ikeda, D. Chen, Y. Yoshinaga, K. Ishibashi, C. Rao, H. Liu, Development of bio-inspired low-noise propeller for a drone. J. Robot. Mechatron. **30**(3), 337–343 (2018)
2. Y. Wu, Y. Ai, W. Ze, T. Jing, X. Song, Y. Chen, A novel aerodynamic noise reduction method based on improving spanwise blade shape for electric propeller aircraft. Int. J. Aerosp. Eng. **4**(2019), 1–10 (2019)
3. N. Isatake, M. Takahashi, H. Hamakawa, E. Kurihara, Aerodynamic Noise radiated from multicopter, in *International Workshop on Environmental Engineering 2019, No.19-206*, (Nago, Okinawa, Japan, 2019)
4. V. Patel, K. Nikam, S. Dikshit, M. Agarwalla, C. Zagade, Propeller design and optimization for drones, in *Recent Advances in Operations Management Applications* (2022), pp. 411–424

Chapter 5
Hydrodynamics of a Partially Slipping Circular Cylinder in a Uniform Flow

K. Nandakumar Chandran, Dijo George Biju, Adarsh R. Nair, and Ranjith S. Kumar

Abstract This article explores numerical modeling of flow past a semi-coated hydrophobic cylinder, aiming for a deep understanding of hydrodynamic instability. Utilizing the finite volume method, the unsteady fluid dynamics and vortex creation are investigated. It evaluates the effectiveness of partial hydrophobic coating in stabilizing high-Reynolds number (*Re*) flows and delves into hydrodynamic features across various Reynolds numbers. Different coating angles θ are considered with a focus on their impact on lift and drag coefficients as well as the Strouhal number. Findings indicate that increasing θ leads to asymmetrical vortex shedding, causing noticeable drag variations. Lift coefficient fluctuations also show notable differences, especially peaking for $\theta < 90°$. The fluctuation of drag coefficients rises until *Re* reaches 80, then decreases.

5.1 Introduction

Investigating flow characteristics around bluff bodies is pivotal in hydraulic engineering, contributing to drag-reducing techniques for energy-efficient systems. Hydrophobic coatings have garnered attention in contexts like autonomous underwater vehicles. The present study aims to integrate attributes of coated and uncoated surfaces for force optimization in aquatic systems and a circular cylinder with partial coating serves as the computational model. Prior research has delved into flow velocities [1] and hydrodynamics influenced by cylinder geometry [2]. As for asymmetric slip, identifying which surface region (aft, fore, or flank) enhances maneuverability is key; [3] ascertained that flank slip notably reduces drag, forming this study's basis. Numerous experiments focus on slip-inducing surfaces to decipher flow separation and wake vortices [4, 5], substantiating drag-reduction merits. Work by [5] revealed

K. N. Chandran · D. G. Biju · A. R. Nair · R. S. Kumar (✉)
Micro/Nano Fluidics Research Laboratory, Department of Mechanical Engineering, College of Engineering Trivandrum (Government of Kerala), Thiruvananthapuram, Kerala 695016, India
e-mail: ranjith@cet.ac.in

A. Suryan et al. (eds.), *Proceedings of the 9th Asian Joint Workshop on Thermophysics and Fluid Science, 27–30 November 2022, Utsunomiya, Japan*, Springer Proceedings in Physics 396, https://doi.org/10.1007/978-981-99-9470-0_5

a 40% contraction in recirculation bubble size and sub-10% drag coefficient diminution for coarse hydrophobic surfaces. Although numerical analyses may yield inflated values compared to experimental findings, some studies are lauded for their precise modeling.

Seo et al. [6] investigated laminar flow around cylinders, demonstrating that increased slip length substantially lowers drag. Mastrokalos et al. [7] probed vortex suppression related to critical slip length. D'Alessio [8] and others reported drag reduction and delayed flow separation at higher Reynolds numbers. Hongjie Huang [9] delineated key factors affecting the drag coefficient, while Choi et al. [10] showed partial slip conditions reduced lift amplitude. Donghyun You [11] found that no-slip conditions elevate wake instability at low speeds, but hydrophobic surfaces mitigate flow separation at higher velocities. Studies extend to rarefied gas flow [12] and magnetized fluid dynamics [13], revealing opportunities for drag reduction under specific viscosity ratios. Research also delved into low Reynolds number gas flows, showing drag coefficients aligning with kinetic theory for $Kn < 1$ [14]. Yancheng Li and Sai Peng [15] applied varying Knudsen numbers to describe how low Reynolds numbers reduce frictional drag, corroborating Hongjie Huang [9]. Other inquiries examined blockage effects [16], wall impacts [17], and aerodynamic drag on NACA0012 airfoils [18]. Current advancements focus on mechanically textured and chemically altered super-hydrophobic surfaces for further drag reduction [19, 20]. This work aims to synthesize slip and no-slip effects for drag and lift analysis, targeting energy-efficient systems via numerical simulation of coated cylindrical forms. The organization of the paper is as follows, the numerical methodology employed is discussed in Sect. 5.2. The results obtained and inferences on account of the physical phenomenon are detailed in Sect. 5.3 before the conclusion in Sect. 5.4.

5.2 Numerical Methodology

This section outlines the computational framework for analyzing uniform laminar flow across a cylinder with selective, asymmetric coatings. The 2D domain includes a cylinder of diameter D, ensconced in a $25D \times 20D$ area, $10D$ distant from the upstream inlet. The fluid traverses the x-axis at an inlet velocity U (Fig. 5.1). A non-uniform structured mesh with finer grids proximate to the cylinder is utilized (Fig. 5.2a). Unsteady simulations are executed across diverse slip lengths (β), Reynolds numbers (Re), and span angles (θ). Studies were undertaken for $\theta = 45°$, 90°, 135°, and 180° to assess their impact on flow separation, drag and lift coefficients, and velocity distribution. Ensuing sections will expound upon the governing equations and validate the model against prior research.

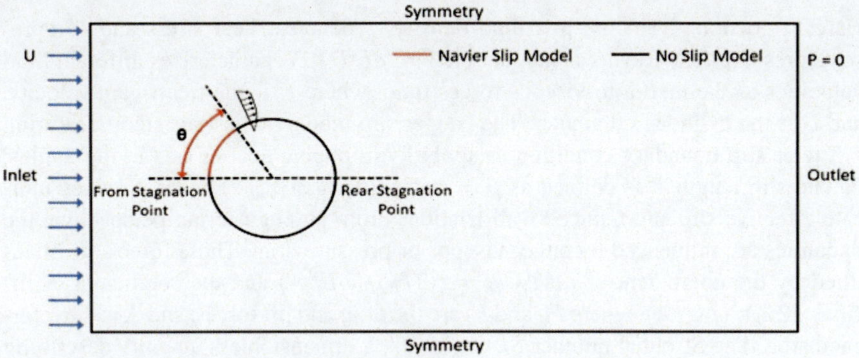

Fig. 5.1 Details of 2D domain and boundary conditions

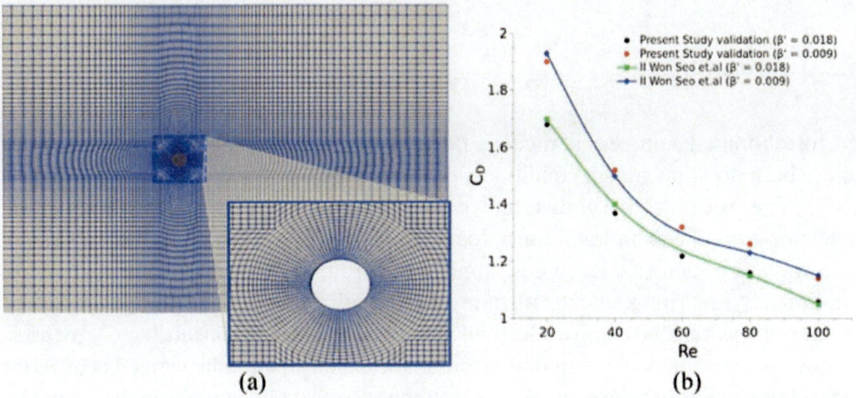

Fig. 5.2 **a** Grid structure **b** Validation of present study employing Navier Slip model with respect to II Won Seo et al.

5.2.1 *Governing Equations and Modelling*

Under the specified Reynolds numbers and stability conditions, the flow reaches a fully developed, laminar state. We employ the Finite Volume Method (FVM) for solving this unsteady issue, focusing on hydrodynamic modeling via the open-source OpenFOAM platform. The fluid flow is modeled as two-dimensional, unsteady, Newtonian, and incompressible. The governing equations for this setup include the continuity equation (Eq. 5.1) and momentum equation (Eq. 5.2):

$$\nabla.\vec{v} = 0 \tag{5.1}$$

$$\frac{\partial \vec{v}}{\partial t} + \nabla.\vec{v}\vec{v} = -\frac{1}{\rho}\nabla P + \frac{1}{\rho}\nabla.\vec{\tau} \tag{5.2}$$

Here, $\vec{v}$ denotes velocity, ρ is fluid density, $\vec{\tau}$ denotes shear stress and P represents pressure. The Reynolds number ($Re = \rho U D/\mu$) characterizes different fluid velocities as the inertia-to-viscous forces ratio, where U is the freestream velocity, and D is the cylinder's diameter. The Navier slip model guides this study, featuring a Navier slip boundary condition to apply hydrophobic coating at varying angles, θ. The slip length β is defined as $\beta = u_{slip}/(\partial u/\partial y)\,|_{wall}$. Higher β values indicate effective slip and reduced wall friction. From an engineering perspective, the dynamics are influenced by either viscous or pressure drag. These forces are quantified by the coefficient of drag $C_D = (2F_d)/\left(\rho U^2 A\right)$ and the coefficient of lift $C_l = (2F_l)/\left(\rho U^2 A\right)$, where F_d and F_l are the drag and lift forces, and A is the reference area. The Strouhal number $St = f D/U$, a dimensionless quantity describing oscillatory flow mechanisms, is also relevant to this study. Throughout the article, the slip length is non-dimensionalized as $\beta' = \beta/D$.

5.2.2 Validation

To corroborate the numerical model's fidelity, a thorough evaluation was carried out under both no-slip and slip conditions, characterized by slip lengths $\beta' = 0.009$ and 0.018. The core objective of this study was to quantitatively scrutinize hydrodynamic behavior around a cylindrical form, focusing particularly on drag coefficients (C_D).

Simulations across a Reynolds number spectrum of 20 to 100 confirmed strong alignment with prior research, affirming the model's validity (Fig. 5.2b). Grid independence was verified using a 9200 cell resolution, primarily concerning C_D metrics. Following an exhaustive temporal discretization analysis, a time step (Δt) of 0.05s was chosen. Furthermore, model convergence was validated through a relative tolerance of less than 10^{-6} per iteration.

5.3 Results and Discussion

This section evaluates numerical outcomes for flow around a cylinder with a selectively and asymmetrically applied hydrophobic layer, concentrating on flow detachment and alterations in lift and drag metrics. The study also examines the Von Karman Vortex Street and associated flow perturbations. Scenarios with variable β' values 0.009, 0.018, and 0.0135 and θ angles 45°, 90°, 135°, and 180° were modeled at Reynolds numbers between 50 and 100, manifesting notable drag mitigation and lift coefficient variability. The research clarifies how Reynolds number, effective slip, and coating span angle influence flow dynamics and vortex-induced instabilities.

5.3.1 *Flow Separation Over Asymmetric- Partially Spanned Slip-On Cylinder Surface*

To deeply understand flow dynamics near solid entities, this research centers on velocity dispersion and separation variation. While it's acknowledged that slip conditions alter the separation angle toward the rear stagnation point, the study scrutinizes the subtle impacts of asymmetrical and partial slip. Particularly, downstream vortex formations and separation angles are assessed in light of the preceding parametric analysis.

Discussions on velocity distributions

Velocity contours around a bluff body elucidate the underlying flow dynamics. Figure 5.3 presents selected contours for varied *Re* and θ settings at β' = 0.0135 and non-dimensionalized time $\tau = 100$ ($\tau = tU/D$). In traditional flow over a cylinder, Karman Street is notable at Reynolds numbers between 40–50; this shifts higher with increased hydrophobicity. Figure 5.3 shows negligible shedding at low *Re*, evolving into prominent vortex streets as velocity rises. Vortex onset is most evident in the *Re* 60–65 range, with increased flow instability thereafter. At θ = 90°, shedding onset shifts to a lower *Re* of 56–60, displaying minor unsteadiness at *Re* = 55. In contrast, for θ = 135° and 180°, vortex initiation occurs at *Re* above 60. These contours reveal how slip impacts velocity distributions, particularly near the wall surface, and inform our understanding of the dynamic forces at play. Figure 5.3 shows negligible shedding at low *Re*, evolving into prominent vortex streets as velocity rises. Vortex onset is most evident in the *Re* 60–65 range, with increased flow instability thereafter. At θ = 90°, shedding onset shifts to a lower *Re* of 56–60, displaying minor unsteadiness at *Re* = 55. In contrast, for θ = 135° and 180°, vortex initiation occurs at *Re* above 60. These contours reveal how slip impacts velocity distributions, particularly near the wall surface, and inform our understanding of the dynamic forces at play.

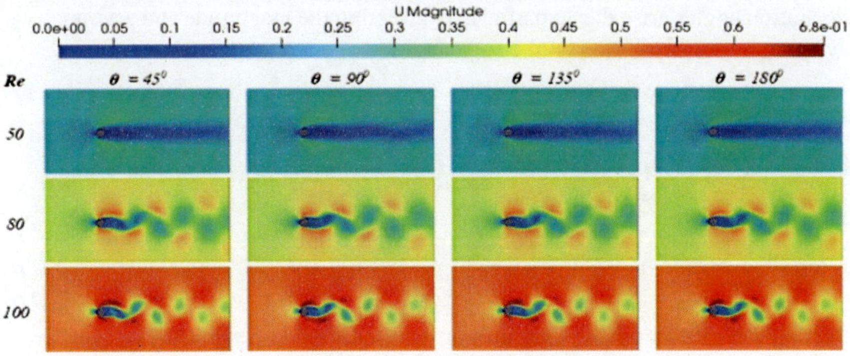

Fig. 5.3 Velocity contour for varying Reynolds number and θ, at β' = 0.0135

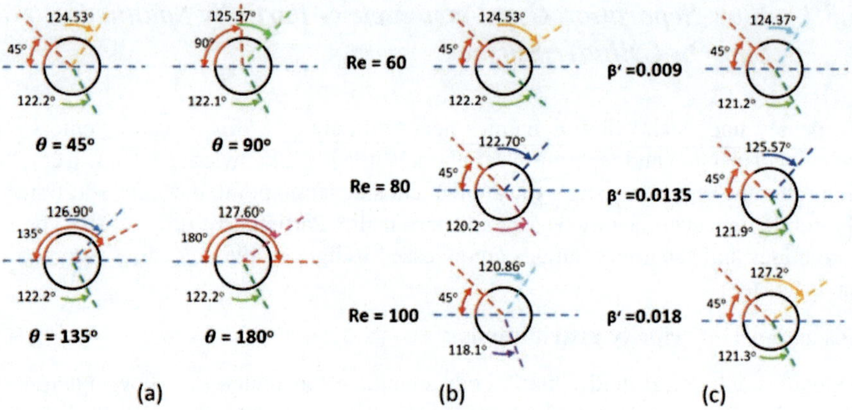

Fig. 5.4 Variations of separation angle for different values of **a** θ, **b** Re **c** β'

Discussions on separation angle

To fully comprehend the complexities of vortex shedding, one must extend the examination beyond mere velocity profiles to include the point of flow separation, represented by separation angle α. While it's recognized that employing slip shifts this angle toward the rear stagnation point, the current study considers partial, non-uniform slip, giving rise to both upper (α_u) and lower (α_l) separation angles, as depicted in Fig. 5.4.

Figure 5.4a illustrates how increasing span angle θ at a fixed β' of 0.0135 and *Re* of 60 alters the upper surface's separation point, highlighting the hydrophobic coating's influence. Figure 5.4b explores separation angle variation at different *Re* levels, holding β' at 0.0135 and θ at 45°, and reveals an earlier onset of separation at higher *Re* values. Concerning α_l, its range remains consistent for a given *Re*, albeit with earlier onset as *Re* increases. Figure 5.4c investigates how varying β' affects α_u and α_l at a constant *Re* of 60 and θ of 45°. The study confirms that asymmetrical separation angles arise due to partial slip, affecting the magnitude and asymmetry of dynamic forces depending on the span angles.

5.3.2 *Effects on Coefficient of Lift*

The lift coefficient, varies temporally for distinct β, θ, and *Re* values. Initially, C_L exhibits negligible fluctuation but stabilizes with increasing τ. Figure 5.5a plots C_L variation against τ at β' = 0.009, 0.0135, and 0.018 for *Re* = 100 and θ = 90°. Beyond τ = 50, amplitude remains constant, but higher β' values cause subtle C_L fluctuations, correlating with increased hydrophobicity. Regarding dependency of C_L on *Re* for β' = 0.009 and θ = 90°, Fig. 5.5b reveals distinct oscillation frequencies

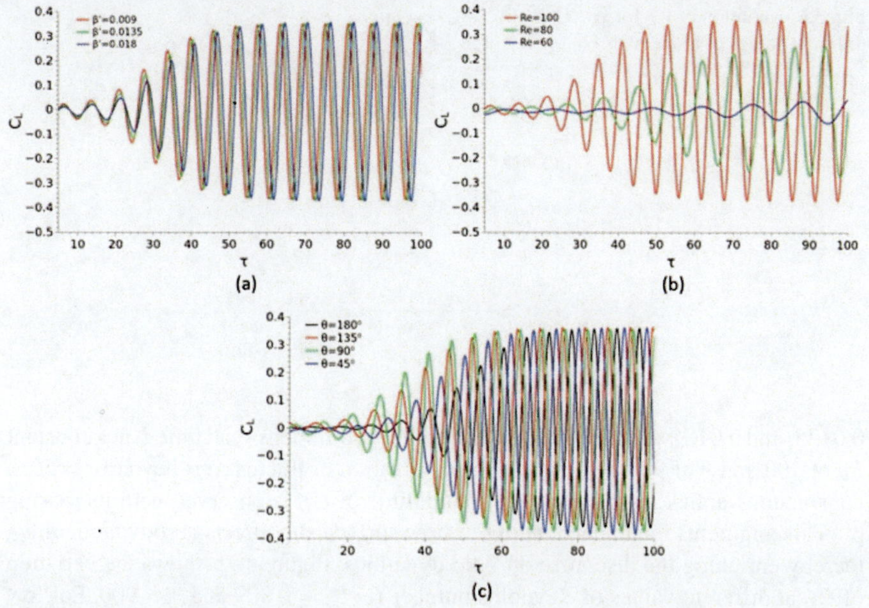

Fig. 5.5 Effect on coefficient of lift by **a** varying β' for θ = 90° and Re = 100 **b** varying Re at β' = 0.009 and θ = 90° **c** varying θ at β' = 0.009 and Re = 100

and significant amplitude changes. An ascending *Re* value results in appreciable C_L elevation.

Additionally, Fig. 5.5c portrays C_L variations for different θ angles at a fixed β' = 0.009 and *Re* = 100. As θ ascends, fluctuation amplitude initially grows until θ = 90°, then diminishes, reaching its nadir at θ = 180°. Beyond τ > 70, upward lift fluctuations converge closely, whereas downward lift shows marked amplitude variation. This discrepancy in C_L fluctuation is attributable to the asymmetric surface texture induced by varying θ. Downward lift fluctuation is more pronounced for θ < 90° and mitigates for θ > 90°. Figure 5.6 elucidates fluctuation amplitude variations in C_L at Re = 70, 80, and 90. While upward lift ($C_{L\text{-}U}$) fluctuations are relatively uniform, downward lift ($C_{L\text{-}l}$) fluctuations are notably more variable. Specifically, $C_{L\text{-}l}$ fluctuation is heightened for θ < 90° and diminishes for θ > 90°.

5.3.3 *Effects on Coefficient of Drag*

The drag coefficient (C_D) manifests as a time-dependent variable under variations of β and θ for assorted Reynolds numbers (*Re*). It's noteworthy that upon the initiation of fluid flow, C_D exhibits elevated values, which subsequently taper off as time elapses, only to ascend again with the onset of vortex shedding. Figure 5.7a offers a compelling visualization of temporal trajectory of C_D for distinct β' values (0.009,

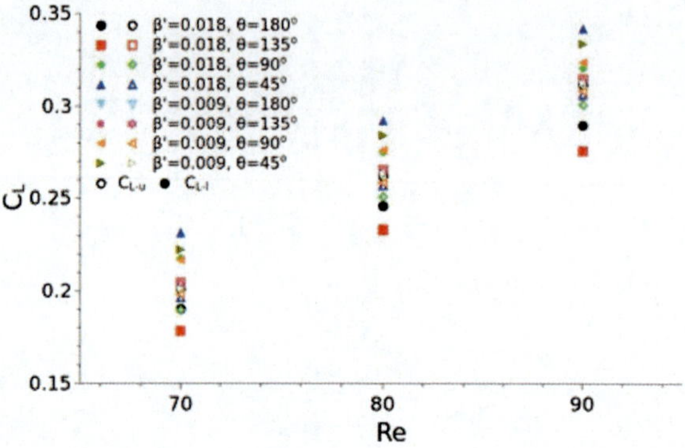

Fig. 5.6 Variation of C_L for different values of β' and θ for Re = 70, 80, 90

0.0135, and 0.018) when scrutinized against the non-dimensional time τ, at a constant *Re* of 100 and θ of 90°. Following the stabilization of fluctuations beyond τ > 50, a conspicuous uptick in the fluctuation amplitude of C_D is observed with increasing β'. This augments the understanding of how surface slip affects aerodynamic drag, thereby enriching the discourse on fluid dynamics. Figure 5.7b shows the variation of C_D at different values of Reynolds number for β' = 0.009 and θ = 900. For low values of *Re*, the value of C_D reduces and reaches a constant value as there is no fluctuation in drag in the absence of vortex shedding. As the value of *Re* increases, the initial reduction in C_D is much higher, but once vortex shedding begins to occur, the value of C_D increases. The fluctuation of the C_D is uniform over the period, and the time average value of C_D gets lowered as the value of *Re* increases. Figure 5.7c shows the variation of C_D at different values of θ for β' = 0.009 and *Re* = 100. As flow initiates the value C_D is higher and it reduces with time, the reduction depends on the value of θ, as the dip in value of C_D is higher for lower θ value (if θ < 90^0) and is higher for greater θ value (if θ > 90°). As for τ > 70, the fluctuation of C_D value is periodic. The time average value of C_D is low for θ > 900 and is higher for θ < 90^0.

It can be clearly observed from Fig. 5.8 the variation of amplitude of fluctuation of C_D for *Re* = 70, 80 and 90. Referencing to the time average value of C_D, the fluctuations due to upward lift is represented using $C_{D\text{-}u}$ and the fluctuations due to downward lift is represented using $C_{D\text{-}l}$. The value of $C_{D\text{-}u}$ decreases as the value of *Re* increases, but the value of $C_{D\text{-}l}$ increases with increase in *Re* value from 70 to 80 and it reduces for *Re* = 90. These variation is fluctuation of $C_{D\text{-}l}$ can be accounted as the effect of non-uniform surface texture, since the variation in the $C_{D\text{-}l}$ depends on the value of θ and shift in separation angle α_u and α_l around the cylinder surface.

5.3.4 *Discussions on Flow Instabilities*

In the context of bluff body dynamics, a pivotal factor for quantifying the frequency of fluctuation resulting from vortex shedding is encapsulated within the Strouhal

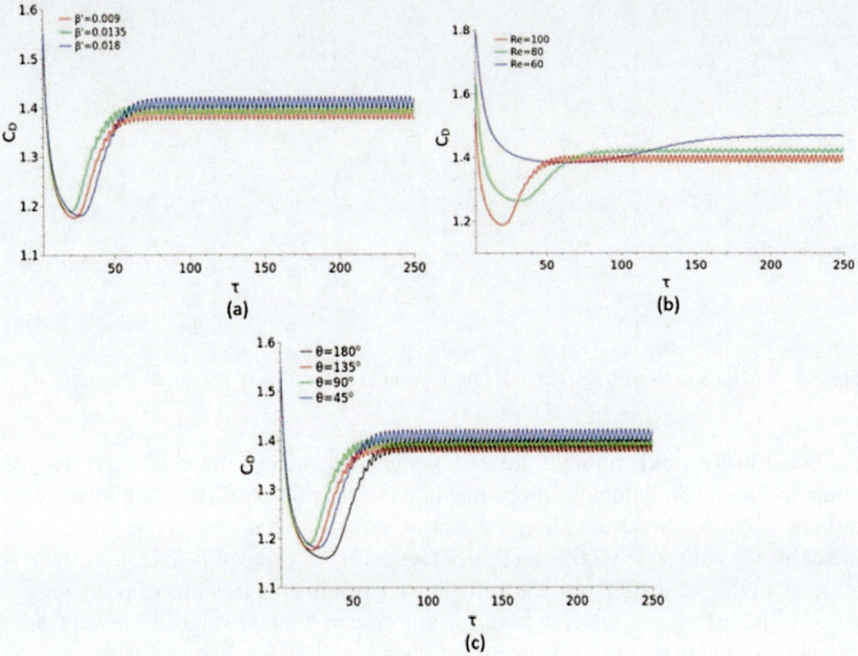

Fig. 5.7 Effect on coefficient of drag by **a** varying β' for θ = 90° and Re = 100 **b** varying Re at β' = 0.009 and θ = 90° **c** varying θ at β' = 0.009 and Re = 100

Fig. 5.8 Variation of C_D for different values of β' and θ for Re = 70, 80, 90

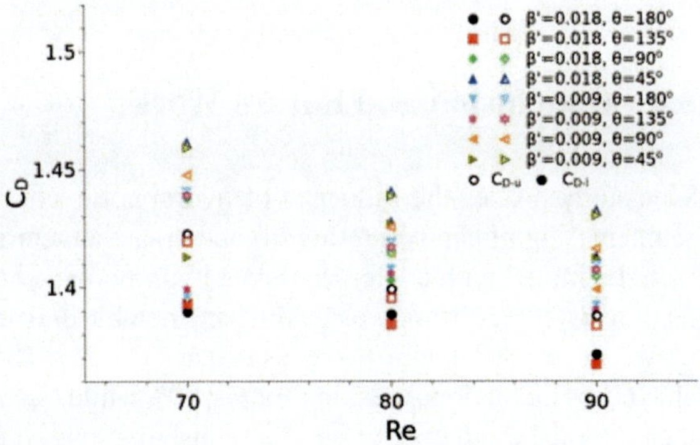

number (*St*). This fundamental parameter, mathematically expressed as, $St = (f \times D)/U$, intricately weaves together the shedding frequency (f), the cylinder's diameter (D), and the free stream velocity (U). This characterization elucidates the intrinsic correlations between shedding phenomena, the geometrical attributes of the cylinder, and the dynamic flow conditions. With the increase in *Re*, the amplification in fluctuations gives rise to substantial instabilities, consequently leading to a hike in the *St*. However, from Fig. 5.9, it becomes evident that the impact on *St* is not solely

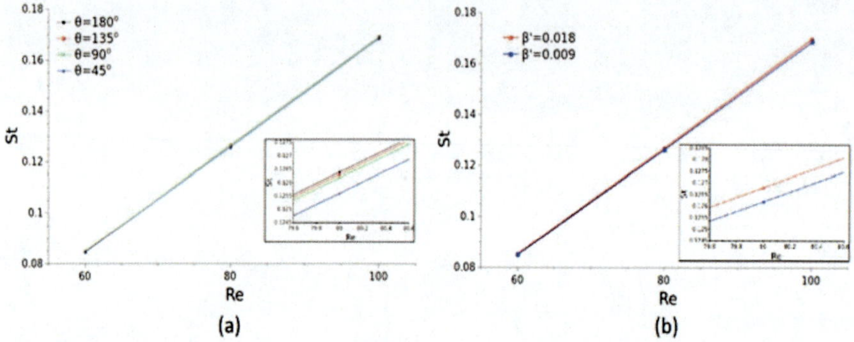

Fig. 5.9 Variation of St with respect to Re for varying a) θ (for β' = 0.009), b) β' (for θ = 45°)

attributed to Reynolds number. Instead, parameters such as θ and β' exert notable influence on the *St*, inducing an upsurge in its value with corresponding increments in both θ and β'. It can be observed that as value of *Re* increases the value of *St* increases for both β' = 0.009 and 0.018. Generally, as value of β' increases there is increase in the value of *St* and the variation gets significant as value of β' increases. Also, as the values of θ increase for a specific case of β' and *Re*, it is observed that *St* increases showing that the asymmetric application of hydrophobic coating over the surface increases the instability. These results can be related to the explanations of changes observed in velocity contour and variations in separation angle as discussed before.

5.4 Conclusion and Future Work

This study probes the influence of hydrophobic coatings on objects in fluid environments, significantly altering lift and drag characteristics. Employing a numerical cylinder model with angle-dependent hydrophobic coating, we find that increased θ leads to asymmetric vortex shedding and notable drag fluctuation. The time-averaged drag for $\theta = 180°$ notably exceeds that at $\theta = 135°$. The textured surface causes variable C_{L-l} fluctuations, peaking for $\theta < 90°$, while C_{D-l} fluctuation initially rises and then diminishes after $Re = 80$. These insights suggest the viability of self-regulating maneuver systems using mixed no-slip and slip surfaces, potentially enabling energy-efficient designs by leveraging existing flow dynamics. Enhanced slip due to coatings could further amplify these effects.

References

1. K. Sarath, K. Manu, An investigation of bluff body flow structures in variable velocity flows. Phys. Fluids **34**, 034102 (2022)
2. W.J.G.S. Pinto, F. Margnat, Influence of cylinder breadth and shape on the onset of flow unsteadiness and the aeolian tone level. Comput. Fluids **228**, 105067 (2021)
3. D. Li, S. Li, Y. Xue, Y. Yang, W. Su, Z. Xia, Y. Shi, H. Lin, H. Duan, The effect of slip distribution on flow past a circular cylinder. J. Fluids Struct. **51**, 211 (2014)
4. J. Wen, H.-B. Hu, Z.-Z. Luo, Z.-Z. Zhang, Experimental investigation of flow past a circular cylinder with hydrophobic coating. J. Hydrodyn. **30**, 992 (2018)
5. N. Kim, H. Kim, H. Park, An experimental study on the effects of rough hydrophobic surfaces on the flow around a circular cylinder. Phys. Fluids **27**, 085113 (2015)
6. I.W. Seo, C.G. Song, Numerical simulation of laminar flow past a circular cylinder with slip conditions. Int. J. Numer. Meth. Fluids **68**, 1538 (2012)
7. M.E. Mastrokalos, C.I. Papadopoulos, L. Kaiktsis, Optimal stabilization of a flow past a partially hydrophobic circular cylinder. Comput. Fluids **107**, 256 (2015)
8. S. D'Alessio, Flow past a slippery cylinder: part 1—circular cylinder. Acta Mech. **229**, 3375 (2018)
9. H. Huang, M. Liu, H. Gu, X. Li, X. Wu, F. Sun, Effect of the slip length on the flow over a hydrophobic circular cylinder. Fluid Dyn. Res. **50**, 025515 (2018)
10. S.D. Choi, T.S. Eum, E.T. Shin, C.G. Song, Numerical investigation of flow around a structure using navier-slip boundary conditions. World J. Eng. (2021)
11. D. You, P. Moin, Effects of hydrophobic surfaces on the drag and lift of a circular cylinder. Phys. Fluids **19**, 081701 (2007)
12. M.A. Gallis, J.R. Torczynski, Effect of slip on vortex shedding from a circular cylinder in a gas flow. Phys. Rev. Fluids **6**, 063402 (2021)
13. S. I. Kamiyama, A. Satoh, Steady flow around a circular cylinder coated with a magnetic fluid film. JSME Int. J. Ser. 2, Fluids Eng. Heat Transfer Power Combust. Thermophys. Properties **31**, 218 (1988)
14. X.-J. Gu, R.W. Barber, B. John, D.R. Emerson, Non-equilibrium effects on flow past a circular cylinder in the slip and early transition regime. J. Fluid Mech. **860**, 654 (2019)
15. Y.C. Li, S. Peng, T. Kouser, Effect of wall slip on laminar flow past a circular cylinder. Acta Mechanica, 1–19 (2022)
16. S. Mittal, S. Prakash Singh, B. Kumar, and R. Kumar, Flow past bluff bodies: effect of blockage, Int. J. Comput. Fluid Dyn. **20**, 163 (2006)
17. A. Ramamurthy, P. Lee, Wall effects on flow past bluff bodies. J. Sound Vib. **31**, 443 (1973)
18. J. Lee, H. Kim, H. Park, Effects of super hydrophobic surfaces on the flow around an naca0012 hydrofoil at low Reynolds numbers. Exp. Fluids **59**, 1 (2018)
19. B.J. Rosenberg, T. Van Buren, M.K. Fu, A.J. Smits, Turbulent drag reduction over air-and liquid-impregnated surfaces. Phys. Fluids **28**, 015103 (2016)
20. P. Muralidhar, N. Ferrer, R. Daniello, J.P. Rothstein, Influence of slip on the flow past super-hydrophobic circular cylinders. J. Fluid Mech. **680**, 459 (2011)

Chapter 6
Effects of Salinity on Carbon Dioxide Microbubble Dissolution into Water

Akimaro Kawahara and Yukihiro Yonemoto

Abstract The present study investigates experimentally characteristics of dissolution of carbon dioxide (CO_2) micro-bubbles generated by a multi-fluid mixer into liquids. As the liquids, artificial seawater and tap water were used to know the effects of salinity on the dissolution. The Sauter mean diameter of the bubbles, concentration of dissolved CO_2 in liquid, volumetric mass transfer coefficient, void fraction, were measured. In addition, the interfacial area concentration was determined from experimental values of the Sauter mean diameter and the void fraction. The volumetric mass transfer coefficient is larger for seawater than tap water because the bubble size becomes smaller for seawater than tap water and then the interfacial area concentration is much larger for seawater than tap water.

6.1 Introduction

Micro bubbles with a diameter of several hundreds of micrometer or less in the liquid have characteristics such as a large gas–liquid interfacial area per unit volume, extremely low rising velocity, and good diffusivity in the horizontal direction [1]. Therefore, the dissolution of the gas inside the bubbles in the liquid is excellent. The application of carbon dioxide microbubbles in seawater is attracting attention. For example, carbon dioxide microbubbles are used as a measure to prevent shellfish from adhering to seawater systems such as heat exchangers and piping in power plants [2]. In addition, the dissolution of carbon dioxide generated in thermal power plants and factories into seawater is an important issue in modern society, and the sequestration/storage of carbon dioxide in ocean [3] and underground [4] are being studied. In addition, one effective use of carbon dioxide in seawater is the cultivation of algae. Carbon dioxide microbubbles are generated in seawater to promote algae photosynthesis [5]. Attempts have also been made to contribute to the prevention of

A. Kawahara · Y. Yonemoto (✉)
Kumamoto University, Kurokami 2-39-1, Chuo-Ku, Kumamoto 860-8555, Japan
e-mail: yonemoto@mech.kumamoto-u.ac.jp

A. Suryan et al. (eds.), *Proceedings of the 9th Asian Joint Workshop on Thermophysics and Fluid Science, 27–30 November 2022, Utsunomiya, Japan*, Springer Proceedings in Physics 396, https://doi.org/10.1007/978-981-99-9470-0_6

global warming by refining oil accumulated in the body of cultured algae and using it as biofuel.

In this study, a microbubble generator (hereinafter abbreviated as MBG) devised by Sadatomi and Kawahara et al. [6, 7] is used to generate carbon dioxide microbubbles in tap water and artificial seawater, thereby the time variations in concentration of dissolved carbon dioxide in those water were measured. The effect of liquid flow rate and carbon dioxide flow rate to MBG and salinity on the dissolution characteristics was investigated experimentally.

6.2 Experiments

Figure 6.1 shows an overview of the MBG [6, 7] used in this experiment. By inserting an orifice with an inner diameter $d_O = 8.4$ mm into a cylindrical pipe with an inner diameter $D = 14.7$ mm to reduce the cross-sectional area of the flow path, the liquid velocity in the reduction area increases and the pressure decreases. When the pressure just downstream the orifice is below atmospheric pressure, the gas is self-sucked from a tube leading to the atmosphere through a porous tube made of a fiber membrane with a pore diameter of ca. 25 μm. The sucked gas is sheared by a high-speed liquid flow, and becomes a lot of fine bubbles and is discharged together with the liquid jet from the generator outlet.

Figure 6.2 shows an outline of the experimental apparatus. A test liquid ($V_L = 100$ L) adjusted to $T_L = 20°$ C was stored in an acrylic cylindrical tank (inner diameter 300 mm, height 2000 mm). For the liquid, tap water and artificial seawater (hereinafter, called as salt water) in which a commercial natural sea salt was dissolved in tap water at a mass concentration of 3.5% were used. The MBG [6, 7] was installed at

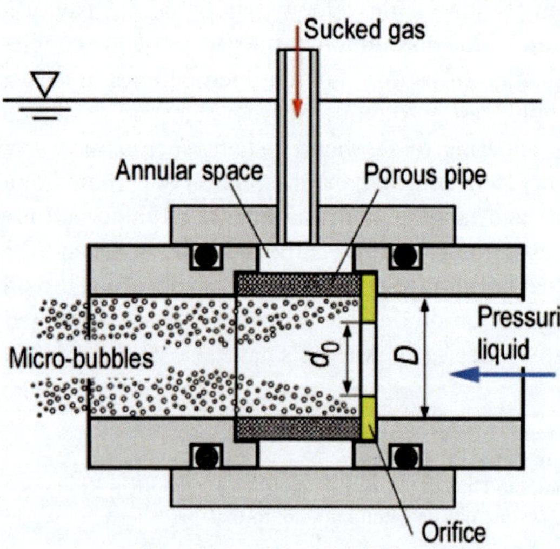

Fig. 6.1 Micro-bubble generator used in the present study (JP Patent 5,103,625)

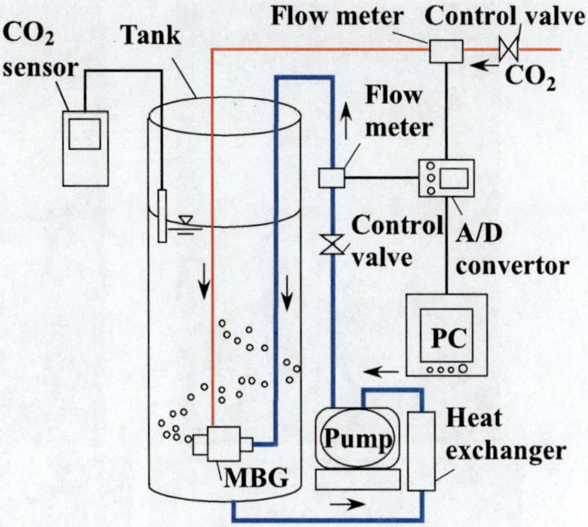

Fig. 6.2 Test apparatus for dissolution of CO2 micro-bubbles

a water depth of 780 mm and aerated with carbon dioxide. The volume flow rates of liquid and gas introduced into the MBG were $Q_L = 30$, 40 and 50 L/min and $Q_G = 0.65$, 1.4 and 3.0 L/min, respectively. Under each condition, microbubbles were generated for 30 min and the time change of dissolved carbon dioxide concentration in the liquid was measured with CO_2 sensor. By substituting the time change into the following equation, the volumetric mass transfer coefficient, $K_L a$ (liquid side mass transfer coefficient $K_L \times$ gas–liquid interfacial area concentration a) was obtained.

$$\frac{dDCO_2}{dt} = K_L a(DCO_{2S} - DCO_2) \tag{6.1}$$

Here, DCO_2 is the carbon dioxide concentration in the liquid, DCO_{2S} is the saturation concentration, and t is the aeration time. Images of bubbles generated in each test liquid were taken with a high-speed video camera. From the images, the diameters d_{Bi} ($i = 1$ to N) of N bubbles were measured with an image analysis [8]. After that, the arithmetic mean diameter d_{Ba} and the Sauter mean diameter d_{BS} were calculated from the measured values. In addition, the void fraction, α, was determined to measure the rise in liquid level during aeration [8]. And then, the interfacial area concentration, a [m^2 / m^3] ($= 6\,\alpha/d_{BS}$), was calculated from the calculated d_{BS} and α.

6.3 Results and Discussion

Figure 6.3 shows aeration images inside the tank for water and salt water case. Lower photos are enlarged images of bubble. Images in tank becomes more cloudy for salt water than tap water, because the bubble size is smaller for salt water than tap water.

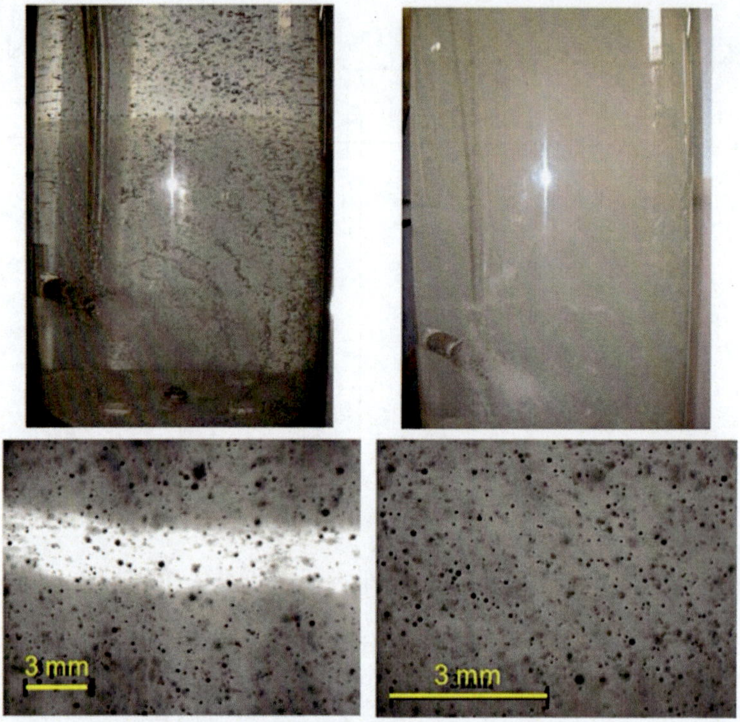

(a) Tap water (b) Seawater

Fig. 6.3 Photos of bubbles in each test liquid ($Q_G = 1.4$ L/min and $Q_L = 30$ L/min)

Figure 6.4 shows the generated bubbles diameters when the carbon dioxide flow rate Q_G is changed under the constant liquid flow rate Q_L; Fig. (a) The arithmetic mean diameter d_{Ba} and Fig. (b) The Sauter mean diameter d_{BS}. In the case of tap water, it can be seen that the d_{BS} value increases especially when Q_G increases. This is because the amount of gas sent into the MBG has increased and the generated bubbles have become easier to merge. In addition, bubble diameter becomes smaller with increasing of Q_L at a fixed Q_G, because the bubbles are broken by high shear and turbulence with high liquid flow. In the case of salt water, even if Q_G is increased, there is no significant change in the d_{BS}, which is less than half that of tap water. The reduction in bubble size with salt addition in water was also reported by Walker et al. [9] and Kawahara et al. [8]. The surface tension of the present salt water ($\sigma = 0.074$ N/m at $T_L = 20$ °C) is similar to that of tap water ($\sigma = 0.072$ N/m), but the bubbles are much smaller. This is thought to be because salt substances and/or ions contained in the salt water adhered to the surface of the bubbles, and coalescence of the bubbles was suppressed.

Figure 6.5 shows the volumetric mass transfer coefficient, K_La with gas volume flow rate, Q_G. It can be seen that under constant liquid flow rate, Q_L, K_La increases with Q_G. When Q_G is constant, K_La increases with Q_L. The cause of the increase of

Fig. 6.4 Variations in bubble diameter with gas volume flow rate

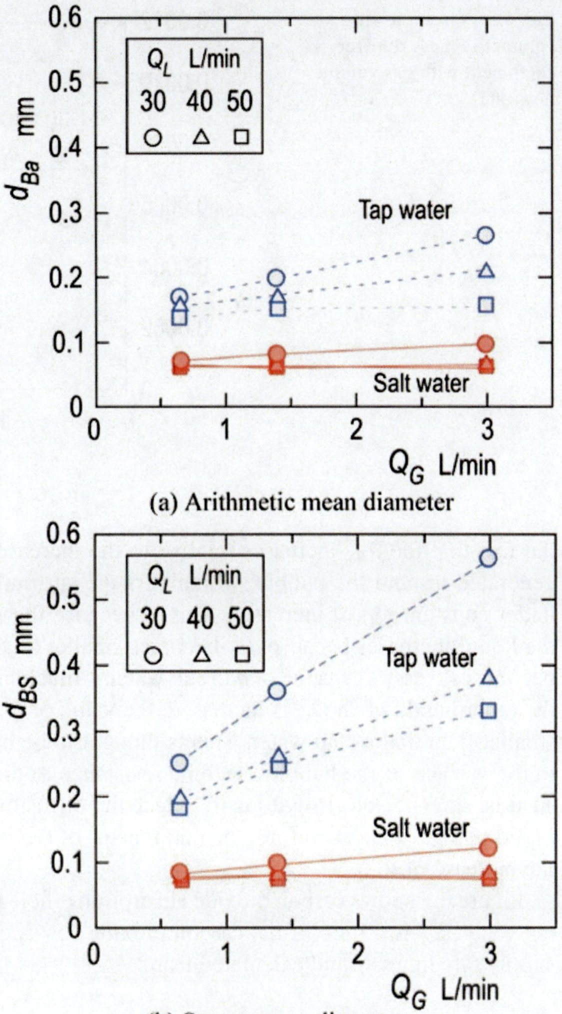

K_La due to the increase of Q_G or Q_L is that the interfacial area concentration a might increase with the increase of gas–liquid flow rates. In addition, K_La of salt water is larger than that of tap water.

Next, K_La is considered by distinguishing between gas–liquid interfacial area concentration a and liquid side mass transfer coefficient, K_L. Figure 6.6a and b show a and K_L, respectively, obtained by experiments. Here, a $(= 6\ \alpha/\ d_{BS})$ was obtained from the measured void fraction α and the Sauter mean diameter d_{BS}. On the other side, K_L was calculated from the measured value of K_La based on the obtained a. For tap water, a increases slightly when Q_G increases under constant Q_L. On the other hand, K_L has increased greatly. This increase in K_L is thought to be due to

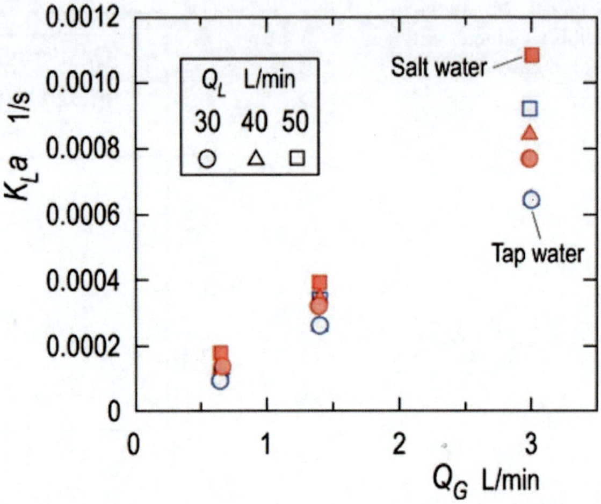

Fig. 6.5 Variations in volumetric mass transfer coefficient with gas volume flow rate

the fact that the d_{BS} increases relative to the increase in Q_G, and a large vortex is generated around the bubble, which stirs the surrounding liquid. As Q_L increases under constant Q_G, a increases. This is because the number of bubbles staying in the liquid increases because d_{BS} becomes smaller and the rising velocity of bubbles becomes slower. The value of a in salt water is much larger than that of tap water. On the other hand, when Q_G is increased, the value of K_L hardly changes and is much smaller than that of tap water. This is thought to be because salt substances adhere to the surface of the bubbles become resistance to diffusion of the substance [10]. That is, salt (i.e., electrolyte) is to reduce the diffusion coefficients of gas phases in liquid phases [11]. Therefore, the main factor of the increase of $K_L a$ in salt water is the increase of a.

Figure 6.7 shows carbon dioxide absorption efficiency, E_A, data for salt water and tap water. E_A is a ratio of the carbon dioxide absorption rate to the carbon dioxide supply rate in the liquid calculated from:

$$E_A = \frac{K_L a V_L DCO_{2S}}{W_{CO2}} \times 100\% \tag{6.2}$$

Here, W_{CO2} is the mass flow rate of carbon dioxide supplied as bubbles in the liquid and V_L is the volume of the liquid filled in the test tank.

E_A increases with increase in gas volume flow rate and/or liquid one for both salt water and tap water, since interfacial area concentration becomes larger with gas and liquid volume flow rate. Regarding the effects of salinity, E_A is larger for salt water than tap water. The larger E_A in salt water is caused by both the longer bubble residence time in liquid and the higher interfacial area concentration than that in tap water.

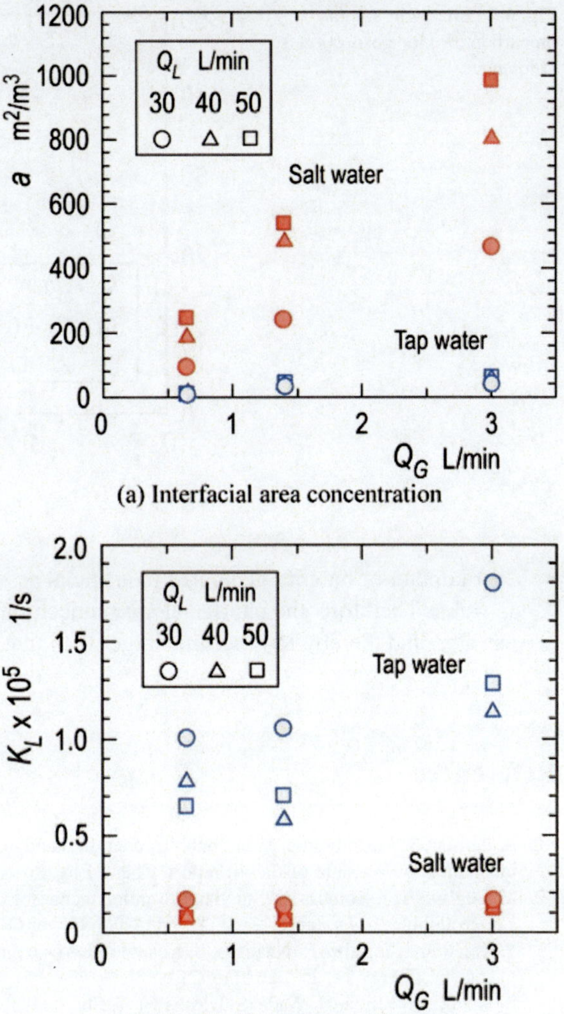

Fig. 6.6 Decomposition of volumetric mass transfer rate

6.4 Conclusions

The following findings were obtained by generating micro-bubbles of carbon dioxide in artificial seawater (salt water) and investigating the dissolution characteristics of the generated bubbles.

- Regardless of the test liquids, as the gas and the liquid flow rates increased, the interfacial area concentration increased, and thereby the volumetric mass transfer coefficient, $K_L a$, increased.

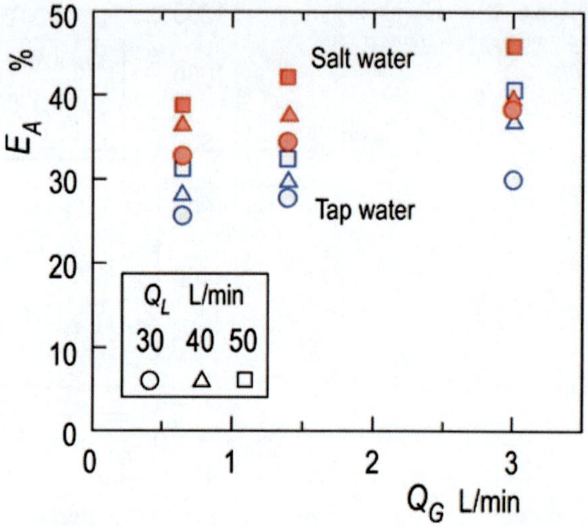

Fig. 6.7 Effects of salinity on carbon dioxide absorption efficiency

- The diameter of bubbles generated in artificial seawater was smaller than that in tap water. Therefore, the interfacial area concentration became larger in artificial seawater, and thereby K_La became larger than that in tap water.

References

1. R. Parmar, S.K. Majumder, Micro bubble generation and micro bubble-aided transport process intensification—a state-of-the-art report. Chem. Eng. Process. **64**, 79–97 (2013)
2. M. Sugimoto, Y. Komura, K. Shioda, Prevention technology of attached marine organism using a microbubble (in Japanese). Proc. Therm. Nucl. Power Gener. Convent., 42–48 (2009)
3. T. Kajishima, T. Saito, R. Nagaosa, S. Kosugi, GLAD: a gas-lift method for CO_2 disposal into the ocean. Energy **22**, 257–262 (1997)
4. S. Suzuki, H. Miida, H. Wada, S. Hirokawa, T. Ebi, K. Inaba, Feasibility study on CO_2 micro-bubble storage (CMS). Energy Proc. **37**, 6002–6009 (2013)
5. K. Ying, M.K.H. Al-Mashhadani, J.O. Hanotu, D.J. Gilmour, W.B. Zimmerman, Enhanced mass transfer in microbubble driven airlift bioreactor for microalgal culture. Engineering **5**, 735–743 (2013)
6. M. Sadatomi, A. Kawahara, Fluids mixer and fluids mixing method, Japanese Patent, JP 5103625
7. M. Sadatomi, A. Kawahara, H. Matsuura, S. Shikatani, Micro-bubble generation rate and bubble dissolution rate into water by a simple multi-fluid mixer with orifice and porous tube. Exp. Therm. Fluid Sci. **41**, 23–30 (2012)
8. A. Kawahara, M. Sadatomi, F. Matsuyama, H. Matsuura, M. Tominaga, M. Noguchi, Prediction of micro-bubble dissolution characteristics in water and seawater. Exp. Thermal Fluid Sci. **33**, 883–894 (2009)
9. A.B. Walker, C. Tsouris, D.W. DePaoli, K.T. Klasson, Ozonation of soluble organics in aqueous solutions using microbubbles. Ozone Sci. Eng. **23**, 77–87 (2001)

10. Y. Hori, K. Hayashi, S. Hosokawa, A. Tomiyama, Mass transfer from single carbon-dioxide bubbles in electrolyte aqueous solutions in vertical pipes. Int. J. Heat Mass Transfer **115**, 663–671 (2017)
11. D.M. Himmelblau, Diffusion of dissolved gases in liquids. Chem. Rev. **64**, 527–545 (1964)

Chapter 7
Numerical Analysis of High-Speed Flow Around a Double-Delta Wing

Yosuke Shiromoto, M. M. A. Alam, and Manabu Takao

Abstract In recent years, supersonic passenger and fighter aircrafts have been receiving an increased interest. These aircrafts usually require thin delta wings with a large leading-edge recession angle. This is because delta wings have an advantage in reducing pressure drag, and that can be obtained without structural difficulties and their small frontal projected area. The flow behavior of a delta wing with a small leading-edge recession angle in low-speed flow has been extensively studied, however, the flow behavior in transonic flow regime has not been well elucidated, so far. In this study, the flow behavior around a double delta wing that was modelled after the Grumman X-47B was investigated by using the computational fluid dynamics (CFD) analysis. 4 (four) different leading-edge shapes, namely 'sharp', 'round', 'triangle' and 'round nosed triangle' were used to investigate the influence of leading-edge shapes on the flow features. Moreover, the mainstream flow velocity was varied from subsonic to transonic to investigate the effect on wing characteristics. From the results, it was found that the round nosed triangle leading-edge shape of the delta wing delays the formation of vortices and the stall, and moreover, it obtained the best lift-to-drag ratio in a wide flow regime.

Y. Shiromoto
Graduate School of Engineering, Osaka Sangyo University, Daito, Osaka, Japan

M. M. A. Alam (✉)
Department of Mechanical Engineering for Transportation, Faculty of Engineering, Osaka Sangyo University, Daito, Osaka, Japan
e-mail: alam@tm.osaka-sandai.ac.jp

M. Takao
National Institute of Technology, Matsue College, Matsue, Shimane, Japan

A. Suryan et al. (eds.), *Proceedings of the 9th Asian Joint Workshop on Thermophysics and Fluid Science, 27–30 November 2022, Utsunomiya, Japan*, Springer Proceedings in Physics 396, https://doi.org/10.1007/978-981-99-9470-0_7

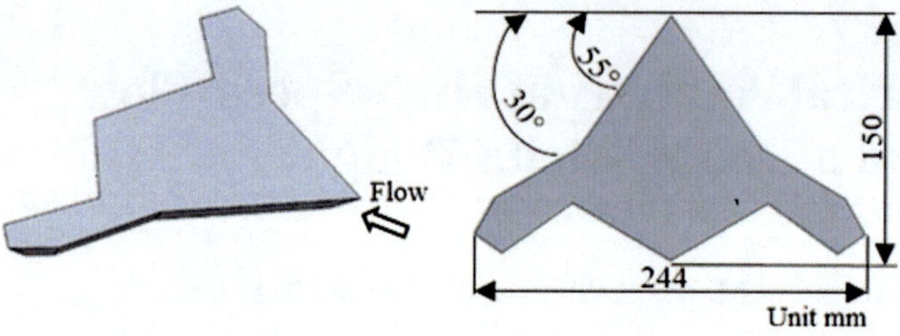

Fig. 7.1 Grumman X-47B double-delta wing

7.1 Intoroduction

At present, the interest of high-speed military and civil aircraft wings is increasing, and these flying wings generally use delta wings or lambda wings. Thus, it is important to understand the flow behavior around these types of wings, and it becomes extremely important in further developing the aerospace industry.

The flow behavior around low leading-edge delta wing at low speed has been extensively studied, however, it is considered that the flow behavior at a high speed has not yet been conducted extensively, so far.

In this study, the Grumman X-47B [1] double-delta wing as shown in Fig. 7.1 was modeled, and the flow around it was investigated by using the computational fluid dynamics (CFD) analysis. The mainstream air velocity and airframe leading edge shape were changed to investigate the flow features around the wing. For analysis, the OpenFOAM [2] was used as the CFD tool.

7.2 Numerical Model and Computational Methods

The Navier–Stokes equations for turbulent, compressible fluid were used to model the high-speed airflow around the delta wing. The *rhoPimpleFoam* [2], a non-stationary solver for laminar or turbulent flow of compressible fluids, which uses the flexible PIMPLE algorithm for time-resolved and pseudo-transient simulations in OpenFOAM was used. Moreover, in this simulation, the standard k-ε model [3, 4] was used as a turbulence model to simulate turbulent stresses.

Figure 7.2 shows the schematics of the computational domain used in this study. As shown, the thickness of the wing is w (10 mm), and the length of the central part of the wing is C (150 mm). As boundary conditions, the no-slip condition was applied to the wing, and the upper and lower ends of the domain and the sides were considered as the atmospheric condition. In this study, 4 (four) different leading-edge shapes, namely 'sharp', 'round', 'triangle' and 'round-nosed triangle', as shown in Fig. 7.3, were used to investigate the influence of leading-edge shapes on the flow features. The mainstream flow Mach number Ma was varied in a range of $0.3 \leq Ma$

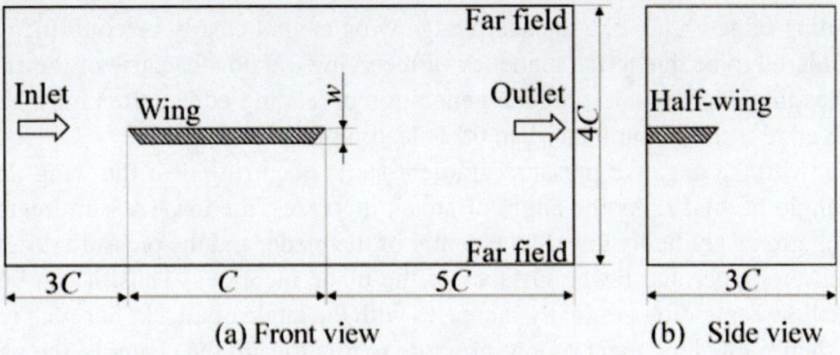

Fig. 7.2 Computational domain and boundary conditions

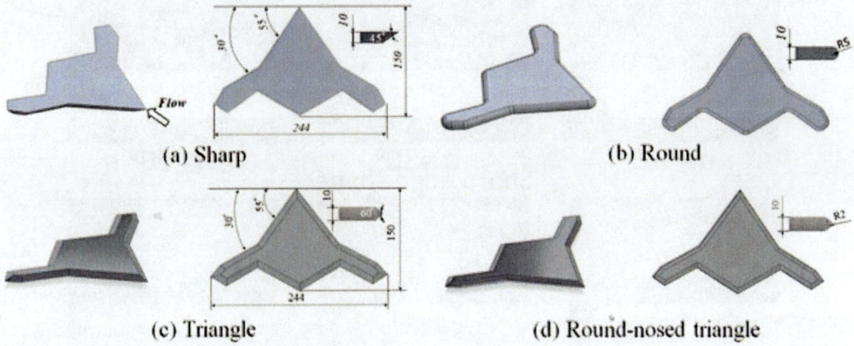

Fig. 7.3 Leading edge shapes of the wing

$\leq$ 1.0. The angle of attack α was changed between 0° and 50°. The SALOME [5], is a free software that provides a generic platform for Pre and Post-Processing for numerical simulation, was used to create mesh for the computational domain.

7.3 Results and Discussion

The flowfield around the delta-wing were visualized and analyzed through the streamlines and pressure distributions. From Fig. 7.4, it was clarified that, for all leading-edge shapes, a slight leading-edge separation vortex is generated near the wing tip when the angle of attack is small. As the angle of attack increases, the size of the leading-edge vortices grows gradually, and moreover, it can be seen that the center of the vortex is moving toward the center of the wing with the increase of angle of attack. Moreover, it was found that a significant difference can be seen in the generation of streamlines on the blade surface with the change of leading-edge shapes. For example, in the case of a round edge wing, the size of the leading-edge vortices rolled up from the leading edge increases as the angle of attack increases, but the center of the leading-edge vortex like the sharp edge wing cannot clearly be confirmed. This is considered to be due to the tendency of receiving the flow because of the smooth round leading edge. In addition, the generation of leading-edge vortex for the sharp leading edge is more dominant than other leading-edge shapes.

Meanwhile, a negative pressure gradient starts occurring near the wing tip at a small angle of attack. As the angle of attack increases, the low-pressure region on the wing grows gradually toward the center of the blade, and the pressure difference between the upper and lower surfaces of the blade increases. Thus, it can be said that the flow separation gradually increases with the angle of attack. For both round-nosed shapes, the increment of low-pressure region toward the center of the wing is

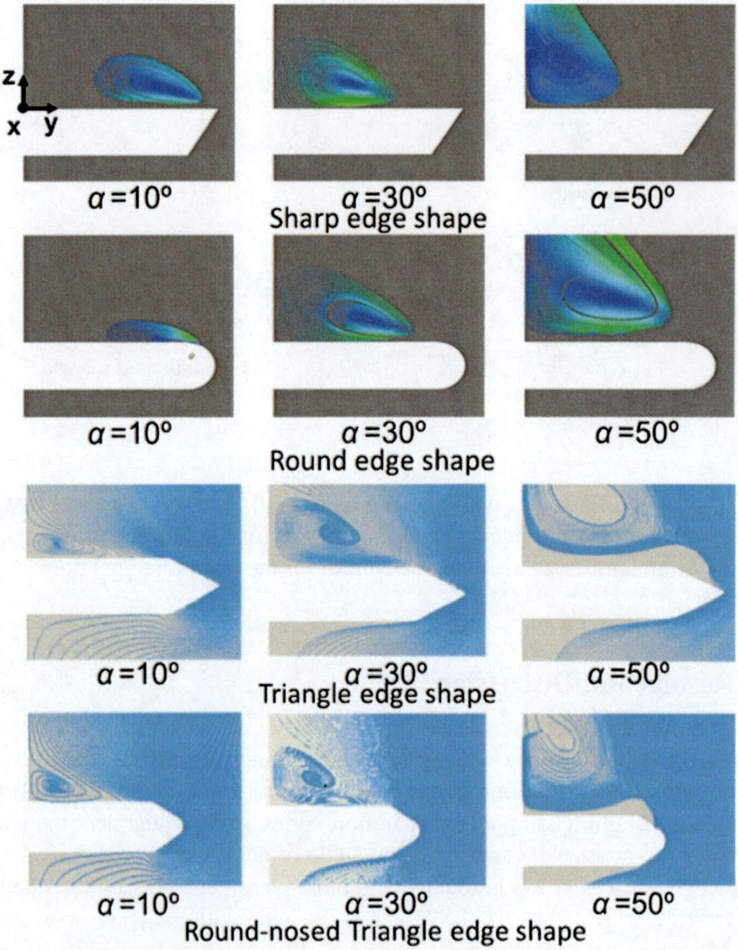

Fig. 7.4 Streamlines on delta wing ($Ma = 0.72$, $x/C = 0.4$)

small. This is considered to be due to the fact that as the angle of attack increases, the low-pressure region becomes wider as the vortex expands.

Moreover, the results for the drag coefficients as shown in Fig. 7.5 suggests that, for all leading-edge shapes, the ratc of increase of the drag coefficient increases as the angle of attack increases, and this tendency was confirmed at all Mach numbers. This can be expected to be due to the increase in the drag due to the increase in the area in contact with the inflowing air when the angle of attack is increased. In addition, it is considered that the leading-edge separation vortices generated on the wing surface also have an influence on the increase in the total drag, therefore, the drag acting on the wing is increased. A comparison shows that the leading edge with round and round-nosed triangle shapes have a lower drag coefficient than other two sharp edge shapes. This is because, in the case of sharp-edged wings, the leading edges are either slanted or triangled, therefore, the downstream air flow rate increases when the incoming air hits. While in the case of round and round-nosed triangle wings, the leading edge is smooth, and it is considered that when the air hits, it disperses in all directions.

In addition, Fig. 7.6 shows a graphical result of the lift-to-drag ratio C_L/C_D against the angle of attack. From the figure, for all leading-edge shapes and Mach numbers, the lift-to-drag ratio approaches the maximum at an angle of attack α of 10°, and for

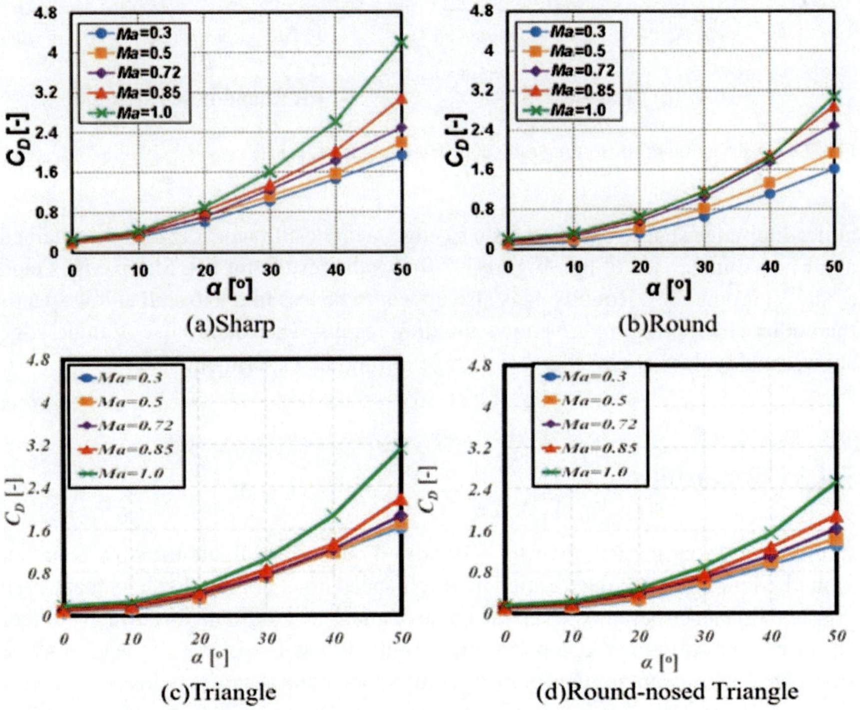

Fig. 7.5 Distribution of drag at different angles of attack

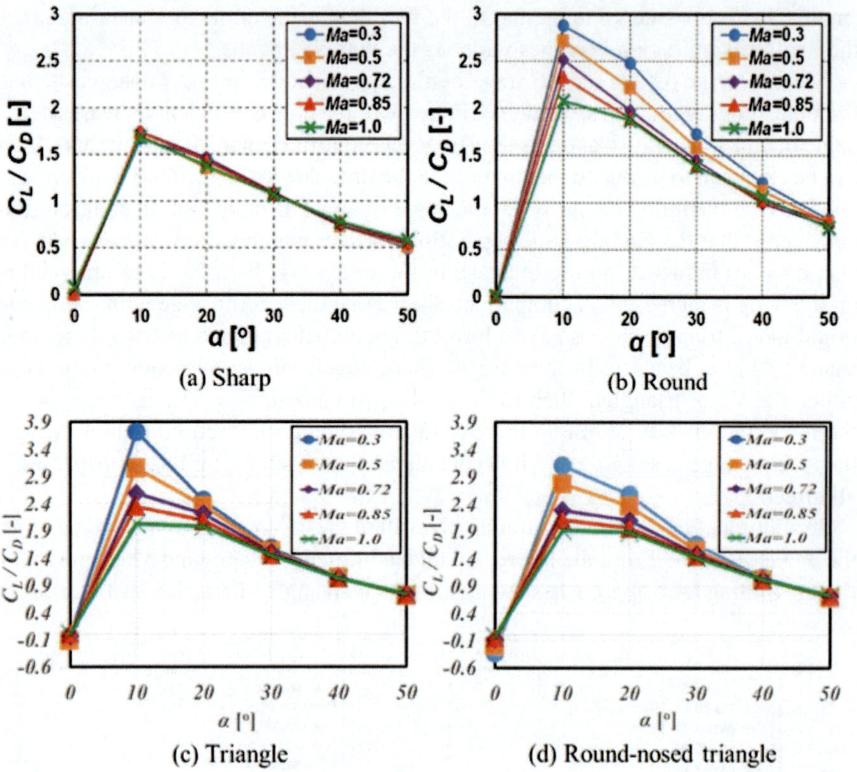

Fig. 7.6 Distribution of lift-to-drag ratio at different angles attack

all leading-edge shapes except the sharp edge, a highest lift-to-drag ratio is obtained at a Mach number *Ma* of 0.3. Figure 7.5 shows the results for lift. Sharp edges tend to stall at around 30°. Round edges also appear to be less likely to stall at low speeds than other shapes. Figure 7.6 shows the drag results. The round-nose triangle edge shows small values at any speed and angle during the experiment.

7.4 Conclusions

In this study, a numerical analysis was carried out to investigate the flow behavior around a double-delta wing traveled in a high-speed flow. The results show that a high drag and lift can be obtained when the angle of attack and the main flow Mach number are increased. Moreover, comparing the results, it was found that the leading-edge shape has a significant influence on the flow velocity and pressure fields on the blade surface, and the sharp edge among four leading edge shapes obtained the lowest lift-to-drag ratio. It was also found that round edges were superior in countering stalls, and round-nose triangles were superior in countering drag.

References

1. M.D. Manshadi, M. Eilbeigi, M.K. Sobhani, M.B. Zadeh, M.A. Vaziry, Experimental study of flow field distribution over a generic cranked double delta wing. Chinese J. Aeronautics **29**(5), 1196–1204 (2016)
2. OpenFOAM, *User guide*, The OpenFOAM Foundation Ltd., UK (2021)
3. W.P. Jones, B.E. Launder, The prediction of laminarization with a two-equation model of turbulence. Int. J. Heat Mass Trans. **15**, 301–314 (1972)
4. B.E. Launder, B.I. Sharma, Application of the energy dissipation model of turbulence to the calculation of flow near a spinning disc. Let. Heat Mass Trans. **1**(2), 131–138 (1974)
5. SALOME (2020). http://www.salome-platform.org/

Chapter 8
Investigation of Cavitation Evolution Through a Butterfly Valve Model at Different Valve Opening Degrees

Haodian Zhang, Zeyong Wu, Kai Wang, Heuy Dong Kim, and Guang Zhang

Abstract As the key control equipment for the transmission of fluid medium, processing valves are widely used in the transmission systems of fluid medium in energy, chemical industry, metallurgy and other fields, which play important roles in the stability and reliability of the system operation. When the flow cross-section is operated in a sudden change, the pressure decreases rapidly at the downstream, which leads to the cavitation in the processing valves. Cavitation makes serious erosion and damage on the valve core and pipeline surface, which leads to the leakage and noise problems in processing valves. This seriously affects the regulation performance and lifetime of processing valves. In this article, numerical simulations were carried out to investigate the transient cavitation in a model butterfly valve. By considering effects of local pressure on formation of cavitation, a modified model for calculating the diameter of cavitation bubbles was derived. Effects of valve opening degree was investigated on the dynamic evolution of cavitation by analyzing formation, development and collapsion of cavitation. The generation, development and collapsion of single cavitation bubble was obtained and discussed in details to state the interaction between vortices and cavitation. Attached and quasi-periodic cavitation were observed and analyzed in detail as well at different valve opening degrees.

8.1 Introduction

Butterfly valve as a regulatory component is always used to achieve the regulation of flow, control the process of medium transmission of pipeline system [1, 2], which is widely applied in aerospace, energy, petrochemical, biomedical and many other

H. Zhang · Z. Wu · K. Wang · G. Zhang (✉)
National-Provincial Joint Engineering Laboratory for Fluid Transmission System, Zhejiang Sci-Tech University, Hangzhou 310018, China
e-mail: zhang1134@zstu.edu.cn

H. D. Kim
Department of Mechanical Engineering, Andong National University, Andong, Korea

A. Suryan et al. (eds.), *Proceedings of the 9th Asian Joint Workshop on Thermophysics and Fluid Science, 27–30 November 2022, Utsunomiya, Japan*, Springer Proceedings in Physics 396, https://doi.org/10.1007/978-981-99-9470-0_8

fields. When designing and using the butterfly valve, the problem caused by cavitation inside it is an important factor to be considered. Cavitation phenomenon could not only cause structural damage of fluid machinery parts, but also leads to hydraulic resonance and noise, giving rise to damage and loss in a wider range. Although many scholars have carried out in-depth research on cavitation phenomenon [3–5], the typical geometry which can generate cavitation, such as nozzles, venturi tubes and hydrofoils, has been thoroughly studied, there is little research on cavitation in butterfly valves. The mechanism and evolution of cavitation inside the butterfly valve is not clear.

In the present studies, numerical simulations are carried out to investigate the cavitation characteristics through a butterfly valve under different regulating conditions. The influence of different inlet flow rates and different opening degrees are explored on the cavitation evolution by using Large Eddy Simulation(LES) and modified Schnerr-Sauer cavitation model [6]. The evolution process of cavitation is analyzed, the relationship between combination coefficient and vapor volume fraction is explored. The influence of cavitation on flow field is reflected by disturbance coefficient, and the correlation between cavitation and vortex is discussed.

8.2 Computation Model and Numerical Validation

8.2.1 Computation Domain and Boundary Conditions

A butterfly valve model is investigated as shown in Fig. 8.1. In order to make sure the full development of the flow, the length of flow zones at the front of the butterfly valve is set as 5 L, and the length of flow zones at the downstream of the butterfly valve is 10 L. In order to ensure the accuracy of numerical calculation, structural mesh is created in the full domain as shown is Fig. 8.2. In this paper, the inlet condition is set according to the inherent flow characteristic curve of butterfly valves. The inherent adjustable ratio of butterfly valve is $R = 30$.

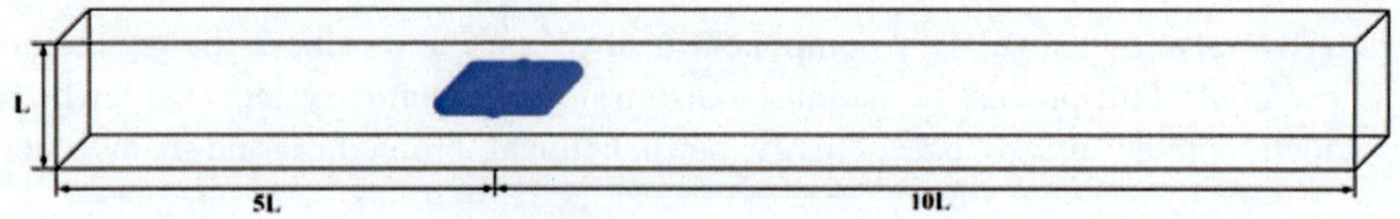

Fig. 8.1 Butterfly valve calculation domain

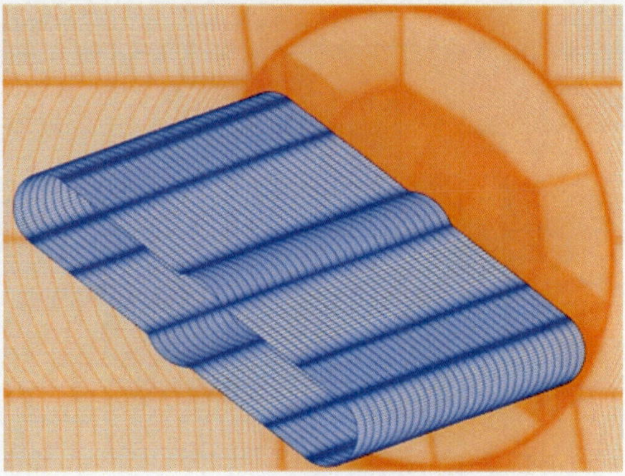

Fig. 8.2 Partial mesh diagram when the opening degree is 60%

8.2.2 Numerical Methods

Since the interphase heat transfer and the compressibility of the fluid are not considered in the solution process, the governing equation mainly includes continuity equation and momentum conservation equation. Large eddy simulation (LES) is adopted as turbulence model in this paper due to the good ability to predict turbulent vortex at different scales. Schnerr-Sauer cavitation model is used for the cavitation model and it considers the vapor–liquid two-phase mixture as a mixture containing a large number of spherical vapor bubbles, and directly calculates the volume fraction of the phase from the net vapor–liquid mass transfer rate to obtain the evaporation source term and condensation source term.

8.2.3 Validation

Five sets of grids of 2.65 million, 3.86 million, 4.66 million, 5.16 million and 5.52 million are created to verify the independence of computational grids as shown in Fig. 8.3. By comparing flow and resistance coefficients of the butterfly valve, with the number of grids increasing, flow and resistance coefficients gradually tend to be stable. When the number of grids reaches 5.16 million, flow and resistance coefficients have already reached a stable value, which is selected for numerical simulations in this paper.

In order to ensure the accuracy of numerical simulation, it is necessary to verify the numerical methods used for present simulations. A three-dimensional Clark-Y hydrofoil is numerically simulated for cavitation flow by using present cavitation model and numerical results are compared with experimental data as shown in Figs. 8.4 and 8.5 [7]. From the quantitative analysis or qualitative analysis, the error between numerical and experimental results is small and acceptable, so it can be considered

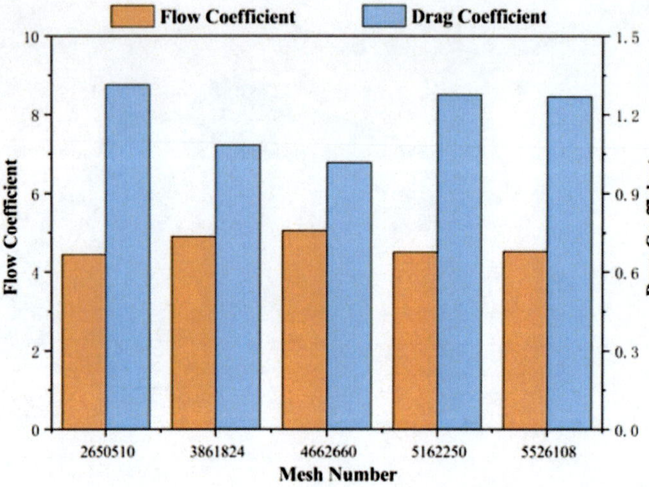

Fig. 8.3 Flow and resistance coefficients at different grid numbers

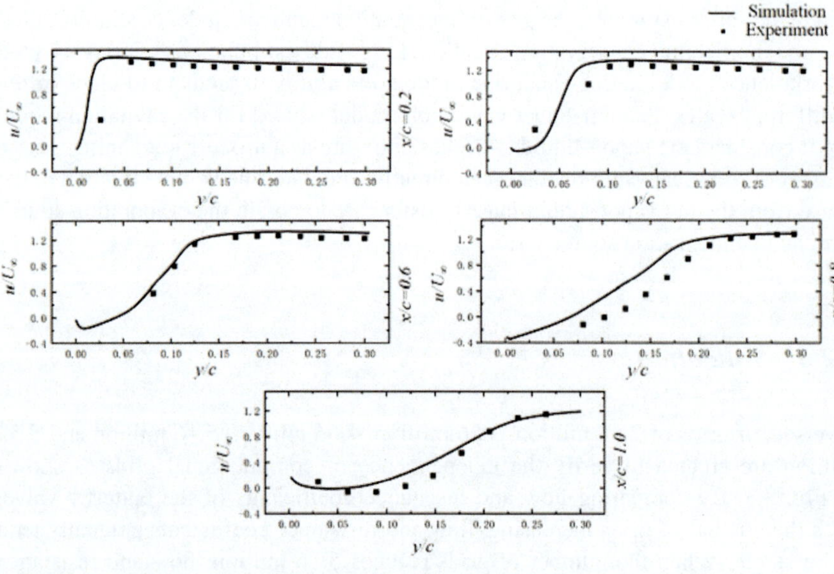

Fig. 8.4 Comparison between numerical and experimental results

that present numerical methods are able to precisely predict cavitation flows through the butterfly valve model.

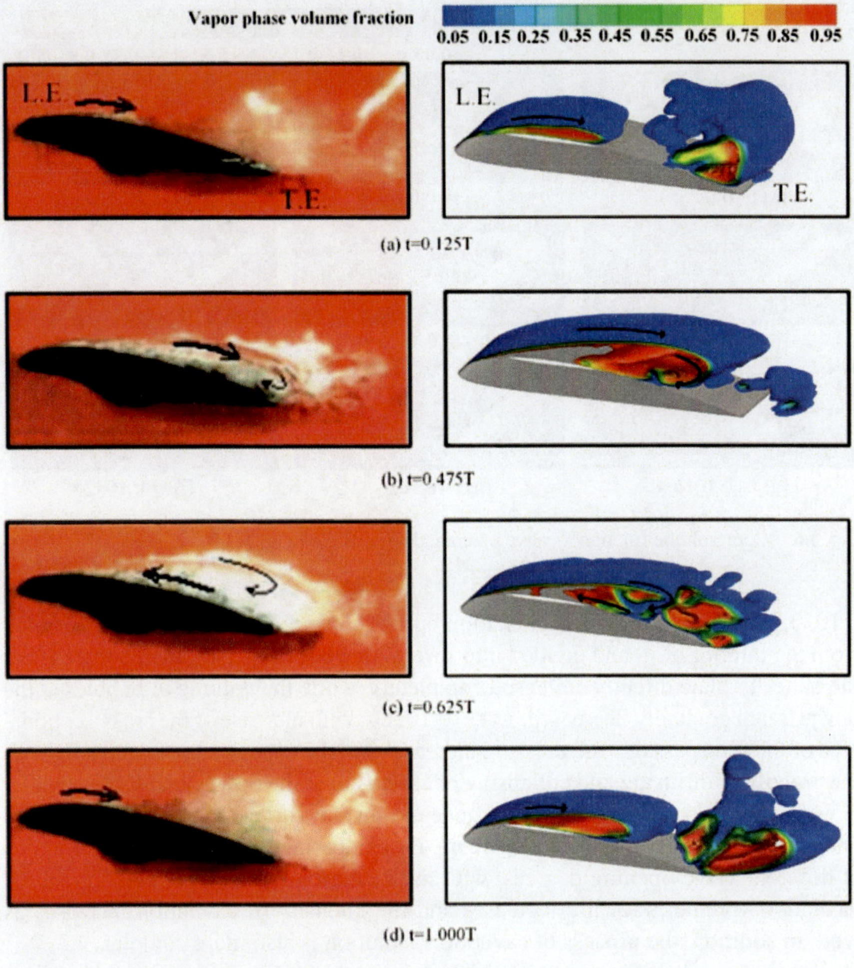

Fig. 8.5 Comparison between numerical and experimental cavitation distributions

8.3 Conclusions and Discussion

8.3.1 Cavitation Distribution

Vapor volume fraction is obtained at valve opening degrees of 60% and 80% as shown in Figs. 8.6, 8.7 and 8.8. It is clearly seen that the evolution of cavitation is obvious periodic. In Fig. 8.6, at t = 0.0992 s, one small bubble is induced at the upper of the butterfly plate. At t = 0.0998 s, a new tiny bubble begins to appear on the upper of the butterfly plate and gradually increases. At t = 0.1004 s, the cavitation in the previous period is gradually squeezed and shrunk, and the cavitation in the current period gradually increases and begins to separate from the upper of the butterfly plate at t =

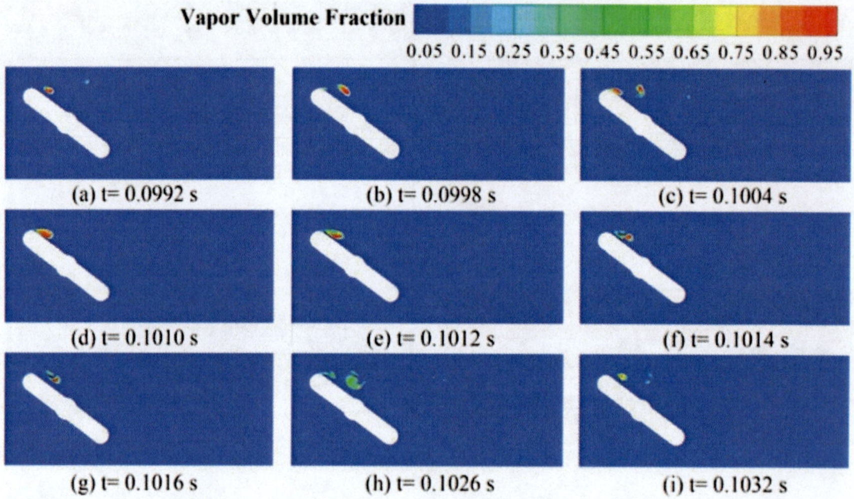

Fig. 8.6 Vapor volume fraction at valve opening degree of 60%

0.1010 s. When t = 0.1012 s, cavitation bubbles start to be fractured and separated from the interior of it, and divided into several tiny bubbles. The small bubbles near the butterfly plate directly collapsed completely, while the volume of bubbles at the downstream gradually decreased. At t = 0.1026 s, with increase of the cross-sectional area of the flow passage, the pressure increases and the vapor volume decreases with the process of diffusing and collapsing of cavitation at the downstream. Meanwhile, new cavitation bubbles continue to generate on the surface of the butterfly plate and gradually expands to the downstream. By comparing the cavitation distributions at different valve opening degrees, with the increase of valve opening degree, the cavitation volume gradually increases and the intensity of cavitation increases as well. In addition, the process of cavitation evolution is also more complex.

8.3.2 *Characteristics of Cavitation Evolution*

The combination coefficient that can reflect the influence of rotation angle and working condition on cavitation is defined as shown in the following equation.

$$\phi = \frac{\sigma}{2(1-\alpha)+\varepsilon} \tag{8.1}$$

where δ represents cavitation number, α represents valve opening, ε is correction factor and $\varepsilon = 0.008$ is fixed in this paper.

Maximum vapor volume fractions and combination coefficients are analyzed at different valve opening degrees as shown in Fig. 8.9. The maximum vapor volume

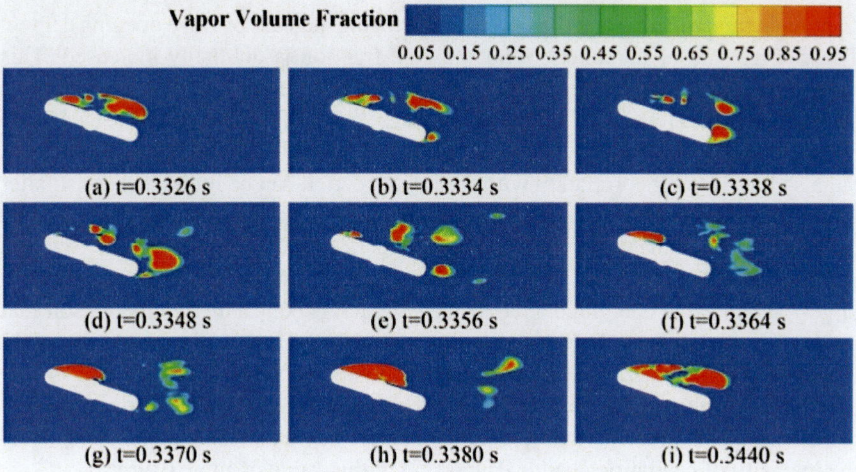

Fig. 8.7 Vapor volume fraction at valve opening degree of 80%

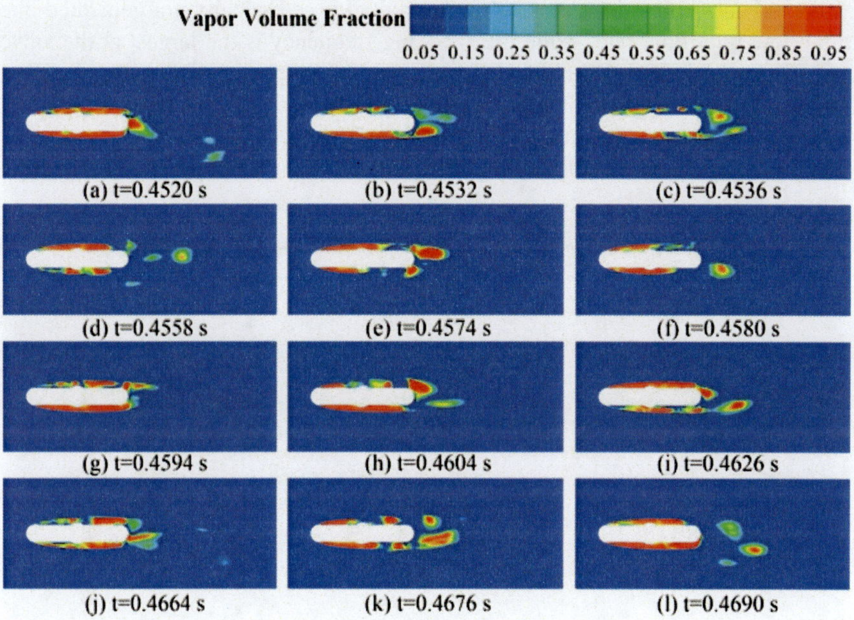

Fig. 8.8 Cavitation development when the opening degree is 100%

fraction increases firstly and then decreases with the increase of valve opening degree. The change rate of maximum vapor volume fraction is gradually increased. This change form is similar to the change trend of the equal percentage flow characteristic curve of the butterfly valve, indicating that the severity of the change of the cavitation phenomenon of the butterfly valve is not only related to the valve opening degree, but also has a strong correlation with the inlet speed. It can be found that the smaller the combination coefficient is, the larger the cavitation volume fraction is. It can also be seen that when the valve opening degree reaches 60%, the vapor volume fraction increases in a significantly faster rate. The maximum cavitation volume reaches the largest at the valve opening degree of 90%, which is contributed to the reason that pressure in the far field does not increase significantly, but the flow rate in far field increases greatly. Therefore, the intensity of cavitation changes increases under these conditions. While the far field pressure has also increased at 100% opening degree, the increase of flow velocity in the far field is not obvious, which inversely leads to the intensity of cavitation change decreasing at this valve opening degree.

The FFT Fourier transform of the vapor volume fraction is calculated to study cavitation shedding frequency at different valve opening degrees as shown in Fig. 8.10. At the valve opening degree of 20%, the cavitation is basically not generated due to the low inlet flow rate, while the shedding frequency is the largest at the valve opening degree is 30%. With the increase of the valve opening degree, the shedding frequency of cavity gradually decreases. At the valve opening degree of 60%, the shedding frequency of cavitation evolution is the lowest. Then it increases at the valve opening degree of 70% and gradually declines at different opening degrees

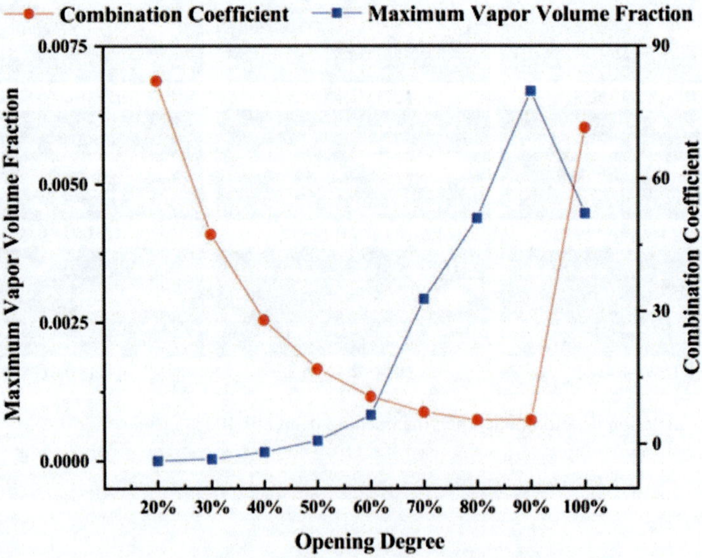

Fig. 8.9 Maximum vapor volume fractions and combination coefficients

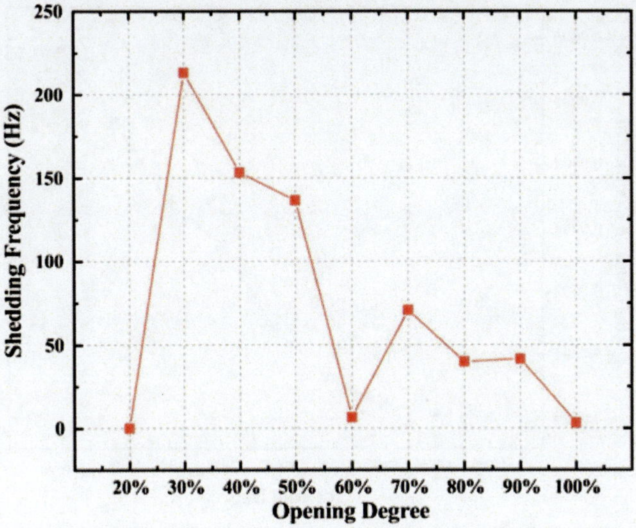

Fig. 8.10 Cavitation shedding frequency at different valve opening degrees

until the shedding frequency of space–time bubbles is the minimum at the valve opening degree of 100%.

The disturbance effect of cavitation evolution at different valve opening degrees is obtained and discussed. The disturbance coefficient ξ is defined to reflect the disturbance degree to the flow field influenced by the cavitation evolution.

$$\xi = \sqrt{\frac{\sum_{i=1}^{n} (V_b^i - \overline{V_b})^2}{n}} \tag{8.2}$$

where, n represents the number of samples needed to calculate the coefficient, V_b^i represents the vapor volume fraction of each sample, and $\overline{V_b}$ represents the average vapor volume fraction of the sample. Through this coefficient, the intensity of cavitation change can be studied and the disturbance degree of the flow field can be obtained.

Figure 8.11 shows the variation of the disturbance coefficient. It can be seen that the disturbance coefficient increases with the increase of the valve opening degree, approximately in a positive proportion. This indicates the flow field becomes more turbulent, which is more greatly affected by the cavitation evolution as the valve opening degree increases.

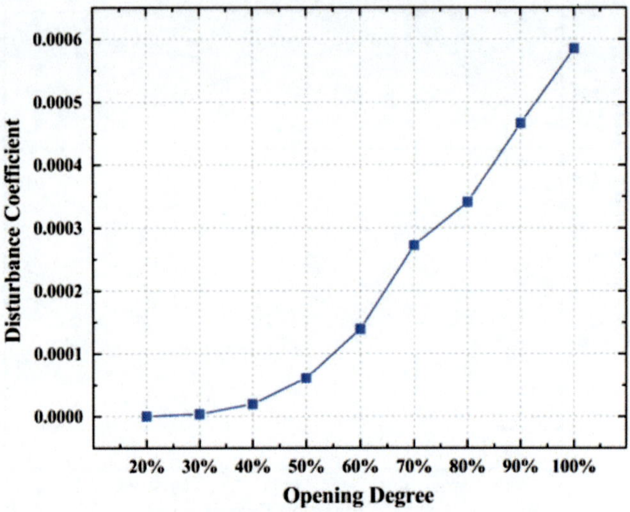

Fig. 8.11 Disturbance coefficients at different valve opening degrees

8.3.3 Correlation Between Cavitation and Vortex

Q criterion is one of the most commonly applied vortex identification methods proposed by Hunt [8]. Q is defined as the residual of the square of the vorticity tensor norm minus the square of the strain rate tensor norm, which can be expressed as:

$$Q = \frac{1}{2}\left(\|\mathbf{B}\|_F^2 - \|\mathbf{A}\|_F^2\right) \tag{8.3}$$

In the above formula, A and B are the symmetric and antisymmetric parts of the velocity gradient tensor respectively.

Cavitation and vortex distributions are obtained at different time under different valve opening degrees as shown in Figs. 8.12 and 8.13. By comparing the structure of cavitation and vortex, it can be seen that there is a good correlation between the position of cavitation and vortex. The shape of the edge position of the cavitation is basically consistent with the position distribution of vortex structures, which shows a certain spatio-temporal synchronization. In Fig. 8.12, at t = 0.0992 s, a part of the cavitation just falls off, and the generated cavitation and fallen cavitation show a certain continuity in location distribution. The cavitation on both sides is still partially attached to the surface of the butterfly plate, and the cavitation in the middle has almost completely fallen off. After that, at t = 0.0998 s, the new generated cavitation at the last moment is gradually grown, but due to the slow generation time of the middle cavitation, the shape of the cavitation is thin in the middle and thick on both sides. At t = 0.1004 s, with the development of time, the generated cavitation periodically starts to be shed and the shed part starts from the edge of one side in the horizontal. The separation of vortex structure was also found when observing vortex structure. At t = 0.1008 s, the cavitation on the other side of the horizontal edge of the butterfly plate also begins to be fractured. With the development of time, the phenomenon of cavitation evolution gradually extends from both sides to the middle of the butterfly plate. At t = 0.1010 s, the cavitation gradually collapses and forms small bubbles at the downstream. At t = 0.1016 s, the cavitation generates in a period eventually fall off completely. The bubbles that fall off keeps in a crescent shape.

By Comparing distributions of cavitation and vortex at valve opening degree of 60% and 80%, it can be seen that the vapor volume fraction is larger and the vortex structure is more complex at the valve opening degree of 80%. The vortex structure basically corresponds to the edge structure of cavitation, which has certain synchronization in time and space development. It indicates that a certain spatio-temporal synchronization exists between cavitation distribution and vortex structure.

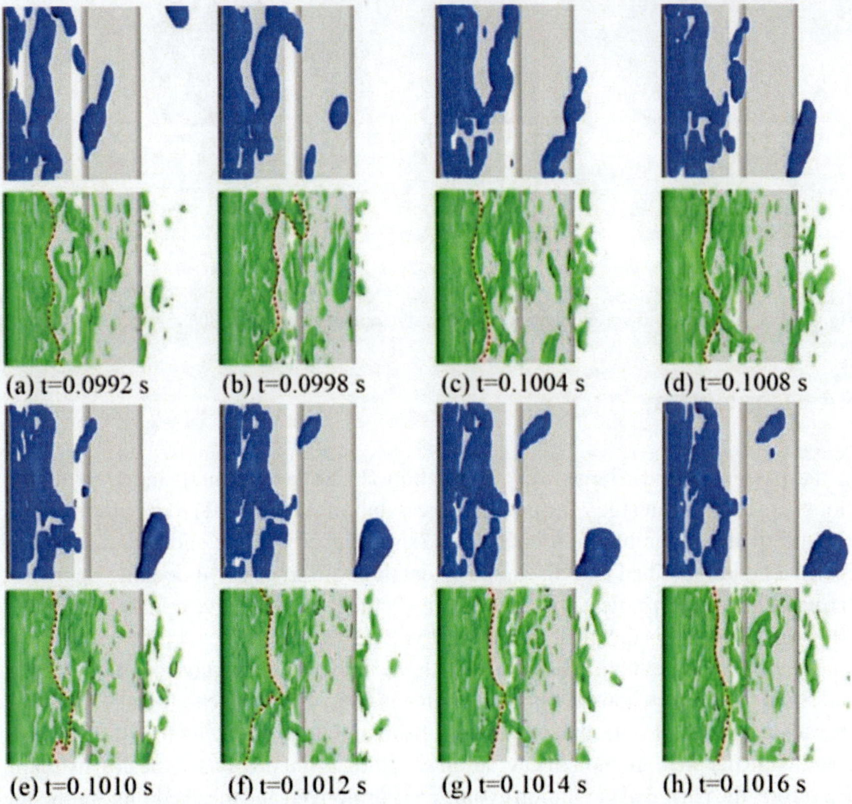

Fig. 8.12 Cavitation and vortex distributions at valve opening degree of 60%

Fig. 8.13 Cavitation and vortex distributions at valve opening degree of 80%

8.4 Conclusion

In the present studies, large eddy simulation (LES) and Schnerr-Sauer cavitation model are carried out study the process of cavitation evolution. The maximum vapor volume fraction, combination coefficient, shedding frequency and the disturbance coefficient are obtained and discussed in detail at different valve opening degree. It is found that although the cavitation volume is the largest at valve opening degree of 90%, the flow field is the most turbulent at valve opening degree of 100%. The shedding frequency of cavitation decreases with the increase of the valve opening degree, and suddenly drops when the opening degree is 60%. This is due to that the shedding characteristics of cavitation have changed from unilateral shedding to bilateral shedding, while the phenomenon of cavitation shedding is not obvious at the downstream. Therefore, the period of cavitation evolution is prolonged and the shedding frequency of cavitation is decreased. It is found that a certain spatio-temporal synchronization exists between cavitation distribution and vortex structure. The evolution process and mechanism of cavitation in butterfly valve are well explored, which provides theoretical basis for the design of anti-cavitation characteristics of butterfly valve.

References

1. J. Tao, Z. Lin, G. Zhang et al., A numerical and experimental study of the time-averaged and transient flow downstream of a butterfly valve. J. Fluids Eng. **144**(5), 051202 (2022)
2. C. Yuan, L. Zhu, S. Liu et al., Numerical study on the cavitating flow through poppet valves concerning the influence of flow instability on cavitation dynamics. J. Mech. Sci. Technol. **36**(2), 761–773 (2022)
3. R. Xue, L. Chen, X. Zhong et al., Unsteady cavitation of liquid nitrogen flow in spray nozzles under fluctuating conditions[J] .Cryogenics **97**, 144–148 (2019)
4. R. Xue, Y. Ruan, X. Liu et al., Numerical study of liquid nitrogen cavitating flow through nozzles of various shapes. Cryogenics **94**, 62–78 (2018)
5. R. Xue, Y. Ruan, X. Liu et al., The influence of cavitation on the flow characteristics of liquid nitrogen through spray nozzles: a CFD study. Cryogenics **86**, 42–56 (2017)
6. G.H. Schnerr, J. Sauer, Physical and numerical modeling of unsteady cavitation dynamics. ICMF International Conference on Multiphase Flow. New Orleans, pp. 1–12 (2001)
7. Y. Long, B. Ji, T. Xing, Verification and validation of large eddy simulation of attached cavitating flow around a Clark-Y hydrofoil. Int. J. Multiph. Flow **115**, 93–107 (2019)
8. J.C.R. Hunt, A.A. Wray, M.P. Eddies, Stream and convergence zones in turbulent flows. Center for Turbulent Research Report CTR-S88, pp. 193–208 (1988)

Chapter 9
Numerical Investigation of Cavitating Flow Around a Hydrofoil Considering Advection–Diffusion and Precipitation of Dissolved Gas in Liquid

Yoshihito Higa, Ryo Iida, Shin-ichi Tsuda, and Satoshi Watanabe

Abstract Hydraulic cavitation is mainly affected by the pressure or velocity field, but the detailed flow structure also depends on other factors such as surface wall roughness or working fluid quality. Among those, the present study has focused on the influence of air content as a non-condensable gas in water, which is one of the factors in water quality. The novel point in this study is that we have applied a simple but physics-based model taking account of the mass transfer of the non-condensable gas based on mass conservation law in the CFD simulations. On the basis of the assumption of a homogeneous bubbly flow, the simulations were conducted by solving five governing equations for cavitation around a hydrofoil of Clark Y-11.7% (angle of attack is 8°) modifying interPhaseChageFoam of OpenFOAM v1706. As a result, the lift coefficient containing air showed higher values than the case without the gas in a certain developed cavitation state showing a periodic change of the hydrofoil lift force, and it may improve the current prediction accuracy of the lift in many CFD simulations. In the case with much air content, an attached cavity around the leading edge due to the influence of the precipitation of air was confirmed even at the time of the minimum lift force, different from the case with little content of air.

Y. Higa · R. Iida
Graduate School of Engineering, Kyushu University, 744 Motooka Nishi-Ku, Fukuka 819-0395, Japan

S. Tsuda (✉) · S. Watanabe
Department of Mechanical Engineering, Kyushu University, 744 Motooka Nishi-Ku, Fukuka 819-0395, Japan
e-mail: tsudashin@mech.kyushu-u.ac.jp

S. Watanabe
e-mail: fmnabe@mech.kyushu-u.ac.jp

A. Suryan et al. (eds.), *Proceedings of the 9th Asian Joint Workshop on Thermophysics and Fluid Science, 27–30 November 2022, Utsunomiya, Japan*, Springer Proceedings in Physics 396, https://doi.org/10.1007/978-981-99-9470-0_9

9.1 Introduction

Cavitation is an important phenomenon in designing hydraulic machines with high-speed rotation. Hence, understanding and predicting cavitation phenomena has been required, and computational fluid dynamics (CFD) is commonly used and becoming a useful tool. Some physical factors affect cavitation such as cavitation number, Reynolds number, surface roughness, and working liquid quality, e.g., the number of bubble nuclei and dissolved gas content. Among those, dissolved gas in water has some influence on cavitation, particularly in industrial working flow conditions without the degassing process. Recently, our group experimentally investigated the influence of the dissolved air in water [1–3] and reported that DO (dissolved oxygen) can change a cavity pressure and the developed state of a cavitating flow. In numerical investigations for hydraulic cavitating flows, a few models have been proposed for cavitating flow considering the effects of dissolved gas [4–6]. However, all the models have assumed a constant concentration or mass of non-condensable gas in the bubbles or employed a coarse model that has not focused on an elementary process of the precipitation or dissolution of the gas in water. Also, most of the previous studies neglected the advection–diffusion of the gas in the liquid phase.

Therefore, in this work, for an improvement of the numerical prediction accuracy of cavitating flow in a more realistic situation, i.e., water with air content, we have developed a physics-based model as an extended model of a well-known homogeneous cavitation model proposed by Schnerr-Sauer (SS model) [7] so that it can properly reflect the mass transfer of air between the liquid water phase and the gas phase as well as advection–diffusion of the dissolved air in water on the basis of an approximation of bubbly flow. The simulations were done for a cavitating flow around the hydrofoil of Clark Y-11.7%, and the effects of the dissolved gas in water were qualitatively investigated.

9.2 Numerical Methods

9.2.1 *Governing Equations with Cavitation Model*

In the present study, the final governing equations were given by the following Eqs. (9.1) to (9.5) based on a homogeneous mixture model considering the mass transfer of non-condensable gas in Eqs. (9.3) to (9.5).

$$\frac{\partial \rho_m}{\partial t} + \frac{\partial \rho_m u_i}{\partial x_i} = 0 \tag{9.1}$$

$$\frac{\partial \rho_m u_i}{\partial t} + \frac{\partial \rho_m u_i u_j}{\partial x_j} - \frac{\partial}{\partial x_j}\left\{\mu_m\left(\frac{\partial u_i}{\partial x_j} + \frac{\partial u_j}{\partial x_i}\right)\right\} = -\frac{\partial p}{\partial x_i} - \gamma_s \kappa \frac{\partial \alpha_l}{\partial x_i} \tag{9.2}$$

$$\frac{\partial \rho_l \alpha_l}{\partial t} + \frac{\partial \rho_l \alpha_l u_i}{\partial x_i} = -(\dot{m}_v + \dot{m}_g) \tag{9.3}$$

$$\frac{\partial \rho_g \alpha_v}{\partial t} + \frac{\partial \rho_g \alpha_v u_i}{\partial x_i} = \dot{m}_g \tag{9.4}$$

$$\frac{\partial c_l \alpha_l}{\partial t} + \frac{\partial c_l \alpha_l u_i}{\partial x_i} - \frac{\partial}{\partial x_i}\left(D\frac{\partial c_l \alpha_l}{\partial x_i}\right) = -\dot{m}_g \tag{9.5}$$

In the above equations, t is the time, u is the velocity and x is the cartesian coordinate, where the subscripts i and j denote the directions of the coordinates. α_l is the liquid phase volume fraction, α_v is the gas phase volume fraction in which both of the vapor and the free non-condensable gas are contained, p is the pressure of the mixture flow field, γ_s is the liquid–vapor surface tension, κ is the curvature of the liquid–vapor surface recognized in each grid scale, ρ_g is the density of the free non-condensable gas inside a bubble, c_l is the concentration of the dissolved non-condensable gas in the liquid phase, D is the diffusion coefficient of the dissolved gas, and R is the bubble radius. ρ_m and μ_m is the homogeneous mixture density and the viscosity, respectively, defined as follows.

$$\rho_m = \rho_l \alpha_l + (\rho_v + \rho_g)\alpha_v \tag{9.6}$$

$$\mu_m = \mu_l \alpha_l + \mu_v \alpha_v + \mu_t \tag{9.7}$$

Here, ρ_l and ρ_v are liquid density and saturation vapor density regarded as constants while ρ_g is variable. μ_t is the turbulent viscosity given by a turbulent model denoted later.

Equation (9.1) is the mass conservation of the homogeneous mixture and Eq. (9.2) is the momentum conservation for the mixture reflecting the surface tension based on the CSF (continuum surface force) model [8] in VOF (volume of fluid) method. Equation (9.3) is the mass conservation of the liquid, where $\dot{m}_v$ and $\dot{m}_g$ represent each source term of the vapor and the non-condensable gas, respectively. In this work, $\dot{m}_v$ was given by Schnerr-Sauer cavitation model, and Eqs. (9.1)–(9.3) were only solved with ignoring $\dot{m}_g$ in Eq. (9.3) if considering vapor cavitation without non-condensable gas content. Equations (9.4) and (9.5) represent the mass conservation of non-condensable gas inside bubbles, and an advection–diffusion equation of the dissolved non-condensable gas in the liquid phase, respectively.

$\dot{m}_v$ and $\dot{m}_g$ respectively represents source terms adding the factor of non-condensable gas using a diffusion flux at a bubble interface $\dot{M}_g$ and the growth rate of bubble radius $\mathrm{D}R/\mathrm{D}t$. In the derivation process of $\dot{m}_v$ and $\dot{m}_g$, we have expressed the material derivative of $\mathrm{D}(\rho_g \alpha_v)/\mathrm{D}t$ as follows referring to the original SS model [7].

$$\frac{\mathrm{D}(\rho_g\alpha_v)}{\mathrm{D}t}=\frac{\mathrm{D}(m_{gb}n)}{\mathrm{D}t}=\frac{\mathrm{D}(m_{gb}n'\alpha_l)}{\mathrm{D}t}=\frac{3\alpha_v}{4\pi R^3}\frac{\mathrm{D}m_{gb}}{\mathrm{D}t}-\frac{\rho_g\alpha_v}{\alpha_l}\frac{\mathrm{D}\alpha_v}{\mathrm{D}t} \tag{9.8}$$

Here, m_{gb} is the mass of the gas inside a single bubble, and n is the number density of bubbles and n' is that defined per unit liquid volume regarded as a constant [7]. Using the relation of $\mathrm{D}(m_{gb})/\mathrm{D}t = 4\pi R^2\dot{M}_g = 3\alpha_v\dot{M}_g/R$, the following expressions on $\dot{m}_v$ and $\dot{m}_g$ have been obtained as the same derivation process of $\dot{m}_v$ in SS model.

$$\dot{m}_v=\frac{3\alpha_v}{\rho_m R}\left\{\rho_l\alpha_l\rho_v\frac{\mathrm{D}R}{\mathrm{D}t}+\alpha_v\left(\rho_g\rho_v\frac{\mathrm{D}R}{\mathrm{D}t}-\rho_v\dot{M}_g\right)\right\} \tag{9.9}$$

$$\dot{m}_g=\frac{3\alpha_v}{\rho_m R}\left\{\rho_l\alpha_l\dot{M}_g+\alpha_v\left(\rho_v\dot{M}_g-\rho_g\rho_v\frac{\mathrm{D}R}{\mathrm{D}t}\right)\right\} \tag{9.10}$$

In the present study, $\dot{M}_g$ was modeled using a nonequilibrium steady solution of the diffusion equation of the gas density around a spherical bubble given by Eq. (9.11).

$$\dot{M}_g=\frac{D}{R}(c_l-c_b)=\frac{D}{R}\left(c_l-\frac{\rho_l}{H/M}p_g\right) \tag{9.11}$$

Here, c_b is the gas density at the liquid side boundary of the bubble surface, and it was given by the gas partial pressure p_g in a bubble using Henry's law, where p_g can be defined from the free gas density ρ_g using the equation of state for the ideal gas. H is the Henry constant between non-condensable gas and water, M is the molar mass ratio of water and non-condensable gas. c_l was given by a solution of the advection–diffusion equation of Eq. (9.5) for the gas in the liquid phase. In fact, the gas diffusion is affected by the flow mixing due to turbulent eddies, and the diffusion coefficient D in this study was calculated assuming the Schmidt number and the turbulent Schmidt number of unity, and it means that D was equal to the kinematic viscosity defined from the sum of molecular and turbulent viscosities. It should be noted that the laminar Schmidt number defined from the molecular properties is about 500, but the influence would be negligible because the flow in a hydraulic cavitation region is turbulent.

$\mathrm{D}R/\mathrm{D}t$ was given by a simplified Rayleigh model for a spherical bubble ignoring the second time derivative of the radius R as follows.

$$\frac{\mathrm{D}R}{\mathrm{D}t}=\mathrm{sign}(p_b-p)\sqrt{\frac{2|p_b-p|}{3\rho_l}} \tag{9.12}$$

Note that the pressure inside the bubble p_b was defined as the sum of the saturation vapor pressure and the partial pressure of the gas, i.e., $p_b = p_v + p_g$.

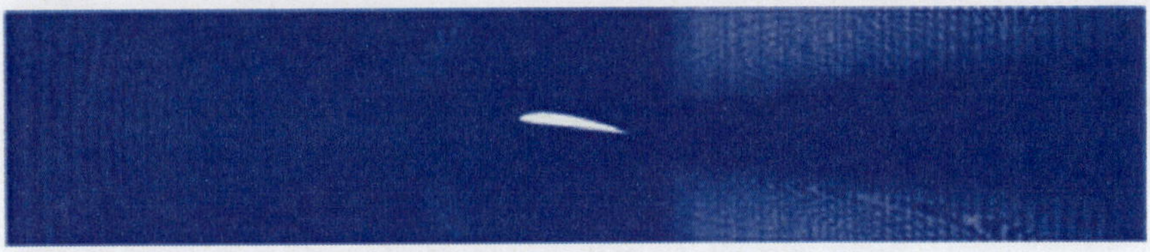

Fig. 9.1 Overview of computational domain

9.2.2 *Computational Setup*

The numerical investigations were conducted for a two-dimensional flow around an isolated Clark Y-11.7% hydrofoil at the angle of attack (AOA) of 8°. OpenFOAM as an open-source software was employed, and interPhaseChangeFoam was applied as the base solver. For the turbulence analysis, an unsteady RANS (URANS) with k-omega SST model was applied for the turbulent closure assuming a fully developed turbulent flow. The computational domain for the two-dimensional simulations is shown in Fig. 9.1, and the chord length was $C = 0.10$ m, the height of domain was $2C$ and the total streamwise length was $10C$. The number of the grid points of the computational domain was approximately 230,000. $U_{in}= 8.1$m/s as the mainstream velocity was imposed at the inlet boundary while the static pressure was imposed at the outlet boundary. The no-slip wall condition on the hydrofoil surface and the slip wall condition on the upper and lower walls were imposed. The time-step was set at 1.0×10^{-6} s and the total computational time was 0.5 s.

9.3 Results and Discussion

In the present study, we applied the following two models, original SS (Schnerr-Sauer) without the non-condensable gas and the extended SS model which considers the mass transfer of the gas with the turbulent diffusion between liquid phase and inside bubbles. In the numerical investigations, the amount of DO in the liquid were set at the following three cases; Low-DO: DO20%, Mid-DO: DO50% and High-DO: DO80% of saturation under atmospheric pressure, which were also given as an inlet boundary condition for c_l. Besides, cavitation number σ was used as a parameter defined as $\sigma = 2(p_{ref} - p_v)/(\rho_l {U_{in}}^2)$, where p_{ref} was a time-averaged reference pressure given at the place $2C$ upstream from the center of the hydrofoil and p_v was the saturation vapor pressure at a room temperature, 293 K.

Figure 9.2 shows the time-averaged lift coefficients against cavitation number. Here, the lift coefficient C_L was defined as $C_L = 2F_L/(\rho_l {U_{in}}^2 C)$, where F_L is the lift force per unit spanwise length. Note that the lift coefficient in DO50% was in advance confirmed to be close to that in DO80%, therefore only the results in SS (DO0%), DO20%, and DO80% are represented in Fig. 9.2.

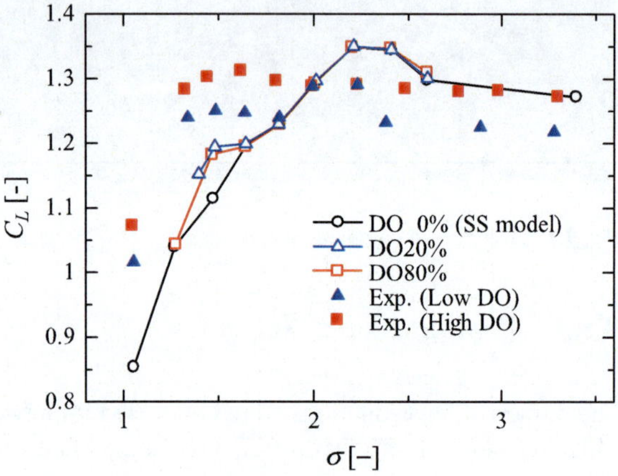

Fig. 9.2 Time-averaged lift coefficient against cavitation number

The computational lift coefficients increase in order of SS, DO80%, DO20% around $\sigma = 1.2$ to $\sigma = 1.6$, which were the conditions under the occurrence of cavitation. In our previous experimental data [3], the higher DO case showed the higher lift also plotted in Fig. 9.2, and the quantitative differences were clearly confirmed, which might be caused not only by the limit of the present gas transfer model but also by the other computational factors such as the homogeneous flow approximation and the space dimension difference. However, according to a previous benchmark test on this hydrofoil with the same AOA, all the simulation results based on the homogeneous approximation underestimated the lift force in the corresponding cavitation numbers [9], but the present homogeneous model reflecting the non-condensable gas influence may improve the prediction accuracy. Below, we focus on the case that σ was between 1.4 and 1.5 where the cavitation developed sufficiently and the difference of the lift coefficient dependent on the air content was clearly visible.

Figures 9.3 and 9.4 show the time fluctuation of the lift coefficient and the cavity structure (void fraction distribution) during one cycle of the lift for DO20% and DO80%, respectively. The numbers from 1 to 5 in the schematic of the lift fluctuation corresponds to those in the void fraction distribution.

In DO20%, the cavity region covering the suction side of the hydrofoil is relatively wide and its void fraction is higher than DO80% at the time of the maximum lift. In addition, at that time, it was confirmed that the sheet cavity length on the suction side was $0.62C$ in DO20% and $0.55C$ in DO80%, respectively. Hence, the suction side was maintained at a lower pressure in DO20%, and it will be a possible reason of the higher lift coefficient. On the other hand, at the time of the minimum lift, an attached cavity is seen on the leading edge in the case of DO80% in Fig. 9.4 but is not in the case of DO20% in Fig. 9.3. The appearance of the attached cavity on the leading edge in DO80% was due to the precipitation of the non-condensable gas in the liquid, and it will be one of the characteristics in such a high DO condition.

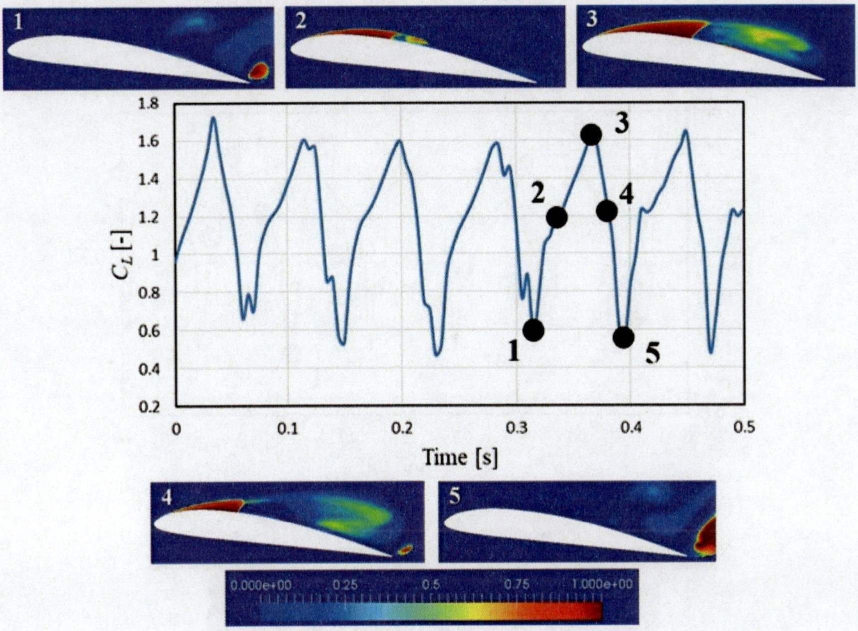

Fig. 9.3 Lift coefficient fluctuation and void fraction during one cycle in DO20% when the cavitation number σ is 1.46

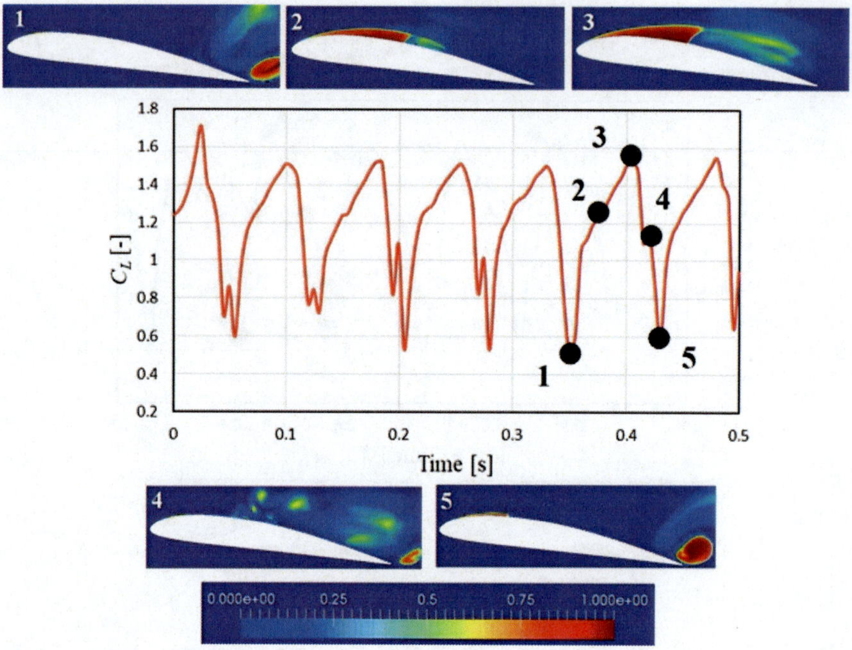

Fig. 9.4 Lift coefficient fluctuation and void fraction during one cycle in DO80% when the cavitation number σ is 1.47

9.4 Conclusion

In the present study, numerical investigations were conducted taking account of the mass transfer of air as a non-condensable gas between liquid water and inside bubbles as well as the advection–diffusion of the dissolved air in the liquid phase. Through the investigation of the effects of the non-condensable gas on a cavitation flow around a hydrofoil of Clark Y-11.7% with the angle of attack of 8 degrees, the following results were obtained.

In the model considering the mass transfer of non-condensable gas based on the mass conservation, some change in the lift coefficients and the cavitation appearance were observed depending on the gas content. In particular, the lift coefficients with the non-condensable gas were generally higher than that of Schnerr-Sauer model without the gas influence, and increases in order of SS, DO80% and DO20% around the cases of cavitation number between $\sigma = 1.6$ and $\sigma = 1.2$. The changes in the lift coefficients and the cavitation appearance mentioned above will be related to the precipitation and the dissolution of non-condensable gas, and the present model may indicate not complete but a partial improvement of the current prediction accuracy of the lift force confirmed in many CFD simulations.

Acknowledgements This work was partially supported by JSPS KAKENHI Grant Number JP20K04269.

References

1. K. Kobayashi, Y. Katayama, S. Watanabe, S. Tsuda, Experimental investigation on cavity pressure inside sheet cavitation. Trans. ASME J. Fluids Eng. **143**(12), 121102 (2021)
2. Y. Matsuura, D. Tsuji, S. Watanabe, S. Tsuda, Effects of dissolved gas on surface pressure of a hydrofoil under cavitation condition. J. Phys. Conf. Ser. **2217**, 012020 (2022)
3. S. Kitamura, Y. Ehara, W. Tsuru, S. Watanabe, S. Tsuda, M. Nohmi, Lift and drag forces around clark Y hydrofoil under cavitating flow at angles of attack with different boundary layer. Turbomacinery **46**(12), 715–722 (2018)
4. A.K. Singhal, M. Athavale, H. Li, Y. Jiang, Mathematical basis and validation of the full cavitation model. Trans. ASME J. Fluids Eng. **124**(3), 617–624 (2002)
5. Y. Iga, Numerical analysis of unsteady cavitating flow byusing a modification based on an assumption of apparent phase equilibrium. IOP Conf. Ser. Earth Environ. Sci. **22**, 052010 (2014)
6. E.H. Malekshah et al., Merging theory-based cavitation model adaptable with non-condensable gas effects in prediction of com-pressible three-phase cavitation flow. Int. J. Heat Mass Transfer **196**(1), 123279 (2022)
7. G.H. Schnerr, J. Sauer, Physical and numerical modeling of unsteady cavitation dynamics. In: Proceedings of 4th international conference on multiphase flow (2001)
8. J.U. Brackbill, D.B. Kothe, C. Zemach, A continuum method for modeling surface tension. J. Comp. Phys. **100**(2), 335–354 (1991)
9. C. Kato, Industry-University Collaborative Project on Numerical Predictions of Cavitating Flows in Hydraulic Machinery: Part 1—Benchmark Test on Cavitating Hydrofoils. In: Proceeding on ASME-JSME-KSME 2011 Joint Fluids Engineering Conference, 445–453 (2011)

Chapter 10
A Brief Review on the Advances of Passive Strut Combustors

Prasanth P. Nair, Abhilash Suryan, and Vinod Narayanan

Abstract Scramjet engines stand as pivotal components of high-speed flight, particularly in demanding hypersonic speeds. Unlike conventional jet engines, scramjets utilize supersonic air for compression and combustion. Achieving effective mixing of fuel and air, is fundamental for optimal combustion. However, it becomes a challenging endeavor due to the limited interaction time. The significance of efficient mixing becomes paramount in scramjet technology, as incomplete combustion due to inadequate mixing undermines engine efficiency. This challenge becomes particularly important in hypersonic conditions where time for interaction is restricted. Consequently, researchers are diligently exploring inventive methodologies to elevate mixing, a crucial factor for increasing combustion efficiency. One of the methods to enhance the mixing is by employing passive struts. Strut-based combustors are a type of intrusive method that enhances mixing and combustion efficiency at the cost of total pressure loss. This paper provides a brief yet comprehensive overview of the ongoing developments in the passive struts used for the scramjet.

P. P. Nair (✉)
Mechanical Engineering Discipline, Indian Institute of Technology Gandhinagar, Gandhinagar 382055, Gujarat, India
e-mail: prasanth.n@iitgn.ac.in

A. Suryan
College of Engineering Trivandrum, Trivandrum 695016, Kerala, India
e-mail: suryan@cet.ac.in

V. Narayanan
Mechanical Engineering Discipline, Indian Institute of Technology Gandhinagar, Gandhinagar 382055, Gujarat, India
e-mail: vinod@iitgn.ac.in

A. Suryan et al. (eds.), *Proceedings of the 9th Asian Joint Workshop on Thermophysics and Fluid Science, 27–30 November 2022, Utsunomiya, Japan*, Springer Proceedings in Physics 396, https://doi.org/10.1007/978-981-99-9470-0_10

10.1 Introduction

Supersonic combustion ramjet engine (SCRAMJET) is a type of air-breathing engine designed to operate at hypersonic speeds, specifically beyond Mach 5. The scramjet finds its application in space launch, hypersonic flight, military, and scientific research. Unlike traditional jet engines, where the inlet to the combustion chamber is subsonic, the incoming air flows into the engine at supersonic speeds. Due to the high-speed nature of the flow, complex flow dynamics, and low retention time, leading to challenges in mixing and combustion of the fuel. Diverse methodologies aimed at augmenting mixing efficiency and ensuring stable combustion within scramjet engines have been thoroughly examined in the works of Seiner et al. [1] and Liu et al. [2]. Amidst the array of techniques investigated, particular emphasis has been placed on the utilization of cavity-based and strut-based configurations for scramjet combustors, captivating the attention of numerous researchers in the field. The plausibility of scramjet in reusable launch vehicle have been discussed by Sam et al. [3]. Extensive investigations, combining both numerical simulations and practical experimentation, have been undertaken concerning cavity-based scramjet combustors [4–10]. In these configurations, the introduction of fuel occurs through a cavity within the combustion chamber's walls, generating a controlled subsonic recirculation zone within the overarching supersonic airflow. Pylon, in combination with a cavity, has also been examined to enhance the mixing performance [11–15].

The present study, however, delves into the intricacies of the strut-based scramjet injector with the passive struts. Within the domain of strut-based injectors [16–24], a distinctive methodology is employed wherein the injection of fuel is orchestrated through strategically positioned struts within the combustor. This strategic placement of struts serves to significantly amplify the intermixing of incoming air and fuel components. The investigation seeks to unravel the manifold complexities associated with enhancing the combustion process and achieving optimal mixing efficiency within hypersonic propulsion systems. Researchers have also studied a combination of cavity-strut configurations to enhance the mixing and combustion [25, 26].

The present paper discusses a configuration featuring three struts. These struts, named "passive" due to the absence of any fuel injection, serve a fundamental role in augmenting the mixing process through their strategic impact on shock distribution, vorticity generation, and the instigation of elevated back-pressure zones. Notably, the struts' presence contributes to the intricate flow dynamics within the system, optimizing the overall mixing efficiency. However, among these struts, a particular one takes on the role of the "active" strut, demarcating it from its passive counterparts. It is within this active strut that the introduction of fuel takes place. This strategic localization of the fuel injection mechanism within the active strut serves as a pivotal aspect of the configuration. By doing so, the researchers aim to exploit the active strut's unique influence on the mixing process, thereby unveiling an intricate interplay between fuel injection, shock dynamics, and induced back pressure.

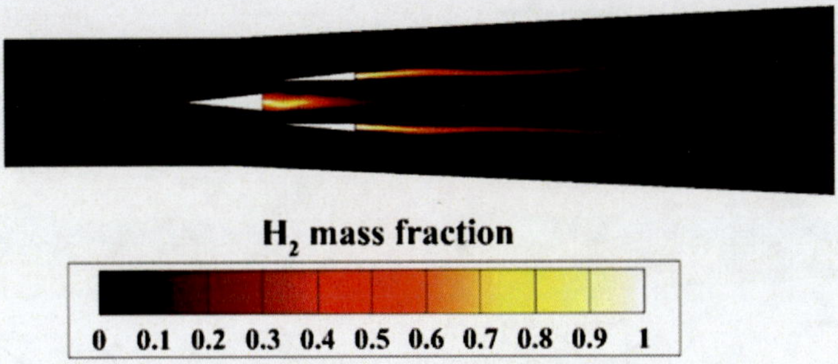

Fig. 10.1 Hydrogen mass fraction contour of reacting flow for three-strut configuration [19]

10.2 Three Strut Configurations

The three-strut configuration for scramjet has been implemented in many previous studies. However, in these studies, the concept of passive struts has not been implemented. The three active strut has the drawback of injecting additional fuel from the two struts in addition to the pressure loss due to the addition of a strut. Figure 10.1 shows the hydrogen mass fraction contour of a three-strut configuration.

It can be observed that injection fuel from all three struts results in significant traces of hydrogen mass fraction downstream of the combustion chamber, signifying inefficient combustion efficiency. An initial evaluation of passive struts was conducted by Nair et al. [19] to compare the combustion efficiency of the three active struts and the combination of one active and two passive struts. They found that in certain cases, the passive strut played a key role in enhancing the combustion performance. It was also noted that the best overall performance was achieved for the configuration having passive struts close to the active strut.

10.3 Shock-Inducing Passive Struts

The configuration of a triple strut having two passive and an active strut can be subdivided into two based on the impacting flow physics on the mixing and combustion. If the shocks are predominant in the flow, the passive struts were named as shock-inducing passive struts, as shown in Fig. 10.2a, b. However, if the mixing and combustion were impacted by the back pressure caused by the passive struts, it was named pressure-inducing passive struts, as shown in Fig. 10.2c, d.

The shock-inducing passive struts enhanced the mixing and combustion efficiency by interacting with the core flow (air/fuel mixture) within and downstream of the passive strut. It was observed in the study [20] that shock-inducing passive struts had marginally performed well in comparison to the single strut configuration. It has less pressure loss in comparison to the pressure-inducing passive struts.

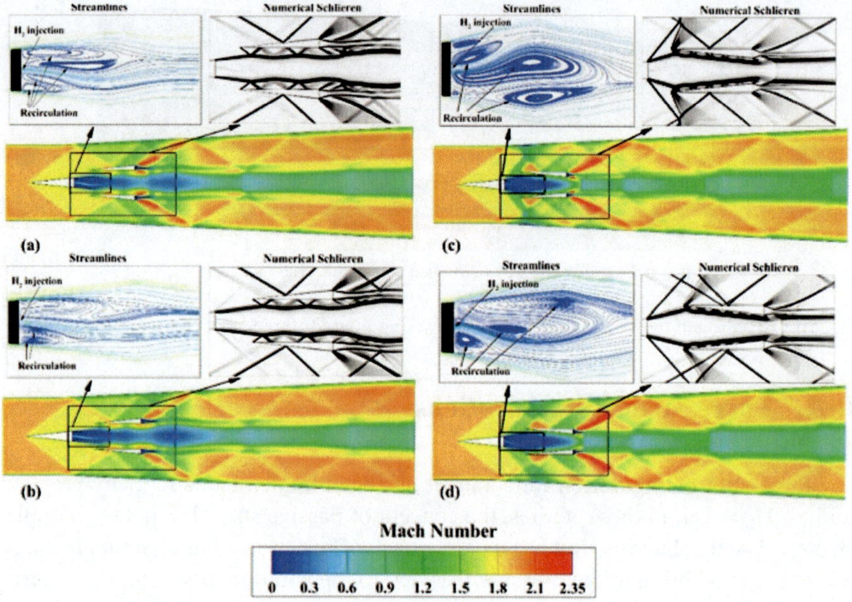

Fig. 10.2 Mach and density gradient contour of reacting flow for shock-inducing and pressure-inducing passive struts [20]

10.4 Pressure-Inducing Passive Struts

The pressure-inducing passive struts had a significant improvement in combustion as well as mixing performance [20]. Illustrated in Fig. 10.3 are the intricate flow dynamics that characterize a multi-strut scramjet combustion chamber, housing a configuration that involves one actively fuel-injecting strut and two passive struts specifically designed to induce back pressure. A non-reacting flow scenario is depicted in Fig. 10.3, and this showcases the complex behavior of gases as they navigate through the combustion chamber.

The supersonic airflow interacts with the active strut, and a sequence of oblique shockwaves manifest at the strut's front tip. These oblique shocks undergo a series of reflections within the confines of the combustion chamber, culminating in the formation of a pattern termed a "shock train." This intricate process is further compounded by secondary shock reflections that arise due to the over-expansion shock, which originates from the base of the passive struts. Notably, these passive struts play a crucial role in generating zones of high pressure within the chamber, promoting the formation of large-scale vortices within the mixing region. The visualization of these vortices is facilitated by the color gradient in Fig. 10.3.

As the core flow continues downstream, an interaction occurs between the shear layer of the flow and the reflected shockwaves. This interaction, aptly termed "shock/shear layer interaction (SSLI)," is characterized by the intricate interplay between

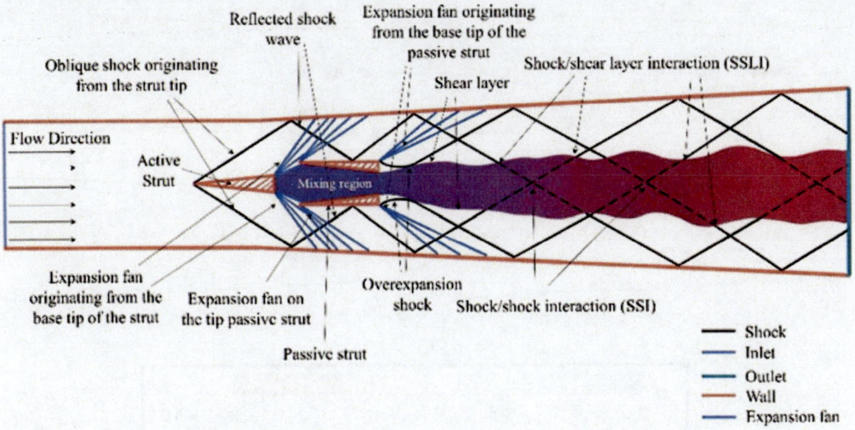

Fig. 10.3 Schematic diagram of the multi-strut scramjet combustion chamber with one active and two pressure-induced passive struts [22]

the shockwaves and the regions of varying flow velocities and directions. The term "shock/shock interaction (SSI)" pertains to the interaction between two distinct shockwaves. It is noteworthy that prior investigations have established that the occurrence of multiple shock interactions and SSLI events serves as a catalyst for enhancing the process of mixing within such configurations.

Thus, the complex interplay of shockwave phenomena, vortical structures, and shear layer interactions within this multi-strut scramjet combustion chamber underscores the profound impact that these intricate dynamics have on the overall mixing efficiency and combustion enhancement. This intricate combination of complex flow and shockwaves sheds light on the sophisticated underlying mechanisms driving the advancement of mixing processes.

In a transient flow scenario, the shock-inducing passive struts create small-scale vortices, and pressure-inducing passive struts create large-scale vortices within and downstream of these struts, as seen in Fig. 10.4. These large-scale vortices are observed to enhance the mixing performance of the scramjet engine [23]. The vortices were observed to be affected by the divergence angle of the passive struts in another study conducted by Nair et al. [22]. It was noted that with the decrease in divergence angle, the large-scale vortices change to small-scale vortices and impact the mixing performance negatively. The total pressure loss also increases with an increase in the divergence angle of the passive strut. The associated acoustics generation was also seen to decrease as the divergence angle of the passive strut decreased.

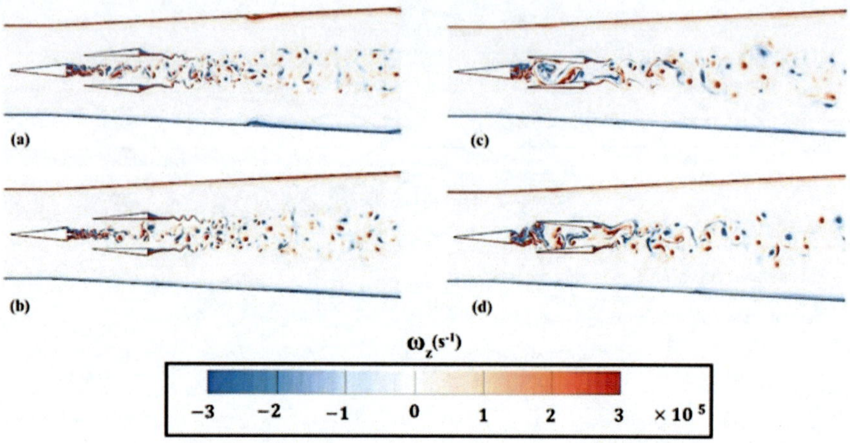

Fig. 10.4 Instantaneous z-vorticity contour for shock-inducing and pressure-inducing passive struts [23]

10.5 Conclusion

A brief review on the passive strut combustors for scramjet has been discussed, exploring the intricate mixing and combustion dynamics. The focus of the study is to elucidate the recent studies and advancements made in the field. Different types of passive strut scramjet combustors have been discussed. Moreover, it is noteworthy that an intriguing trade-off exists between achieving enhanced mixing and incurring pressure losses. This implies that while optimizing the mixing process can yield substantial benefits, it might necessitate accepting a certain level of pressure loss within the system. The underlying principle here is that augmenting the rate of mixing stands to greatly enhance the overall performance of the scramjet engine. This, in turn, holds the potential to reduce the length of the combustor, leading to a simultaneous reduction in skin friction drag. In essence, this research not only delves into the recent advancements within the realm of passive struts in scramjet combustors but also underscores the intricate balance between mixing efficiency, combustion efficiency, shock, and pressure dynamics.

References

1. J.M. Seiner, S.M. Dash, D.C. Kenzakowski, Historical survey on enhanced mixing in scramjet engines. J. Propul. Power **17**, 1273–1286 (2001). https://doi.org/10.2514/2.5876
2. Q. Liu, D. Baccarella, T. Lee, Review of combustion stabilization for hypersonic airbreathing propulsion. Prog. Aerosp. Sci. **119**, 100636 (2020). https://doi.org/10.1016/j.paerosci.2020.100636

3. L. Sam, P.T. Idithsaj, P.P. Nair, A. Suryan, V. Narayanan, Prospects for scramjet engines in reusable launch applications: a review. Int. J. Hydrogen Energy (2023).https://doi.org/10.1016/j.ijhydene.2023.05.341
4. H. Wang, Z. Wang, M. Sun, Experimental study of oscillations in a scramjet combustor with cavity flameholders. Exp. Thermal Fluid Sci. **45**, 259–263 (2013). https://doi.org/10.1016/j.expthermflusci.2012.10.013
5. H. Wang, Z. Wang, M. Sun, N. Qin, Large-Eddy/Reynolds-averaged Navier-Stokes simulation of combustion oscillations in a cavity-based supersonic combustor. Int. J. Hydrogen Energy **38**, 5918–5927 (2013). https://doi.org/10.1016/j.ijhydene.2013.02.100
6. M.-B. Sun, X.-D. Cui, H.-B. Wang, V. Bychkov, Flame flashback in a supersonic combustor fueled by ethylene with cavity flameholder. J. Propul. Power **31**, 976–981 (2015). https://doi.org/10.2514/1.B35580
7. M. Sun, Z. Zhong, J. Liang, H. Wang, Experimental investigation on combustion performance of cavity-strut injection of supercritical kerosene in supersonic model combustor. Acta Astronaut. **127**, 112–119 (2016). https://doi.org/10.1016/j.actaastro.2016.05.035
8. Z. Wang, Z. Cai, M. Sun, H. Wang, Y. Zhang, Large eddy simulation of the flame stabilization process in a scramjet combustor with rearwall-expansion cavity. Int. J. Hydrogen Energy **41**, 19278–19288 (2016). https://doi.org/10.1016/j.ijhydene.2016.09.012
9. Z. Cai, X. Zhu, M. Sun, Z. Wang, Experiments on flame stabilization in a scramjet combustor with a rear-wall-expansion cavity. Int. J. Hydrogen Energy **42**, 26752–26761 (2017). https://doi.org/10.1016/j.ijhydene.2017.09.059
10. F. Li, T. Wang, K. Yang, J. Zhang, H. Wang, M. Sun, Z. Wang, P. Li, Effect of fuel temperature on mixing characteristics of a kerosene jet injected into a cavity-based supersonic combustor. Phys. Fluids **35**, 043307 (2023). https://doi.org/10.1063/5.0145494
11. A.B. Freeborn, P.I. King, M.R. Gruber, Swept-leading-edge pylon effects on a scramjet pylon-cavity flameholder flowfield. J. Propul. Power **25**, 571–582 (2009). https://doi.org/10.2514/1.39546
12. A. Oamjee, R. Sadanandan, Effects of fuel injection angle on mixing performance of scramjet pylon-cavity flameholder. Phys. Fluids **32**, 116108 (2020). https://doi.org/10.1063/5.0026125
13. N. Verma, A. Vaidyanathan, Liquid jet breakup behind a pylon in supersonic flow. Exp. Thermal Fluid Sci. **113**, 109984 (2020). https://doi.org/10.1016/j.expthermflusci.2019.109984
14. P.P. Nair, V. Narayanan, Numerical study on the effect of the pylon downstream of the cavity-based scramjet combustor. Presented at the 8th Thermal and Fluids Engineering Conference (TFEC) (2023). https://doi.org/10.1615/TFEC2023.aer.046452
15. P.P. Nair, V. Narayanan, Effect of injection method on the pylon placed downstream of the cavity, in AIAA AVIATION 2023 Forum. American Institute of Aeronautics and Astronautics (2023). https://doi.org/10.2514/6.2023-4137
16. M. Oevermann, Numerical investigation of turbulent hydrogen combustion in a SCRAMJET using flamelet modeling. Aerosp. Sci. Technol. **4**, 463–480 (2000). https://doi.org/10.1016/S1270-9638(00)01070-1
17. F. Génin, S. Menon, Simulation of turbulent mixing behind a strut injector in supersonic flow. AIAA J. **48**, 526–539 (2010). https://doi.org/10.2514/1.43647
18. A. Mura, J.-F. Izard, Numerical simulation of supersonic nonpremixed turbulent combustion in a scramjet combustor model. J. Propul. Power **26**, 858–868 (2010). https://doi.org/10.2514/1.48074
19. P.P. Nair, S.A., A. Suryan, S. Nizetic, Investigation of flow characteristics in supersonic combustion ramjet combustor toward improvement of combustion efficiency. Int. J. Energy Res. **45**, 231–253 (2021). https://doi.org/10.1002/er.5257
20. P.P. Nair, V. Narayanan, A. Suryan, Combustion efficiency improvement for scramjet combustor with strut based flame stabilizer using passive techniques. Int. J. Hydrogen Energy **46**, 40054–40072 (2021). https://doi.org/10.1016/j.ijhydene.2021.09.224
21. P.P. Nair, V. Narayanan, Numerical study on the effect of equivalence ratio in passive strut configuration, in AIAA SCITECH 2023 Forum. American Institute of Aeronautics and Astronautics (2023). https://doi.org/10.2514/6.2023-1645

22. P.P. Nair, A. Suryan, V. Narayanan, Effect of passive strut angle on the vortical structures and mixing characteristics of scramjet combustor. Phys. Fluids **35**, 066111 (2023). https://doi.org/10.1063/5.0151676
23. P.P. Nair, A. Suryan, V. Narayanan, Modal analysis of mixing characteristics in scramjet combustor with passive struts. Int. J. Hydrogen Energy **47**, 34656–34675 (2022). https://doi.org/10.1016/j.ijhydene.2022.08.061
24. A. Singh, S. Mukhopadhyay, Effect of strut profile, angled injection, and blunting of leading edge on mixing of supersonic hydrogen-air streams. Int. J. Hydrogen Energy **48**, 1175–1188 (2023). https://doi.org/10.1016/j.ijhydene.2022.09.240
25. A.C. Rajesh, S. Jeyakumar, K. Jayaraman, M. Karaca, Steady and unsteady flow characteristics of dual cavity in strut injection scramjet combustor. Int. J. Hydrogen Energy **48**, 28174–28186 (2023). https://doi.org/10.1016/j.ijhydene.2023.04.017
26. S.S. Yarasai, D. Ravi, S. Yoganand, T.K. Raj Rajagopal, Numerical investigation on the performance and combustion characteristics of a cavity based scramjet combustor with novel strut injectors. Int. J. Hydrogen Energy **48**, 5681–5695 (2023). https://doi.org/10.1016/j.ijhydene.2022.11.150

Chapter 11
Relation Between Cellular Structure of Underexpanded Radial Jet and Curvature Effect of Slit Nozzle

Tatsuya Ohkawa, Yoko Sakakibara, Masaki Endo, and Hiromasa Suzuki

Abstract A radial underexpanded jet is a supersonic jet issuing from a small slit between two circular tubes which are placed face to face with each other. The small slit comprised of these two tubes is referred to as a 'slit nozzle' in this study. Such flow is a simple model of the wall jet of jets impinging on an object or the flow downstream of pressure reducing valves. In this study, the jet structure and the oscillation phenomena of the radial jet issuing from the slit nozzle was numerically investigated using the symmetric TVD (Total Variation Diminishing) scheme and the frequency of oscillation was analyzed by FFT analysis. Computational results showed that the jet had a shock cell structure. Such structure is well known to be the typical structure of underexpanded rectangular or axisymmetric jet. The most remarkable difference between the radial jet and those jets was the decreasing ratio of the cell size, and the difference becomes conspicuous if the nozzle diameter D is smaller. The oscillation behavior of the radial jet also depends on D because of its curvature effect. The frequency of oscillation decreases gradually as the nozzle diameter increases and suddenly drops at about $D/b = 20$, where b is the width of the slit between two tubes.

11.1 Introduction

When high pressure gas issues from a convergent nozzle, the jet often becomes underexpanded. It is well-known that an underexpanded jet has a cellular structure which consists of the expansion region and the compression region alternately and emits high frequency acoustic wave called 'screech tone'. This jet has been used in various industrial scenes.

T. Ohkawa (✉) · Y. Sakakibara · M. Endo
Tokyo Denki University, Ishizaka, Hiki, Saitama, Hatoyama 350-0394, Japan
e-mail: 21udm01@ms.dendai.ac.jp

H. Suzuki
Tokyo Metropolitan College of Industrial Technology, 1-10-40 Higashi-Ohi, Shinagawa, Tokyo, Japan

A. Suryan et al. (eds.), *Proceedings of the 9th Asian Joint Workshop on Thermophysics and Fluid Science, 27–30 November 2022, Utsunomiya, Japan*, Springer Proceedings in Physics 396, https://doi.org/10.1007/978-981-99-9470-0_11

For example, the underexpanded jet is used as the assist gas in laser processing to remove molten metal (debris) generated by the laser processing [1]. Another example of industrial applications of the jet is the cooling jet for glass tempering process. The effectiveness of using an underexpanded jet for this process was studied by Tamai [2].

In such industrial applications, when the underexpanded jet impinges on a workpiece, the jet spreads radially along the wall surface of workpiece. This wall jet emits acoustic waves, and the acoustic waves interact with the main jet issuing from the nozzle. As a result, the acoustic waves affect the unsteady behavior of the main jet, and the jet strongly oscillates and radiates high frequency sound.

Because such phenomena reduce the processing accuracy and cause fatigue failure of the nozzle in the above applications, many studies have been conducted on the axisymmetric jet [3–5] or on the rectangular jet [6, 7] concerning the oscillation phenomena and radiated sound.

In this study, we focus the radially spreading underexpanded jet like the wall jet, which also appears downstream of the poppet valve and the pressure reducing valve [8, 9]. When these valves are partially open, the high frequency oscillation of valve is derived by the underexpanded radial jet, which causes the fatigue failure of valve.

We investigate the basic characteristics of the underexpanded radial jet using a flow model shown in Fig. 11.1. This nozzle consists of two circular tubes placed face to face each other with a small gap between them. The jet issues radially from this small gap into the atmosphere. The nozzle made by the two tubes is named a 'slit nozzle'. In this study, the jet structure and the oscillation phenomena of the radial jet issuing from the slit nozzle was numerically investigated and the frequency of oscillation were analyzed by FFT analysis.

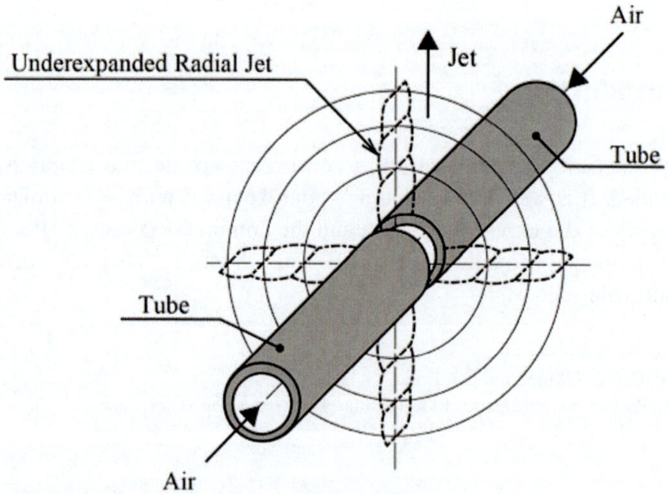

Fig. 11.1 Underexpanded radial jet issuing from slit nozzle

11.2 Computational Model for Underexpanded Radial Jet

Figure 11.2 shows the computational model for underexpanded radial jet issuing from the slit nozzle. The flow field is axially symmetric against the central axis of the tube that is in the direction of x-axis, and symmetric against the central axis of the jet that is in the direction of y-axis. The computational grid is fine near the jet, and gradually becomes coarser as the distance from the nozzle increases. The atmospheric conditions are applied on the top and the right boundaries to avoid undesirable wave reflection from the boundary. The slip wall conditions are applied on the nozzle wall and the symmetric conditions on the left boundary. The small gap between two tubes b is 2 mm, and the nozzle exit is assumed to be the critical conditions.

This flow field is analysed by means of Yee's symmetric TVD scheme for the axisymmetric Euler equations [10]. The ratio of atmospheric pressure p_a to the stagnation pressure in high pressure chamber p_0, that defined as *NPR*, is changed between $NPR = 4.4$ to 4.8. The tube diameter of the slit nozzle is changed in the range from $D/b = 6.0$ to 90.0, and the effects of D/b on the flow field were examined.

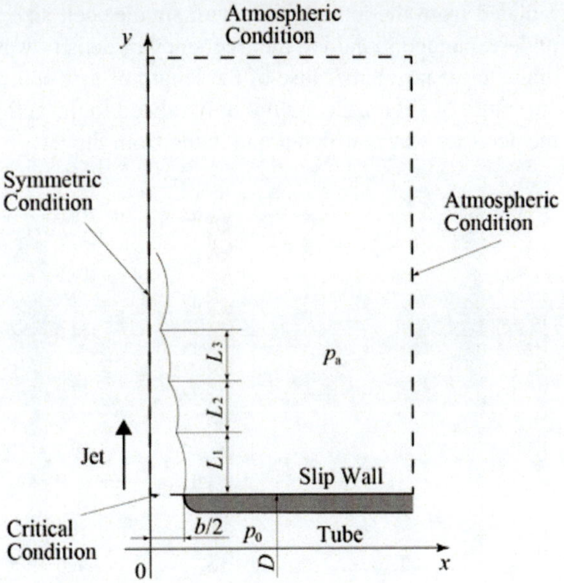

Fig. 11.2 Computational model for underexpanded radial jet

11.3 Results and Discussion

11.3.1 Cellular Structure of Radial Jet

The cellular structure similar to the axisymmetric and the two-dimensional rectangular jet is formed in the underexpanded radial jet. However, the difference from the case of those jets is the decreasing ratio of cell size. Figure 11.3 shows the flow field of the underexpanded radial jet obtained by the computation. The cell size at $D/b = 6.0$ is smaller than those at $D/b = 12.0$, and this characteristic is especially noticeable for cells further downstream. Thus, the smaller the tube diameter, the smaller the decreasing ratio of cell size in underexpanded radial jet.

Furthermore, this cell structure transitions to a vortex region in the downstream and generates periodic density waves around the jet. Focusing on this density waves, the pattern of the radiation of density waves are changed depending on the tube diameter. Specifically, the wavelength of the density wave in Fig. 11.3b is longer than that in Fig. 11.3a.

In case of two-dimensional rectangular jet, the frequency of the acoustic waves radiated from the jet is higher with smaller cell size. Therefore, in the case of the underexpanded radial jet, the frequency of density wave is higher with smaller tube diameter of nozzle because of the length of first cell. This shows that the effects of curvature of slit nozzle are not only related to the cellular structure of jet but also to the acoustic waves which are radiated from the jet.

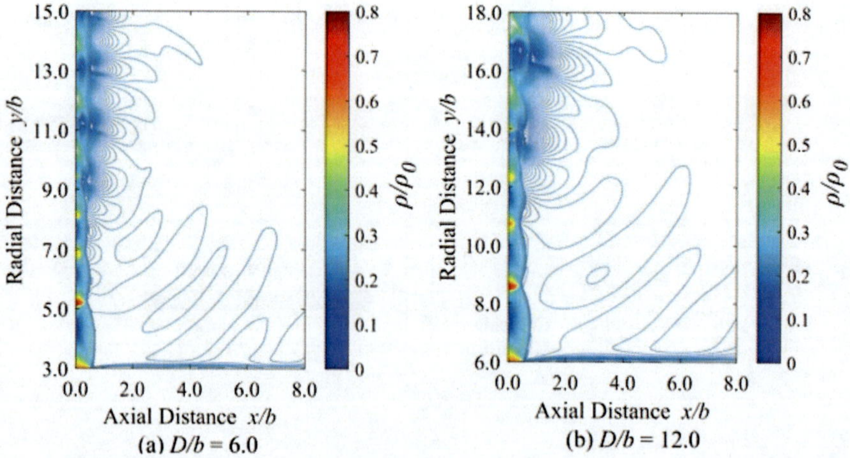

Fig. 11.3 Density contours of flow field

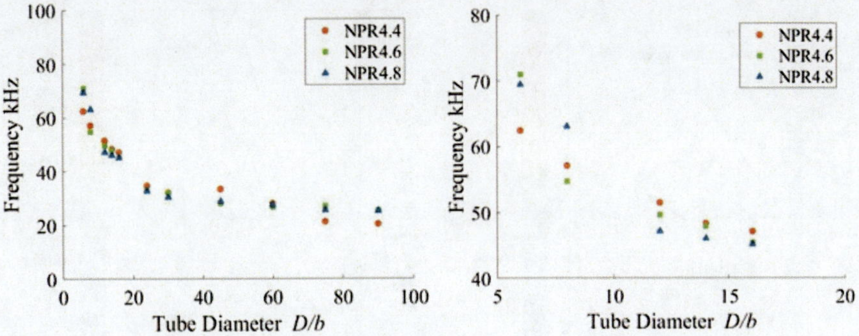

(a) $D/b = 6.0$ to 90.0 (b) $D/b = 6.0$ to 16.0

Fig. 11.4 Oscillation of underexpanded radial jet

11.3.2 The Oscillation of Radial Jet

Figure 11.4a shows the frequency of the oscillation against the tube diameter of slit nozzle and Fig. 11.4b is the enlargement of the range from $D/b = 6.0$ to 16.0 of Fig. 11.4a. In Fig. 11.4a, with increase in the tube diameter, the frequency of the oscillation decreases in the range of $D/b < 16.0$. Between $D/b = 16.0$ and 24.0, it drops sharply and then decreases gradually again. Overall trends indicate that the frequency of oscillation also decreases with increasing pressure ratio. However, the frequency of oscillation is clearly high at $NPR = 4.6$ for $D/b = 6.0$ and at $NPR = 4.8$ for $D/b = 6.0$ to 8.0 as shown in Fig. 11.4b, and the decreasing ratio of frequency is large when D/b is small.

Figure 11.5 shows a rate of change of density $\partial\rho/\partial x$ at $NPR = 4.6$ for $D/b = 6.0$ and 8.0, which is similar to Schlieren visualization. At $D/b = 6.0$, we can see three groups of density waves around the jet. The first group propagates toward the upstream nozzle, the second one propagates with vortices which move downstream, and the third one propagates in the x direction between these two waves. The last density waves clearly have a higher frequency than the others. However, at $D/b = 8.0$, these waves are difficult to distinguish. These density waves are generated due to the interference between the upstream propagating wave and the downstream propagating wave, and its frequency double the others. When these high frequency density waves clearly appear, the frequency of the jet oscillation includes this frequency. Therefore, it is considered that the oscillation behaviors of the jet are affected by the density waves generated around the jet.

In addition, for these waves, the amplitude of the upstream propagating density wave is larger than that of the downstream propagating wave for $D/b = 6.0$. However, for $D/b = 8.0$, this characteristic is opposite, namely the downstream propagating wave is larger than that of the upstream propagating wave. The upstream propagating density wave is a part of the feedback loop of screech, and affects the oscillation of

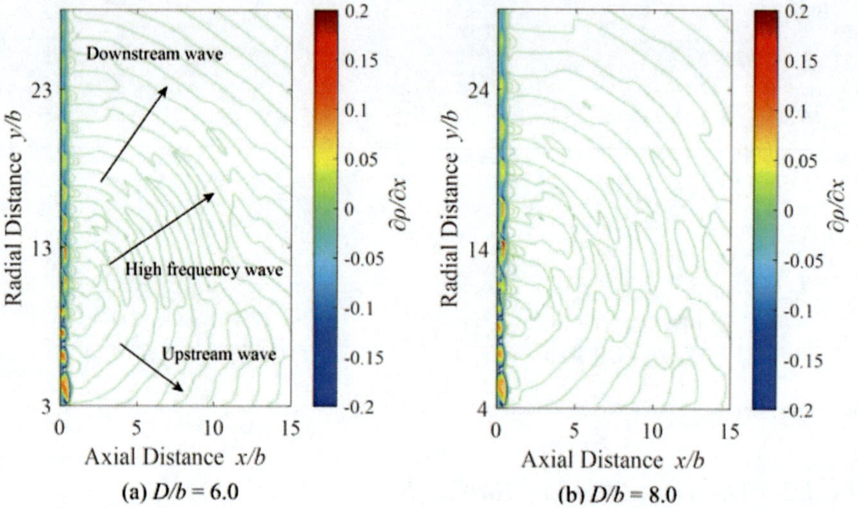

Fig. 11.5 Rate of change of density

the jet. It can be considered that the change in amplitude of the upstream propagating density wave caused the sudden change in the oscillation of the jet structure.

11.4 Conclusion

The underexpanded radial jet issuing from the slit nozzle was numerically simulated, the jet structure and the oscillation behavior was analyzed, and the following results were drawn.

(1) The frequency of oscillation of radial jet is high when *D*/*b* is small.
(2) Three groups of density wave (upstream propagating waves, downstream propagating waves with vortices, and the waves between them) appear around the jet.
(3) The frequency of waves between two waves is almost double that of other waves and, when these waves appear, the frequency of jet oscillation includes this frequency.

References

1. M. Kanaoka, S. Kotou, Study of the relationship between the accuracy in Co_2 laser cutting and the assist gas. Japan Soc. Mech. Eng. **59**, 1952–1959 (1993). ((in Japanese))

2. K. Tamai, S. Aratani, H. Ojima, K. Takayama, Experimental investigation for 2.0mm tempered glass produced by air quenching method. J. Ceramic Soc. Japan **110**, 299–303 (2002). (in Japanese)
3. J. Panda, Shock Oscillation in underexpanded Screeching Jets. J. Fluid Mech. **363**, 173–198 (1998)
4. D. Edgington-Mitchell, Aeroacoustic resonance and self-excitation in screeching and impinging supersonic jets-a review. Int. J. Aeroacoustics **18**, 118–188 (2019)
5. Y. Ching, Loh, Lennart, S. Hultgren, Jet Screech Noise Computation, NASA TM-2003–212626 (2003)
6. C.K.W. Tam, The shock-cell structures and screech tone frequencies of rectangular and non-axisymmetric supersonic jets. J. Sound Vib. **121**, 135–147 (1988)
7. T. Yamagishi, S. Kaji, Numerical simulation on screech tone generated by two-dimensional supersonic jetsProceedings 5th International Congress on Sound and Vibration, 691–698 (1997)
8. M. Nakano, E. Outa, K. Tajima, Aerodynamic study on noise and vibration generated in high pressure gas valves bulletin of the JSME **173**, 691–698 (1979)
9. Y. Duan, A. Revell, J. Sinha, W., Hahn A Computational fluid dynamics (CFD) analysis of fluid excitations on the spindle in a high-pressure valve. Int. J. Pressure Vessels Piping, 175 (2019)
10. H.C. Yee, Upwind and Symmetric Shock Capturing Schemes NASA TM-89464 (1987)

Chapter 12
Investigation of Flow and Fuel/Air Mixing in a Heavy-Duty Gas Turbine DLN Combustor Using Laser Diagnostics and Numerical Simulation

Yuan Feng, Xuesong Li, Xiaodong Ren, Chunwei Gu, Xuan Lv, Shanshan Li, and Ziye Wang

Abstract Cold flow and mixing characteristics inside a 300 MW class F-class heavy-duty DLN combustor are thoroughly explored. Employing PIV, PLIF experiments, and numerical simulations, the research evaluates velocity distribution and mixing within the combustion chamber. The experimental phase introduces innovative optical schemes, enhancing experiment execution. Various turbulence models are assessed for suitability. LES simulations closely replicates experimental features, while RANS models like RNG k-epsilon and Realizable k-epsilon capture the combustion chamber flow characteristics, aiding structure selection. Mixing analysis reveals effective fuel–air premixing within the chamber, maintaining uniformity despite experimental concentration differences. LES agrees well with experiments in mixing simulations, while RANS models show less precision in predicting fuel mixing rates, highlighting a research area for improvement. This study contributes insights into DLN combustor flow and mixing dynamics, harmonizing experimentation and numerical simulation. Implications include turbulence model selection and experiment guidance for similar contexts.

12.1 Introduction

In the context of carbon reduction and the integration of renewable energy, there is an increasing significance placed on the role of heavy-duty gas turbines, in which the combustor system holds a crucial position of energy conversion. In contemporary heavy-duty gas turbines, the can-annular DLN combustion system

Y. Feng · X. Li · X. Ren (✉) · C. Gu
Tsinghua University, Beijing 100084, China
e-mail: rxd@mail.tsinghua.edu.cn

X. Lv · S. Li · Z. Wang
China United Gas Turbine Technology, Beijing 100084, China

A. Suryan et al. (eds.), *Proceedings of the 9th Asian Joint Workshop on Thermophysics and Fluid Science, 27–30 November 2022, Utsunomiya, Japan*, Springer Proceedings in Physics 396, https://doi.org/10.1007/978-981-99-9470-0_12

is commonly employed to achieve efficient energy conversion. This system generates lean-premixed combustion of fuel and air, which reduces flame temperature and subsequently lowers NOx emissions [1]. Multiple swirlers are typically employed in DLN combustors to ensure stable combustion. These swirlers generate recirculation zones within the combustion chamber, promoting the recirculation of gases towards the flame's root, facilitating the continuous ignition process by mixing with unburned gas. Due to the unique working principles, the flow structure and fuel–air mixing significantly impact the combustor's operational characteristics. Thus, it is essential to design a well-optimized flow structure to anchor the flame ideally, while achieving uniform fuel–air mixing to prevent thermal acoustic oscillations and minimize pollutant emissions.

Numerical tools play a crucial role in the combustion chamber design process, reducing costs significantly. Hence, numerous model combustors have been investigated both experimentally and numerically in previous studies, with turbulence models being a decisive factor in obtaining accurate results. For instance, Li Yajun analyzed the performance of different turbulent models in an intense swirling flow combustor, finding that the Realizable k-ε turbulent model, incorporating swirl flow modification, outperformed others in simulating intense swirling flow combustion fields [2]. W. J. S. Ramaekers studied non-reacting flow in a gas turbine combustor and observed good agreement between LES (Large Eddy Simulation) outcomes and experimental results [3]. P. A. Strakey investigated a swirl-stabilized premixed combustor under cold-flow conditions and found good agreement between experimental data and LES simulations, whereas the results from RANS simulations were less encouraging [4]. Gilles Reichling compared the k-ω-SST and k-ω-SST-SAS turbulence models, highlighting that the URANS (Unsteady Reynolds-averaged Navier–Stokes) calculations were unable to capture the full unsteadiness of the flow field within the combustor [5]. In the realm of mixing field studies, Fujun Sun examined the effect of swirl intensity on non-reacting flow fields and fuel–air premixing characteristics for lean premixed combustors, utilizing the standard k-ε model and wall functions [6]. Tom Tanneberger found that steady-state simulations using the realizable k-ε turbulence model were unable to accurately represent fuel–air mixing in the main flow [7]. Festus validated a CFD (Computational Fluid Dynamics) model against experimental data from a generic Siemens DLE combustion system and determined that the SST turbulence model provided the best agreement [8]. In the research conducted by Phil T. King et al., a radial swirl DLE combustion system's gaseous fuel–air mixing performance was investigated using various RANS turbulence models. However, none of the models were able to accurately predict fuel jet penetration or combustor fuel–air distribution, although improved fuel jet penetration was achieved by employing an extremely low value(0.1) for the turbulent Schmidt number [9]. In general, These studies recommend different turbulence models based on the specific characteristics within various model combustors.

However, most of the experiments and simulation studies focused on laboratory-level single nozzle models or sectors of annular combustors, rather than the can-annular combustors used in industrial heavy-duty gas turbines, which hold a more significant position in the energy sector. Model combustors possess simple structures, small geometric dimensions, and low flow rates, while DLN combustors are larger, more complex, and typically have multiple nozzles. Due to these substantial differences in research objects, the turbulence models verified in simple burners must be reevaluated for their applicability in DLN combustors. Studying Industrial combustors with laboratory skills like laser diagnostics are challenging and uncommon. Given these circumstances, there is a clear need to conduct experimental and simulation studies on flow and mixing phenomena in heavy-duty DLN combustors. This article focuses on PIV and PLIF experiments conducted on a 300 MW F-class heavy-duty DLN combustor and performs numerical simulations of the cold flow field and mixing field. The study compares and analyzes different flow and mixing results obtained using various turbulence models. Notably, the successful implementation of optical testing experiments on industrial DLN combustor flow and mixing fields in this study is the first time in China. The research team overcame numerous challenges to obtain high signal-to-noise ratio results, providing robust support for the validation of numerical tools.

12.2 Method

The research in this experiment focuses on a specific module of the DLN combustor found in a 300 MW F-class gas turbine. This module consists of a headend, a combustion chamber, and a corresponding pressure plenum sector. The headend in this study contains 7 nozzles, and the directions of nozzles 2 and 5 are clockwise, opposite to other nozzles, as shown in Fig. 12.1.

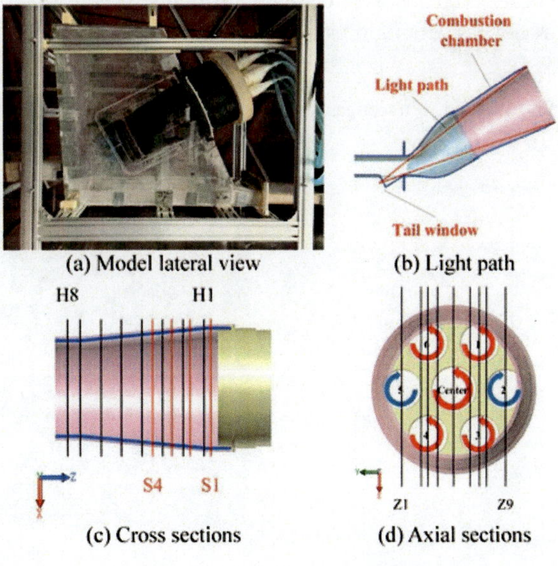

Fig. 12.1 Optical design of the experiment

To optimize the optical scheme design, the goal is to capture a larger area of the measurement section. Previous PLIF experiments conducted by Siemens and Mitsubishi [10–13] were unable to capture most of the section without using image stitching due to the obstruction caused by the transition piece. In contrast, the research team in this study designed a tail window on the transition piece to maximize the shooting area of the horizontal and vertical cross-sections inside the combustion chamber.

In order to gather extensive flow and mixing information within the combustion chamber, 2D PIV experiments photographed eight cross-sections (H1-H8) and nine longitudinal sections (Z1-Z9), while PLIF experiments captured four cross-sections (S1-S4). The positions of these measured sections are indicated in Fig. 12.1c, d. For instrumentation, the TSI system was used for PIV measurement experiments, and the Dantec system employing image intensification was used for PLIF measurement experiments.

Acetone was used as a tracer in the PLIF experiments due to its stable properties, strong signal, and resistance to quenching. In this study, a mixture of acetone and nitrogen was used as the tracer gas, requiring a substantial flow rate of up to 4300 NL/min. Since the tracer gas mixes with air in the premixing tube, its concentration decreases to approximately 1/10 of the original. Thus, a high acetone concentration is necessary in the tracer gas to ensure a detectable fluorescence signal even after dilution.

The traditional Laskin nozzle method and homemade bubble seeder [14] were not suitable for this experiment. Siemens' tracer gas generation system, which measures and controls the flow rate of liquid acetone entering the system and uses high-temperature water vapor as the heat source, presented challenges due to safety requirements and increased system complexity [15]. After careful consideration, the published plans were deemed unsuitable. Therefore, the research team redesigned a tracer gas generation system, as depicted in Fig. 12.2.

Fig. 12.2 PLIF tracer gas system

In this study, computational fluid dynamics (CFD) simulations were conducted using Starccm + software for mesh generation and FLUENT software for solving. The pressure-based SIMPLE method was utilized, and cold flow field simulations were performed using various turbulence models, including Standard k-ε, RNG k-ε, Realizable k-ε, SST k-ω, RSM, and LES.

Based on the simulation results of the flow field, the RNG k-ε, Realizable k-ε, RSM, and LES turbulence models were selected to simulate the mixing field. The Enhanced Wall Treatment was employed for models with wall functions, while the WALE sub-grid stress model was chosen for the LES model, following recommendations from previous studies [16]. Considering the large scale of the experimental model, certain simplifications were made based on the geometry design of the combustion chamber. The suitability of the simplified model had been established in previous studies [17].

Regarding the mesh, a tetrahedral mesh was utilized, and 5 layers of prism layers were added as a boundary layer to ensure that the y + of the first mesh layer met the requirements for LES calculations. As the combustion chamber contained structures of different scales, the mesh was refined in the areas of swirl blades, fuel holes, and other relevant positions during the mesh design process. The mesh with 51.86 million elements was employed for subsequent flow and mixing simulations. Mesh independence results has been shown in previous study of the team [17].

12.3 Results and Discussion

In the DLN combustion chamber, the interaction between the swirling flows generated by the 7 nozzles leads to a unique flow phenomenon. Figure 12.3 illustrates the flow changes inside the combustion chamber, highlighting the coupling effect between the swirling flows. At cross-section H1, the swirling flows generated by each nozzle are relatively independent. However, as the flow progresses to cross-section H3, it becomes evident that two clockwise swirling flows merge with the surrounding counterclockwise swirling flow. Nozzles 1, 4, and the central nozzle produce swirling flows that are mostly separated and converge into two high-speed regions on the left and right sides.

Accurately simulating the interaction between vortices and the overall flow changes within a combustion chamber is crucial for validating the suitability of turbulence models. Due to background scattering, some experimental data may be missing during the PIV recording process, particularly for sections H11-H3, which are closer to the head background. To calibrate the turbulence model, data from sections H4 to H7 are selected, as shown in Fig. 12.4. According to the PIV experimental results, the main flow characteristics inside the combustion chamber include the formation of two high-speed regions and the transition of the low-speed region from an X-shaped to a strip-shaped pattern. A comparison of the simulation results reveals that the standard k-ε model fails to capture the features of the high-speed and low-speed regions. The RNG k-ε model closely matches the experimental results for

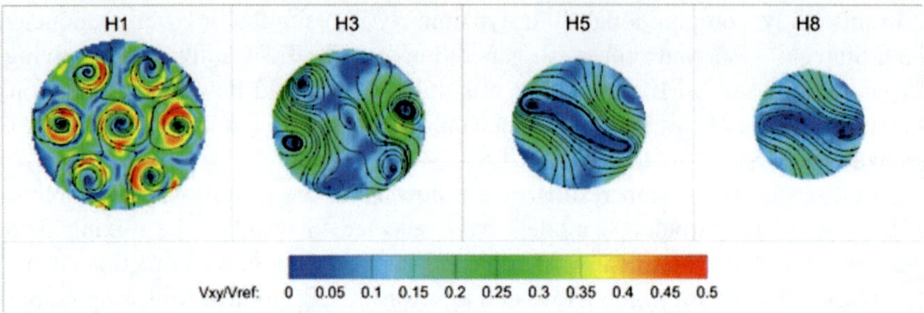

Fig. 12.3 Contour and streamline of H1, H3, H5 and H8

section H4, but as the flow develops, the rotation angle of the low-speed region differs from the experimental value. The Realizable k-ε model exhibits better agreement with the experiment, capturing the position and rotation angle of the high-speed and low-speed regions, albeit with an overall rotation angle exceeding the experimental value. The SST k-ω model shows some discrepancies in the coupled flow features. While the RSM model reproduces the primary flow characteristics observed in the experiment, the flow bend is sharper, and a new low-speed region appears at different positions. The LES simulation not only exhibits similar trends to the experiment but also matches the flow details more closely, making it the best match overall.

To quantitatively compare the differences between the simulation results, data was extracted from a diameter based on the position changes in the high-speed zone for sections H4-H7, as shown in Fig. 12.5. With the exception of the standard k-ε model, all other models were able to capture two velocity peaks, resembling the shape observed in the experiment. The Vxy calculation results from the LES model

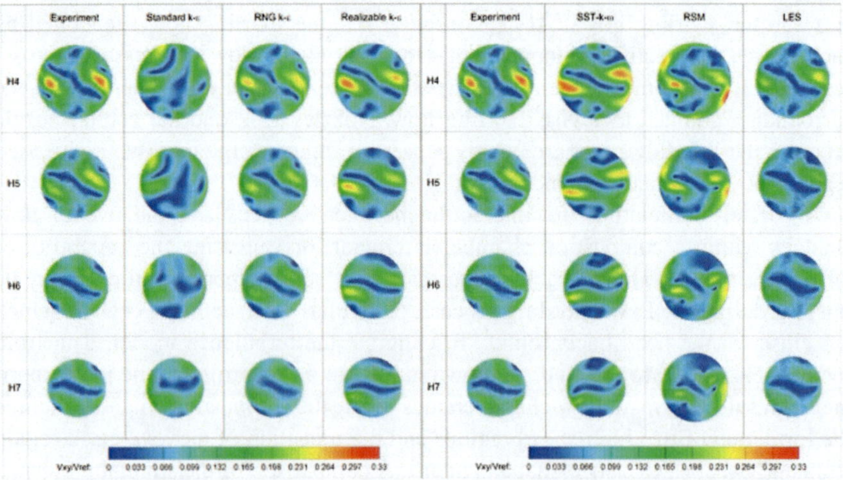

Fig. 12.4 Vxy contour comparison

were slightly lower than the experimental results, while the other models yielded higher values. The RSM model showed the largest deviation from the experimental results, while the RNG k-ε and Realizable k-ε models demonstrated good agreement. The LES model exhibited the closest agreement with the experimental results.

Comparison between the results obtained from the PLIF experiment and the simulation results is shown in Fig. 12.6. Based on the experimental findings, the tracer gas is already well mixed with air in the premixing tube. In section S1, there is a concentration difference between regions due to varying fuel flow rates from the nozzles. However, as the flow progresses through sections S1-S4, the concentration difference gradually diminishes, and the fuel spreads evenly. It is important to note

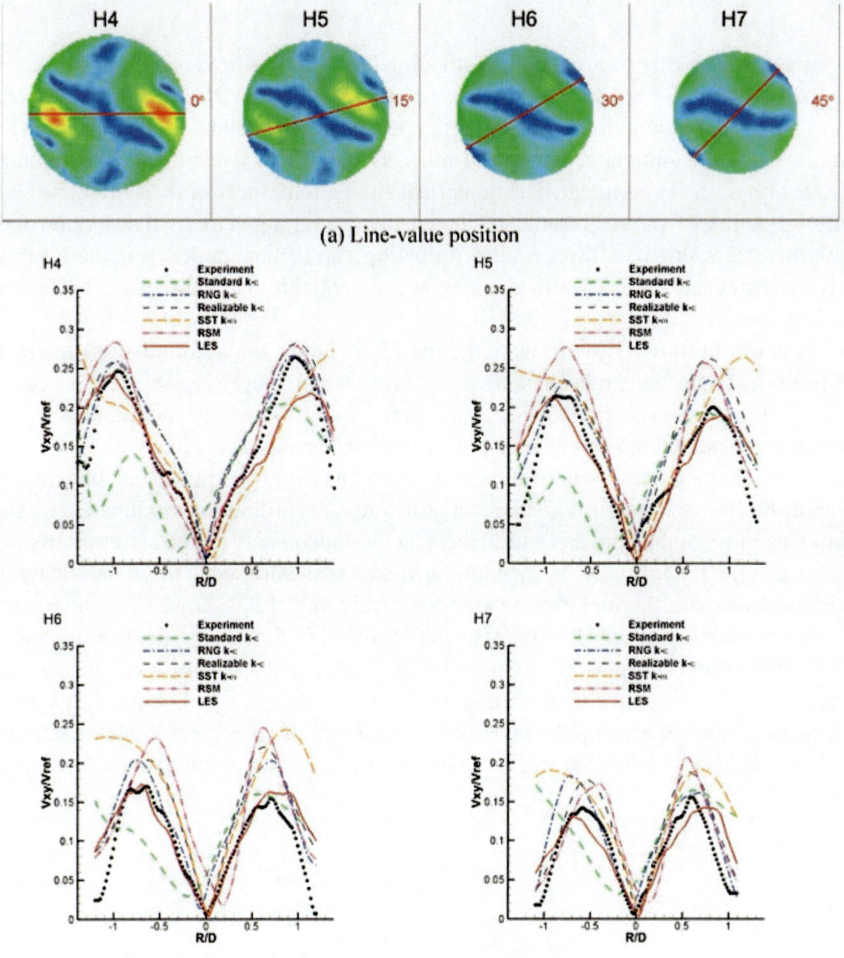

(a) Line-value position

(b) Line diagram comparison on H4- H7

Fig. 12.5 Vxy line diagram comparison

that the PLIF experiment was conducted under test conditions that differed from actual operation conditions. Different fuel quantities were set to better verify the numerical tools, imposing higher demands on the simulation. The results obtained from the LES simulation exhibited good agreement with the experimental results. Conversely, the RANS calculation results encountered an issue with a slow mixing rate, only capturing the general trend of fuel diffusion. There was still a significant disparity in concentration absolute value compared to the experiment, similar to previous research studies [9, 18, 19].

12.4 Conclusion

This study primarily focused on investigating the cold flow field and mixing field of a 300 MW class F-class heavy-duty DLN combustor through PIV, PLIF experiments and numerical simulations. The experimental phase involved introducing innovative designs for optical schemes and other systems, which facilitated the smooth execution of the experiment. The numerical simulation aspect of the study assessed the suitability of various turbulence models by examining velocity distribution and mixing characteristics. Based on the evaluation, turbulence models suitable for this type of combustor were identified, along with potential research directions for future investigations.

Among the flow field simulations, the LES (Large Eddy Simulation) method closely matched the experimental characteristics, accurately reproducing velocity distribution. The RANS models, particularly RNG k-epsilon and Realizable k-epsilon, displayed similarity to the experimental results and effectively captured the flow characteristics within the combustion chamber, making them suitable for rapid selection of combustion chamber structures. Regarding the mixing field of the combustion chamber, the fuel and air were already adequately mixed in the premixing tube. Despite the substantial concentration differences imposed in the experiment for different nozzles, the fuel and air were uniformly mixed S4, showcasing the robust mixing capability of the DLN combustion chamber. In the mixing field simulations, the LES method also closely aligned with the experimental results, while the RANS methods exhibit less accurate simulation of fuel mixing rates and only rough agreement with the experimental trend. Achieving more precise predictions of mixing rates in RANS simulations in DLN combustors is an important avenue for future research.

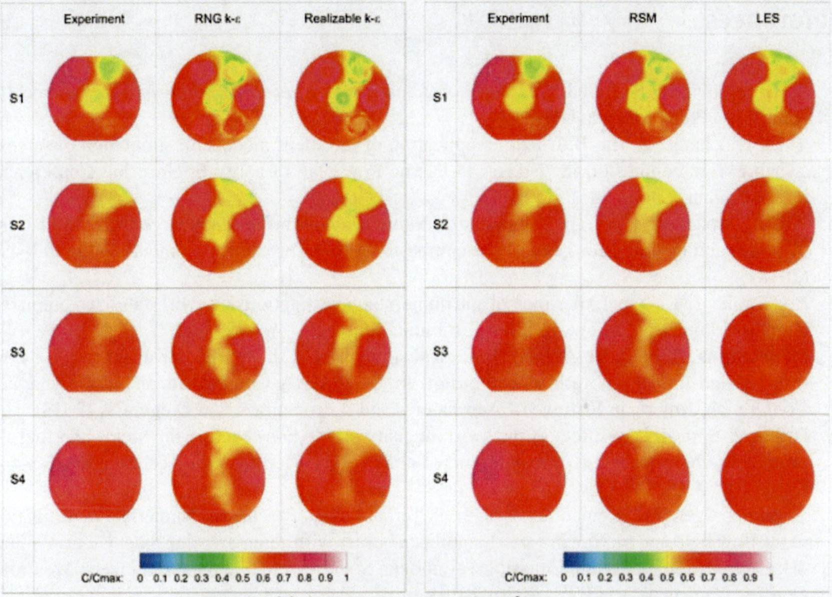

(a) Acetone contour comparison

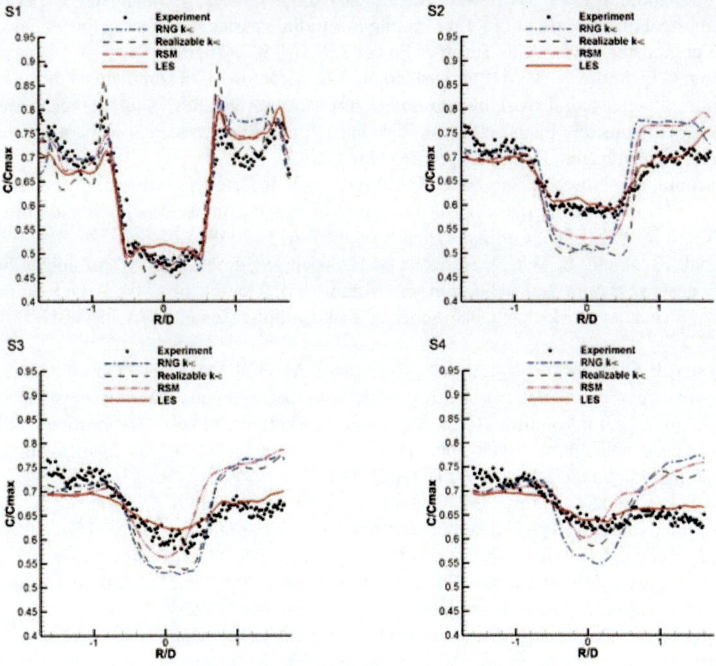

(b) Line diagram comparison on S1-S4

Fig. 12.6 Mixing field comparison

References

1. A.H. Lefebvre, D.R. Ballal, *Gas Turbine Combustion: Alternative Fuels and Emissions*, 3rd edn. (CRC Press, Florida, US, 2010)
2. Y. Li, H. Zheng, Y. Mu, B. Cheng, Application of turbulent models on simulation of intense swirling flow combustor, in: 2011 Asia-Pacific Power and Energy Engineering Conference, IEEE, Wuhan, China, pp. 1–4 (2011)
3. W.J.S. Ramaekers, F.A. Tap, B.J. Boersma, Subgrid model influence in large eddy simulations of non-reacting flow in a gas turbine combustor. Flow Turbulence Combust. **100**, 457–479 (2018)
4. P.A. Strakey, M.J. Yip, Experimental and numerical investigation of a swirl stabilized premixed combustor under cold-flow conditions. J. Fluids Eng. **129**, 942–953 (2007)
5. G. Reichling, B. Noll, M. Aigner, Numerical simulation of the non-reactive and reactive flow in a swirled model gas turbine combustor, in: 21st AIAA Computational Fluid Dynamics Conference, American Institute of Aeronautics and Astronautics, San Diego, CA (2013)
6. F. Sun, J. Suo, Z. Liu, Effect of the swirl intensity on the non-reacting flowfields and fuel-air premixing characteristics for lean premixed combustors. J. Mech. Sci. Technol. **36**, 433–444 (2022)
7. T. Tanneberger, T.G. Reichel, O. Krüger, S. Terhaar, C.O. Paschereit, Numerical investigation of the flow field and mixing in a swirl-stabilized burner with a non-swirling axial jet, in: Volume 4B: Combustion, Fuels and Emissions, American Society of Mechanical Engineers, Montreal, Quebec, Canada, p. V04BT04A026 (2015)
8. F.E. Agbonzikilo, I. Owen, J. Stewart, S.K. Sadasivuni, M. Riley, V. Sanderson, Experimental and numerical investigation of fuel-air mixing in a radial swirler slot of a dry low emission gas turbine combustor. J. Eng. Gas Turbines Power **138**, 061502 (2016)
9. P.T. King, G.E. Andrews, M.M. Pourkashanian, A.C. McIntosh, CFD predictions of isothermal fuel-air mixing in a radial swirl low nox combustor using various RANS turbulence models, in: Volume 2: Combustion, Fuels and Emissions, Parts A and B, American Society of Mechanical Engineers, Copenhagen, Denmark, pp. 973–983 (2012)
10. S. Tanimura, M. Nose, K. Ishizaka, S. Takiguchi, J. Rodriguez, Advanced dry low NOx combustor for mitsubishi g class gas turbines, in: Volume 3: Combustion, Fuels and Emissions, Parts A and B, ASMEDC, Berlin, Germany, pp. 607–615 (2008)
11. B.F. Kock, B. Prade, B. Witzel, H. Streb, M.H. Koenig, Combustion system update SGT5–4000F: design, testing and validation, in: Volume 1B: Combustion, Fuels and Emissions, American Society of Mechanical Engineers, San Antonio, Texas, USA, p. V01BT04A046 (2013)
12. H. Krämer, F. Dinkelacker, A. Leipertz, G. Poeschl, M. Huth, M. Lenze, Optimization of the mixing quality of a real size gas turbine burner with instantaneous planar laser-induced fluorescence imaging, in: Volume 2: Coal, Biomass and Alternative Fuels; Combustion and Fuels; Oil and Gas Applications; Cycle Innovations, American Society of Mechanical Engineers, Indianapolis, Indiana, USA, p. V002T02A027 (1999)
13. B. Janus, J. Bigalk, L. Helmers, B. Witzel, Y. Ghermay, M. Huth, C. Johnson, K. Landry, R. Sunshine, successfully validated combustion system upgrade for the SGT5/6–8000H gas turbines: technical features and test results, in: Volume 3A: Coal, Biomass and Alternative Fuels; Cycle Innovations; Electric Power; Industrial and Cogeneration, American Society of Mechanical Engineers, Düsseldorf, Germany, p. V03AT09A009 (2014)
14. J. Hermann, M. Greifenstein, B. Boehm, A. Dreizler, Experimental investigation of global combustion characteristics in an effusion cooled single sector model gas turbine combustor. Flow Turbulence Combust. **102**, 1025–1052 (2019)
15. B. Witzel, Application of Optical Diagnostics to Support the Development of Industrial Gas Turbine Combustors, University of Duisburg-Essen (2015)
16. R.D. Shah, H. Patil, J. Banerjee, A relative assessment of sub grid scale models for large eddy simulation of co-axial combustor. J. Mech. Sci. Technol. **26**, 1753–1763 (2012)

17. Y. Feng, X. Li, X. Ren, C. Gu, X. Lv, S. Li, Z. Wang, Experimental and numerical investigation of the non-reacting flow in a high-fidelity heavy-duty gas turbine DLN combustor. Energies **15**, 9551 (2022)
18. G. He, Y. Guo, A.T. Hsu, The effect of Schmidt number on turbulent scalar mixing in a jet-in-crossflow, in: Volume 2: Coal, Biomass and Alternative Fuels; Combustion and Fuels; Oil and Gas Applications; Cycle Innovations, American Society of Mechanical Engineers, Indianapolis, Indiana, USA, p. V002T02A029 (1999)
19. A. Innocenti, A. Andreini, A. Giusti, B. Facchini, M. Cerutti, G. Ceccherini, G. Riccio, Numerical investigations of NOx emissions of a partially premixed burner for natural gas operations in industrial gas turbine, in: Volume 4B: Combustion, Fuels and Emissions, American Society of Mechanical Engineers, Düsseldorf, Germany, p. V04BT04A045 (2014)

Chapter 13
Measurement of Fluid Friction Acting on Flat Plate

Takamitsu Fujino, Masaki Endo, Hiromasa Suzuki, and Yoko Sakakibara

Abstract There are several types of air resistance acting on an aircraft. One of them, frictional resistance, accounts for half of the air resistance, and thus, measurement of the frictional resistance plays an important role in engineering practice. Oil film interferometry is one of the methods for measuring the frictional resistance. Oil dropped on an object is thinly stretched by the frictional resistance. Interference fringes are generated when monochromatic light is applied to the oil. The spacing between neighboring fringes changes depending on the gradient of thickness of the oil. The air was sent to a wind tunnel using a blower and passed through several rectifying grids for a uniform flow field to be formed. The wind speed at the centre of the wind tunnel exit, i.e., the mainstream speed, was measured using a pitot tube. A flat plate, onto which a drop of oil was dropped, was placed along the stream at the centre of the wind tunnel exit. The change in fringe spacing was measured and the coefficient of surface frictional resistance was estimated. The relation between the Reynolds number based on the distance from the front edge of the flat plate and the coefficient of surface friction resistance is summarized. The results show good agreement between the experimental and the theoretical trends.

13.1 Introduction

Aircraft are subject to aerodynamic drag when flying. Therefore, reducing aerodynamic drag improves fuel efficiency. Surface friction drag accounts for the largest portion of the drag. At subsonic speeds, the ratio is about half [1]. Therefore, reducing surface friction drag is one way to improve aircraft performance. Surface frictional resistance causes shear stress on the wall surface. Methods for measuring shear stress

T. Fujino (✉) · M. Endo · Y. Sakakibara
Tokyo Denki University, Ishizaka 350-0394, Hatoyama, Hiki, Saitama, Japan
e-mail: 22rmm17@ms.dendai.ac.jp

H. Suzuki
Tokyo Metropolitan College of Industrial Technology, 1-10-40 Higashi-Ohi, Shinagawa 140-0011, Tokyo, Japan

A. Suryan et al. (eds.), *Proceedings of the 9th Asian Joint Workshop on Thermophysics and Fluid Science, 27–30 November 2022, Utsunomiya, Japan*, Springer Proceedings in Physics 396, https://doi.org/10.1007/978-981-99-9470-0_13

on walls have been studied for more than 20 years. However, a database to discuss the universality of this method has not yet been established. In this study, oil film interferometry is used to measure shear stresses in a flow field parallel to a flat plate, and the results are compared with theoretical values.

13.2 Oil Film Interferometry

Oil film interferometry is one of methods of measuring shear stress. When a wedge-shaped oil film is irradiated with monochromatic light, interference fringes are generated due to optical path differences. A drop of oil is dropped onto a flat plate, and the shear stress of the air causes the oil to spread and the distance between the interference fringes to widen gradually. The shear stress is calculated from the speed at which the interference fringes move. Figure 13.1 illustrates the optical path difference of light in the thin film. The light passing through point A is reflected from the contact surface, point B, between the oil and the flat plate and goes to point C. The light in the oil going to the point C and the light reflected at the point C produce the light path difference DBC shown as the red line in the figure. This optical path difference DBC causes alternating bright and dark lines to appear. The difference between the thicknesses of the oil film on the adjacent dark lines is given by the following equation.

$$h_k = \frac{(k+\frac{1}{2})\lambda_0}{2\sqrt{{n_{oil}}^2-\sin^2\alpha}} \tag{13.1}$$

$$\Delta h = \frac{\lambda_0}{2\sqrt{{n_{oil}}^2-\sin^2\alpha}} \tag{13.2}$$

where h_k, λ_0, n_{oil}, and α are the thickness of the oil, the wavelength of the sodium lamp, the refractive index of the oil, and the incident angle of the sodium lamp, respectively. The sodium lamp is used as a light source for visualization, Considering the oil as a Newtonian fluid, the velocity u of the oil at a height y from the flat plate can be expressed by the following equation.

$$u_k = \frac{\tau}{\mu} y = \frac{\tau}{\rho\nu} y \tag{13.3}$$

where τ, ρ, and ν are the shear stress, the oil density, and the kinematic viscosity coefficient, respectively. Figure 13.2 shows a schematic of interference fringes generated in the oil. S_k, u_k, and h_k in the figure indicate the wavelength, the shear rate, and the height of the kth interference fringe, respectively. A wedge-shaped film of oil on a flat plate gradually spreads downstream due to shear stress τ, and the apex angle of the wedge becomes smaller. Assuming that the shear stress τ does not vary locally, the distance between neighboring dark lines, i.e., the wavelength of the interference fringes, can be expressed by the following equation using Eq. (13.3), because the difference in velocity between neighboring dark lines is proportional to the decrease in the wedge top angle [2].

$$\frac{dS_k}{dt} = u_k - u_{k-1} = \frac{\tau}{\mu}(h_k - h_{k-1}) = \frac{\tau}{\mu}\Delta h \tag{13.4}$$

Substituting Eq. (13.2) into Eq. (13.4) and solving for shear stress τ yields the following equation.

$$\tau = \frac{2\rho\nu\sqrt{n_{oil}{}^2 - \sin^2\alpha}}{\lambda_0} \cdot \frac{dS}{dt} \tag{13.5}$$

From Eq. (13.5), the shear stress τ can be obtained from the temporal variation of the interference fringe wavelength *dS/dt*.

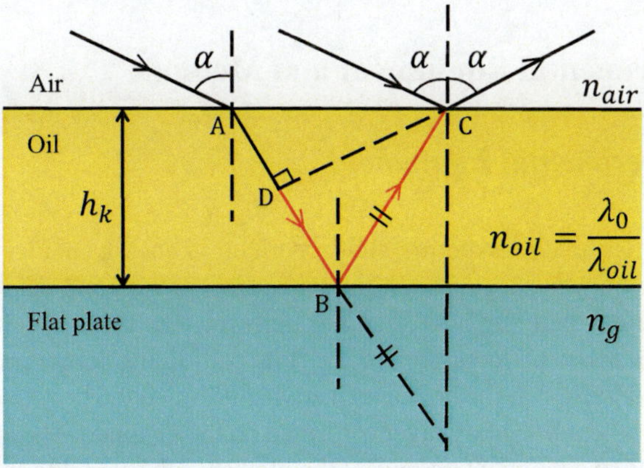

Fig. 13.1 Optical path

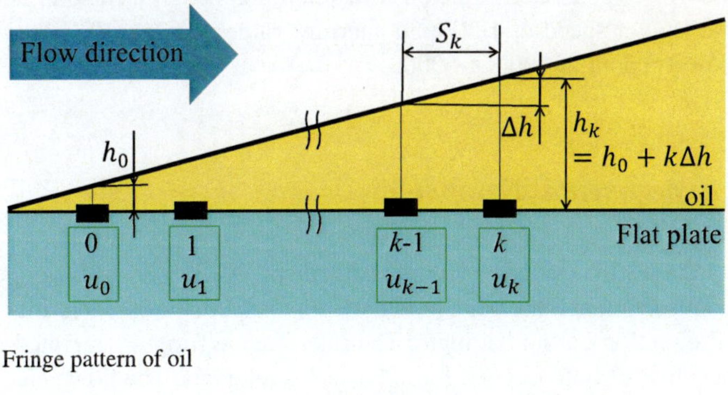

Fig. 13.2 Fringe pattern of oil

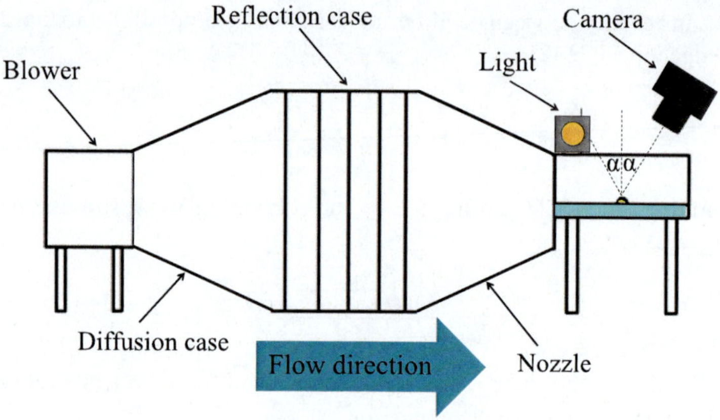

Fig. 13.3 Schematic view of experimental equipment

13.3 Experimental Equipment and Methods

13.3.1 Experimental Equipment

A schematic view of the experimental apparatus is shown in Fig. 13.3. The air supplied by the blower goes through the expansion section and passes through the rectifying grating, where the flow becomes uniform. The flow velocity is increased by reducing the cross-sectional area of the flow path in the compression section. A tripping wire is attached at the nozzle exit so that it forced the flow to become turbulent, and the turbulent boundary layer is formed on the flat plate. The air velocity was changed by adjusting the input voltage of the blower with a variable autotransformer. The flat plate was installed at the center of the nozzle exit in the height direction. The oil film on the flat plate was illuminated with light from a sodium lamp, and the interference fringes generated in the oil were photographed by a camera. The camera was set to a shutter speed of 1/200, an aperture value of F9, an ISO sensitivity of 250, and a pixel count of 4000 × 6000.

13.3.2 Experimental Equipment

Figure 13.4 shows the optical system arrangement. A drop of oil is applied to the measurement section and the camera is fixed on a tripod at $\alpha = 30°$ with respect to the vertical direction on the flat plate. The interference fringes generated in the oil were consecutively photographed for 500 s at 5-s intervals. The flow velocity at the center of the nozzle exit was measured using a pitot tube and used as the main flow velocity. The input voltage applied to the blower was varied in 10 V steps from 60 to 100 V using a variable autotransformer, and measurements were carried out three

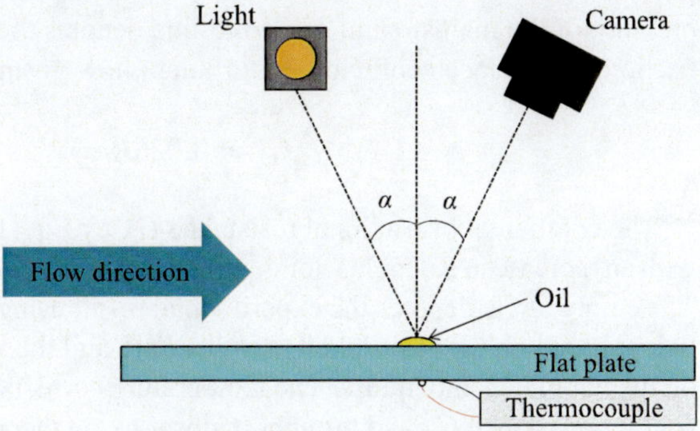

Fig. 13.4 Schematic of the test section

times at each input voltage. Since the viscosity of the oil depends on the temperature of the oil [3, 4], the temperature of the oil was measured using thermocouples under the assumption that the temperature of the measurement plate is equal to that of the oil.

13.4 Results of Experiments

Figure 13.5 shows a binarized image of the interference fringes observed in the experiment. The air flows from top to bottom. Figure 13.5a is an image taken at 200 s after the start of photographing, and Fig. 13.5b is an image at 500 s. Both images visualize the same region at a flow velocity of 14.9 m/s. The oil is stretched by the shear stress with the air, and the oil thickness becomes thinner. As a result, the wavelength of the interference fringes increases with time. To estimate the typical fringe spacing between neighboring dark lines, FFT analysis was performed on the image shown in Fig. 13.5. The dominant frequency components was obtained and the wavelength *S* in pixel. The relation between the wavelength *S* of the interference fringes and elapsed time *t* is shown in Fig. 13.6. The solid line is an approximately straight line obtained by the least square method. The wavelength *S* is proportional to the elapsed time *t*. The red lines shown in Fig. 13.5a are drawn at 17 equally spaced locations in the y-direction of the image. Figure 13.7 shows the results of *dS/dt* on each red line shown in Fig. 13.5a. The abscissa denotes the distance from the left edge of the image. Since the values in the vicinity of the edge are lower than those in the center, the average is calculated using the values at 13 locations in the center of the image. In the same way, the average of *dS/dt* at the different flow velocities was calculated. Figure 13.8 shows the values of *dS/dt* against Reynolds number. As the Reynolds number *Rex* increases, *dS/dt* tends to increase. The shear stress was obtained by substituting the value of *dS/dt* into Eq. (13.5). In this study, however, the values of ρ, n_{oil}, α, and λ_0 are constants. Figure 13.9 shows variation of the coefficient of frictional resistance C_f against the Reynolds number. The circle denotes the experimental results, where the shear stress τ is divided by the dynamic pressure of the mainstream. The solid line denotes the theoretical value of the local frictional resistance coefficient in turbulent flow given by white [5].

$$C_f = 0.026Rex^{-1/7} \tag{13.6}$$

The coefficient of frictional resistance C_f by Eq. (13.6) ends to decrease slightly with an increase in Reynolds number. The experimental results have a same tendency. The empty circle denotes the experimental result using the kinematic viscosity coefficient based on the atmospheric temperature and the solid circle denotes that based on the flat plate temperature. The temperature correction increased the coefficient of friction resistance C_f and brought it closer to the theoretical value. The temperature correction improved the accuracy by up to 7%.

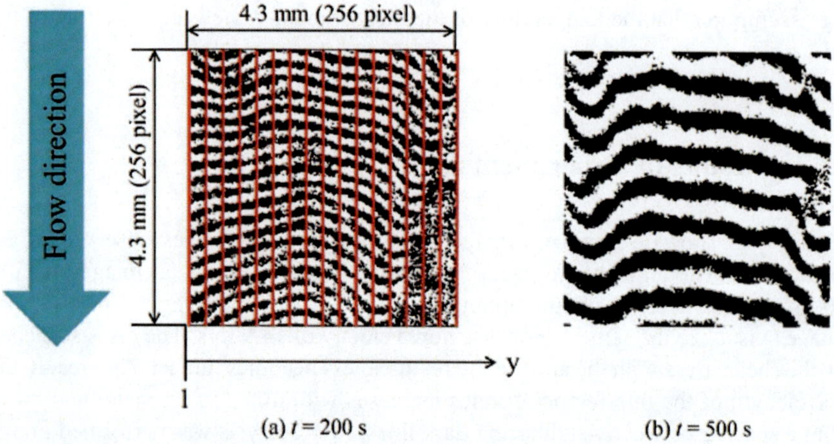

Fig. 13.5 Binarized image

Fig. 13.6 Relation between wavelength and time

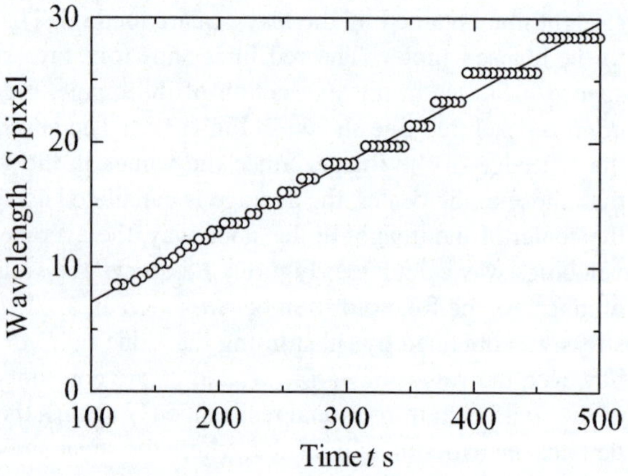

Fig. 13.7 Variation of *dS/dt* against position in *y* direction

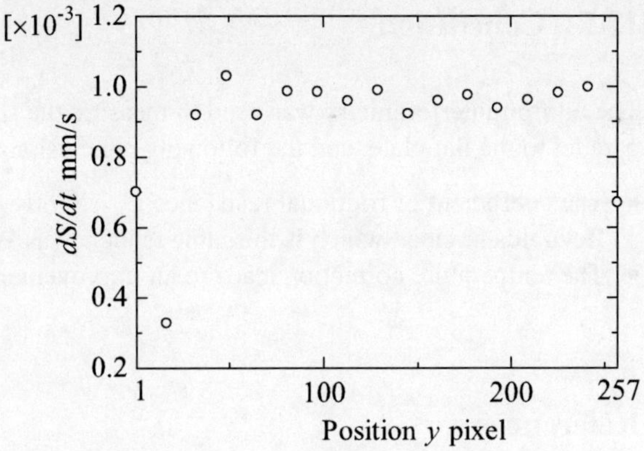

Fig. 13.8 Variation of *dS/dt* against *Rex*

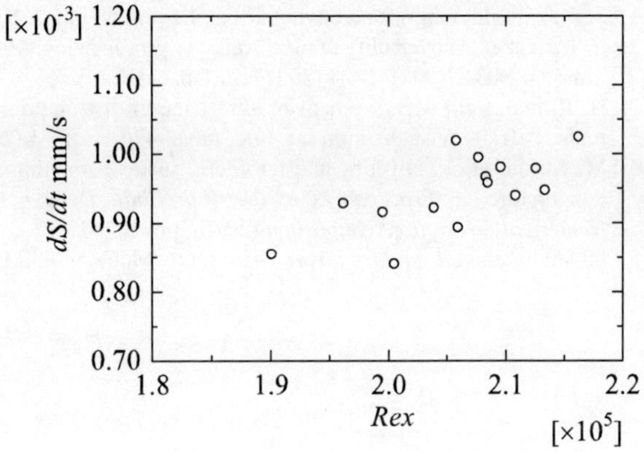

Fig. 13.9 Friction resistance coefficient measured by OFI

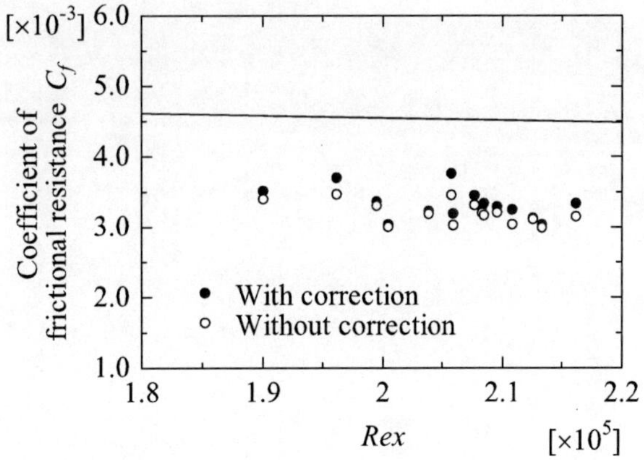

13.5 Conclusion

The oil film interferometry was used to measure the shear stress from the flow field parallel to the flat plate, and the following conclusions were obtained.

- The coefficient of frictional resistance C_f slightly decreases with increasing the Reynolds number, which is the same tendency as White's equation.
- The temperature correction leads to an improvement in accuracy.

References

1. K. Nakahashi, Aircraft aerodynamics and optimization. J. Nagare **26**, 259–265 (2007)
2. Y. Tsuji et al., Universality of mean velocity profile in high Re-number turbulent boundary layer. Trans. JSME **87**(904), 1–24 (2021) (in Japanese)
3. H. IIjima et al., Measurement of skin friction distribution using oil film for industrial wind tunnels. Trans. Visualization Inf. Soc. Japan **41**(9), 28–34 (2021)
4. M. Kurita, et al., Oil-film interferometry using correction for variation of oil-viscosity with temperature, in *Proceedings of the 48th Fluid Dynamics Conference/the 34th Aerospace Numerical Simulation Symposium* (2016), pp. 207–211
5. H. Schlichting, *Boundary Layer Theory*, ed., McGraw-Hill Book Co, (1978)

Chapter 14
Application of Pressure-Sensitive Paint Measurement Technology in Pressure Measurement of Cascade Blade Surface

Xiangfu Lei, Limin Gao, Longrui Chang, Guanhua Yang, and Xiaochen Mao

Abstract As a contactless, high-resolution flow visualization technology, pressure sensitive paint (PSP) measurement technology has the problems of limited visual angle and difficult optical layout and others in internal flow measurement. In order to explore the usability of PSP measurement technology on high solidity and complex three-dimensional blade, optical layout and PSP measurement has been carried out for a linear cascade blade with a solidity of 1.53 and a camber angle of 39.61°, Through three-dimensional reconfiguration, the experimental results are restored to three-dimensional blades. The PSP results are compared with the measurement results of pressure taps. The results show that the pressure resolution of PSP measurement is about 59/mm^2 pressure measuring points; The results measured by PSP under different working conditions accord with the aerodynamic law well, and the whole domain pressure distribution on the suction surface of twisted cascade under a wide range of working conditions is obtained; The error between PSP measurement results and pressure tap measurement results is within 1.673%.

14.1 Introduction

In wind tunnel and flight experiments, surface pressure measurements play a crucial role in uncovering information regarding surface flow phenomena, such as shock wave, flow separation, and boundary layer transition [1]. These measurements serve as a key verification tool for turbomachine design and performance prediction. Traditionally, pressure measurements are obtained by installing an array of pressure sensors on the surface of the experimental model through pressure taps [2]. However, this method is limited by the model's strength and installation constraints,

X. Lei · L. Gao (✉) · L. Chang · G. Yang · X. Mao
School of Power and Energy, Northwestern Polytechnical University, Xi'an 710072, China
e-mail: gaolm@nwpu.edu.cn

National Key Laboratory of Aerodynamic Design and Research, Northwestern Polytechnical University, Xi'an 710072, China

A. Suryan et al. (eds.), *Proceedings of the 9th Asian Joint Workshop on Thermophysics and Fluid Science, 27–30 November 2022, Utsunomiya, Japan*, Springer Proceedings in Physics 396, https://doi.org/10.1007/978-981-99-9470-0_14

resulting in discrete pressure data from the pressure taps. Drilling thinner blade leading edges can be challenging, and the presence of pressure taps may damage the blade model's surface. Consequently, these limitations restrict the application of pressure measurement technology in turbomachine experiments.

Since the 1980s, pressure-sensitive paint (PSP) measurement technology has gained popularity among aerodynamic researchers due to its "non-invasive" characteristics, full-field pressure measurement, high resolution, and other advantages that traditional pressure tap measurement lacks. This technology has been extensively applied to the aerodynamics of aircraft [3], airfoils [4], missiles [5], and other outflow research. However, due to the complex structure and restricted optical path, it was not until 2002 that NASA Lewis Research Center [6] successfully applied the PSP measurement technique to a transonic cascade wind tunnel and captured images of shock waves on the transonic cascade. Gao et al. [7] successfully utilized the PSP measurement technique to measure pressure on the suction surface of a linear cascade with low solidity, yielding results that were consistent with those of pressure taps. To enable quantitative pressure analysis and overcome the limitations of narrow field of vision, they also developed the PSP three-dimensional reconstruction technique [8, 9] and optical path layout method based on Cartesian coordinate system [10]. These techniques were then applied to perform full-domain surface pressure measurement on a linear cascade with a solidity of 1.73 [11] and a linear cascade with a solidity of up to 2.3 [12], but even with the aforementioned methods implemented, in order to capture a comprehensive image of the blade, a dual-camera light source system is still required. In order to obtain full-domain pressure distribution of cascade, Cheng et al. [13] utilized two optical path layouts with camera light sources. One layout was positioned at the wind tunnel exit and the other was on the side of the endwall. The PSP measurement technique was employed to obtain these results. Wang et al. [14] developed an endoscopic PSP technique to measure the air film cooling efficiency of a transonic turbine cascade with a solidity of 0.86, and the measurement error was within 1.7%. However, the insertion of an endoscope into the flow field inevitably introduces disturbances to the flow field.

The narrow spatial field of view poses challenges in capturing comprehensive images in blade experiments. Existing solutions have increased experimental complexity and reduced efficiency. To address this, this paper propose a novel optical path layout method based on spherical coordinates. This method enables rapid positioning of optimal locations for cameras and light sources, eliminating the need for a dual-camera system. This technique is applied to a compressor cascade with a solidity of 1.53, and the surface pressure distribution of the three-dimensional blade is reconstructed using three-dimensional reconstruction technology. The experimental results serve as valuable references for blade design and the analysis of flow mechanisms.

14.2 Experimental Subjects

14.2.1 Test Linear Cascade

The basic blade profile used in this experiment is the controlled diffusion blade airfoil (CDA) NPU-A1 [15]. Figure 14.1 shows the blade to be tested painted with PSP, and the test cascade consists of seven blades and grid tray, as shown in Fig. 14.2, with chord length of 69.5 mm, solidity of 48 mm, blade height of 100 mm, apsect ratio of 1.36, and other specific geometric parameters shown in Table 14.1. The blade to be measured is located in the second channel with better total pressure loss periodicity, and 9 pressure taps are set at 20 and 80% blade height positions, and 10 pressure taps are set at 50% blade height for verification and comparison of measurement results. In order to avoid camera shots and light path obstruction from the light source as much as possible, highly translucent quartz glass was used for both sides of grid tray.

Considering the influence of blade curvature on the image 3D reconstruction, and with the help of existing surface pressure taps, a total of 26 marker points were set on the straight blade surface as shown in Fig. 14.3, in which the red points are marker points and the rest points are blade surface pressure taps. The marker points are set more intensively in the area with larger curvature.

Fig. 14.1 Cascade model

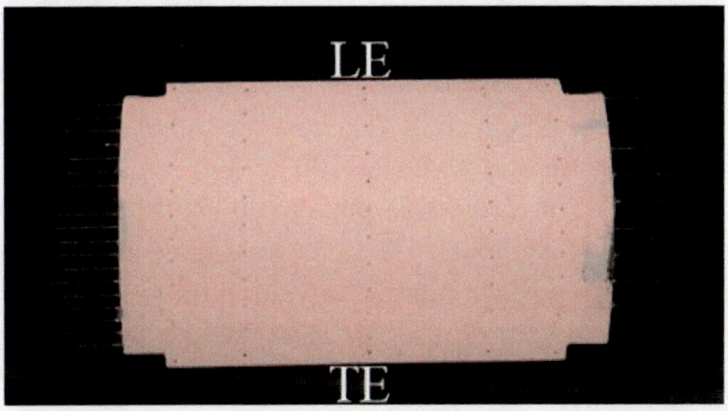

Fig. 14.2 Test cascade model

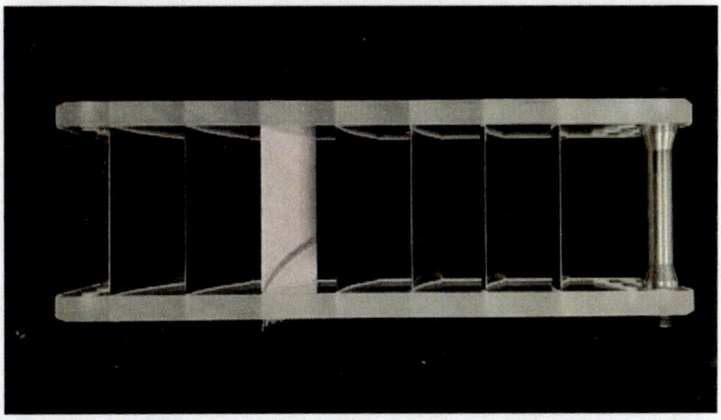

Table 14.1 Geometric parameters of cascade

Parameters	Value	Parameters	Value
Chord	73.6 mm	Max. thickness	3.51 mm
Height	100 mm	Solidity	1.53
Aspect ratio	1.36	Stagger angle	26.56°

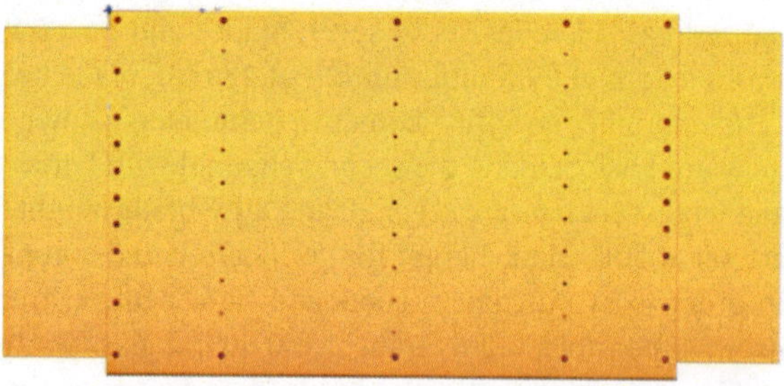

Fig. 14.3 Positions of marking points and pressure taps

14.3 Experimental Method

14.3.1 Linear Cascade Tunnel and Measurement System

The experiments were conducted in a high subsonic linear airfoil wind tunnel at the National Key Laboratory of Airfoil Cascade Aerodynamics of Northwestern Polytechnical University. The test section of the wind tunnel is a rectangular section with a length and width of 100 × 300 mm, as shown in Fig. 14.4; the test piece is mounted on a rotatable plate to adjust the inlet flow incidence continuously. The wind tunnel was extensively commissioned and measured in the laboratory [16, 17]: the incoming Mach number deviation did not exceed ±0.0005, the inlet flow angle deviation did not exceed ±1°, and the turbulence did not exceed 2.0%.

Fig. 14.4 Test section of linear cascade tunnel

14.3.2 PSP Measurement System and Paint

The spectrogram of the the PSP paint is shown in Fig. 14.5, and the sprayed blade is shown in Fig. 14.6. Under ambient light, the pressure-sensitive paint is pink in color. The previous steady-state calibration results showed that the paint can maintain very good linearity in the range of 20–180 kPa [12].

The experiments were conducted with an independently assembled PSP internal flow measurement system, which consists of a domestic high stability and high-power UV-LED array light source and a 14-bit high-sensitivity scientific-grade CCD camera (2048 × 2048) camera, respectively. Among them, the light source stability

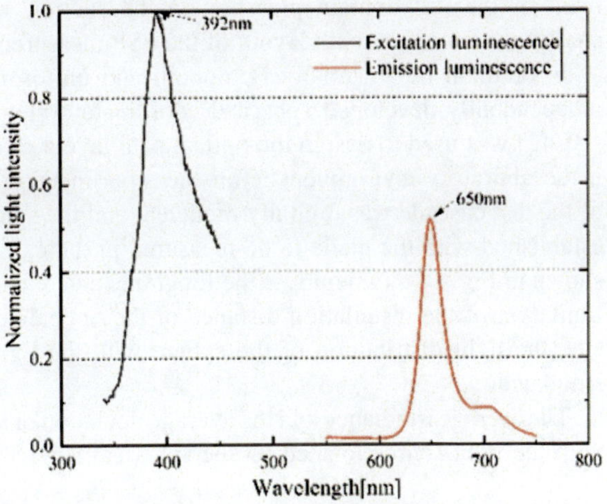

Fig. 14.5 Excitation and emission spectra of PSP

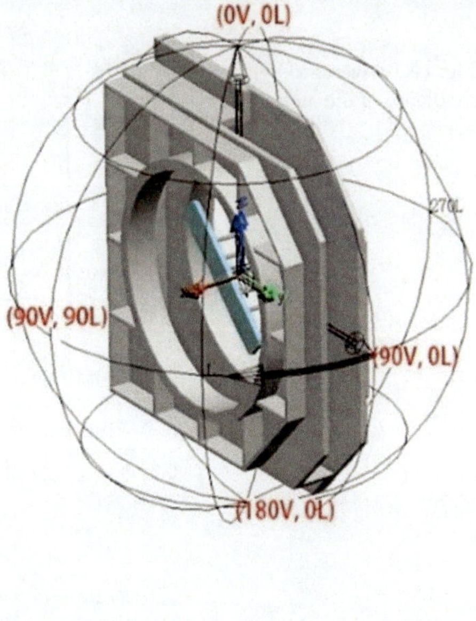

Fig. 14.6 Setting of spherical coordinates

is not less than 99.5%, and the camera is equipped with a band-pass filter with a wavelength of 650 ± 20 nm.

14.4 Experimental Scheme

14.4.1 Optical Arrangement Layout

The measured blade is located in the middle channel of test cascade and the passage of the cascade is unusually narrow, which, together with the demand for the measurement of the full domain pressure on the suction surface, undoubtedly poses a challenge to the light path layout of the PSP measurement system.

In order to obtain relatively smooth and uniform light and photo effects, the independently developed spherical coordinate system optical path layout method [10, 11] was used to design the optical path layout of the PSP measurement system in the laboratory environment before the experiment: first, the wind tunnel test section of the test cascade was digitally modeled, and the spherical coordinate system was established with the blade to be measured in the cascade channel as the center, as shown in Fig. 14.6 (where L is the longitude and V is the latitude); Considering the limitation of the installation distance of the optical equipment at the experimental site, the light distribution of the sphere with the light source at R = 300 mm is simulated.

The average irradiance and the average deviation of irradiance on the blade surface with the light source located on the spherical surface are shown in Figs. 14.7 and 14.8.

The results show that the maximum average irradiance is obtained on the surface of the blade suction surface with the light source located at ($L = 77°$, $V = 109°$)

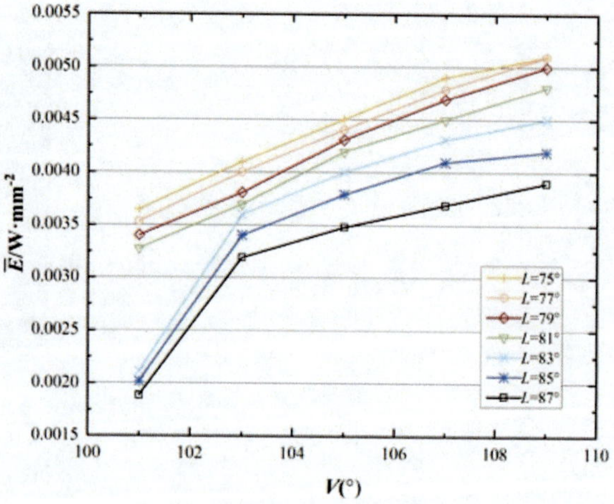

Fig. 14.7 Averaged light irradiance of the measured blade

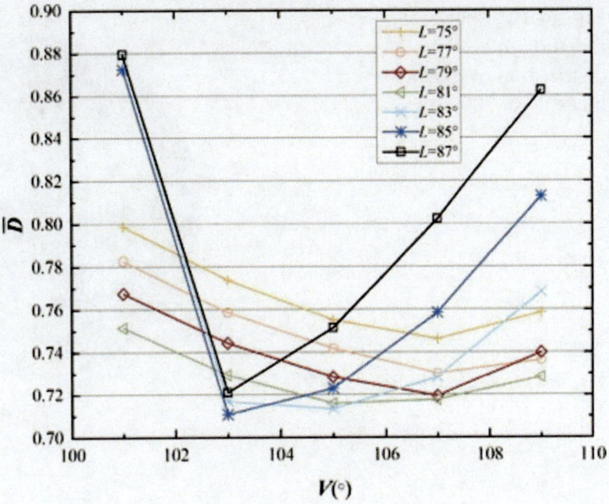

Fig. 14.8 Averaged light irradiance deviation of the measured blade

on a spherical surface with $R = 300$ mm and the average deviation is close to the lowest value. This means that when the light source is located at this coordinate, the probe molecules inside the PSP coating of the suction surface of the blade will receive more uniform and certain intensity of excitation light. At this time, the light irradiance distribution of the suction surface of the blade is shown in Fig. 14.9: except for the leading edge of the blade endwall, the light distribution uniformity is good, and the light intensity is high enough, which meets the light efficiency requirements of PSP measurement experiments on light source irradiation.

Using the blade as the light source, the light irradiation intensity of the camera located in the $R = 300$ mm sphere is simulated in reverse (Fig. 14.10). It can be

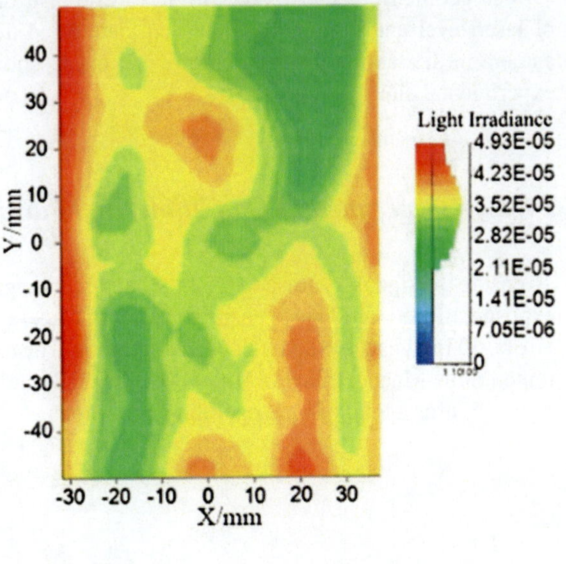

Fig. 14.9 Light irradiance distribution of the measured blade at different longitude and latitude

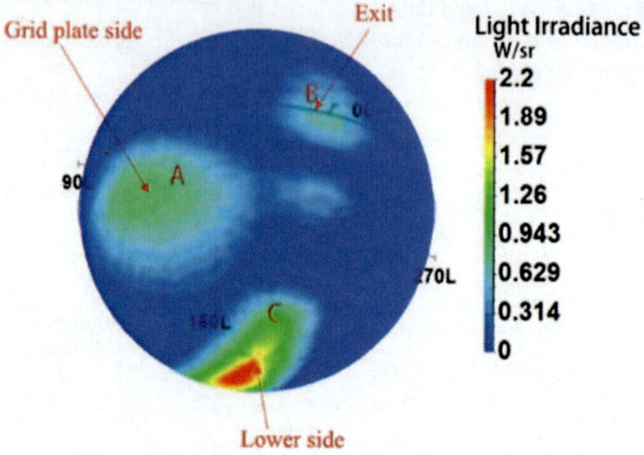

Fig. 14.10 Spherical radiation intensity distribution of the measured blade ($R = 300$ mm)

seen that the area with higher light radiation intensity has three areas A, B and C: area B corresponds to the downstream of the wind tunnel exit, which will not only cause camera vibration but also affect camera safety; area C corresponds to the lower side of the leaf grid wind tunnel, which cannot provide enough space for camera installation; area A corresponds to the end wall side of the leaf grid, which can ensure the smoothness and safety of the camera light path through quartz glass.

By converting the spherical coordinates obtained from the light path layout into right-angle coordinates, the specific position of the light source camera can be precisely determined: the spatial coordinates of the light source are (391.0, 104.0, −139.6 mm) and the spatial coordinates of the camera are (306.2, 76.3, −96.5 mm). Camera in the lower left side of the light source, the height of the two devices are close, almost all are positive to the measured blade channel, the specific optical path layout is shown in Fig. 14.11. in the experimental environment, the measured blade surface geometric center as the origin to establish the coordinate system, the use of laser level and laser rangefinder can determine the excitation light source and camera in the laboratory installation position, as shown in Fig. 14.12 for the final experimental site.

14.4.2 Experimental Conditions and Parameter Settings

To ensure the signal-to-noise ratio of the captured images, the aperture $f = 2.8$ and the exposure time $t = 100$ ms to ensure maximum camera light intake; to reduce random errors, 30 images were taken in each condtions. The experiments were conducted at three inflow Mach numbers ($Ma_1 = 0.3, 0.5, 0.7$) and three incidence angle ($i = -5°, 0°, 5°$) for a total of nine operating conditions.

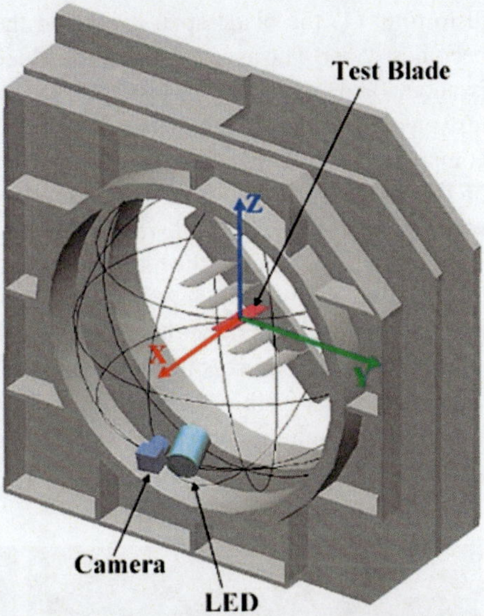

Fig. 14.11 Schematic diagram of the simulated layout

Fig. 14.12 The site layout of the camera and light source

14.5 Results and Discussion

14.5.1 Original PSP Image Analysis

Figures 14.13 and 14.14 show the reference image and blow image when the incidence angle $i = -5°$ and $i = 5°$ under the condition of $Ma_1 = 0.7$. It can be seen that due to the limitation of the photographing angle, all images have obvious image distortion: (1) the blade span height on the image decreases and the chord length becomes wider. The aspect ratio of the blade image calculated according to the pixel is about 0.84, which is far less than the actual aspect ratio of the swept cascade test piece; When the incidence angle increases, the blade image will be further compressed. (2) The spanwise compression of the leading edge line of the blade on the image is more serious than that of the trailing edge line, showing a trapezoid with narrow lower edge and wide upper edge.

Since the irradiance of the light source and camera has been fully considered before the experiment, the luminance distribution of the reference images of Fig. 14.13a and Fig. 14.14a for the two incidence angles ($Ma_1 = 0.7$) is relatively uniform, and the marking points and pressure measuring holes on the blade surface of the blowing image are clearly visible, and the luminance of the leading edge of the blade is slightly lower, which is almost consistent with the irradiance distribution on the blade surface simulated by the light path layout.

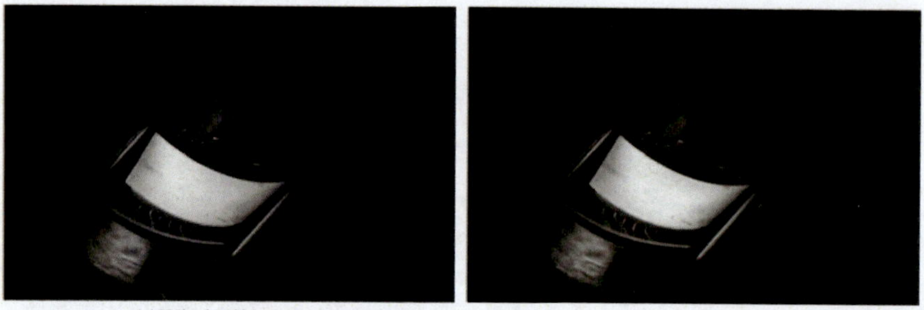

(a)Wind-off image (b)Wind-on image

Fig. 14.13 Wind-off and wind-on images of blade suction surface ($i = -5°$)

(a)Wind-off image (b)Wind-on image

Fig. 14.14 Wind-off and wind-on images of blade suction surface ($i = 5°$)

14.5.2 PSP Versus Pressure Tap Data

The self-developed 3D image reconstruction algorithm [8, 10] and PSP image processing [9] are used to conduct 3D reconstruction, ratio processing and field calibration of PSP experimental results under all working conditions, and the pressure distribution of the whole area of the blade surface of the cascade can be obtained.

The static pressure measured by the PSP at 50% blade height of the cascade at the inlet Mach number of 0.7 and three incidence angles is quantitatively extracted and compared with the measurement results of the pressure tap. In Fig. 14.15, it is the local atmospheric pressure. It can be seen from the figure that the static pressure measured by PSP at three incidence angles is highly consistent with the results measured by the pressure measuring taps, no matter in terms of the static pressure size or the chord distribution of static pressure: when the incidence angle of incoming flow is −5° and 0°, the incoming flow direction is close to be axial, and the air flow on the blade suction surface first increases and decreases along the leading edge of the blade, then decelerates and diffuses until the trailing edge of the blade; As the position of the pressure measuring taps remains unchanged, the minimum static pressure measured at the two incidence angles of $i = -5° \sim 0°$ is 32.143%, but PSP has the advantage of high spatial resolution, accurately capturing the peak position of surface isentropic Mach number from 29.5% chord length to 30.2% chord length due to the change of incidence angle.

However, when the angle of incidence of the incoming flow is 5°, the front stagnation point of the airflow entering the cascade moves to the pressure surface. During the flow around the leading edge, the airflow first decelerates and pressurizes on the suction surface, and then increases and depressurizes after the chord length of 31.3%. The static pressure on the rear surface of the blade suction surface is almost unchanged. It can be inferred that the airflow separation occurred behind the fifth pressure measuring point at the 5° angle of incidence.

From the measurement data volume and range of both, the PSP measurement data can reach up to 200 points along the full chord length direction, while the traditional pressure tap pressure measurement method is limited by the limit of blade thickness. Only nine fixed chord lengths (16.67 ~ 83.92%) of 50% blade height and the surface static pressure at eight positions (16.67 ~ 91%) of 20 and 80% blade height are

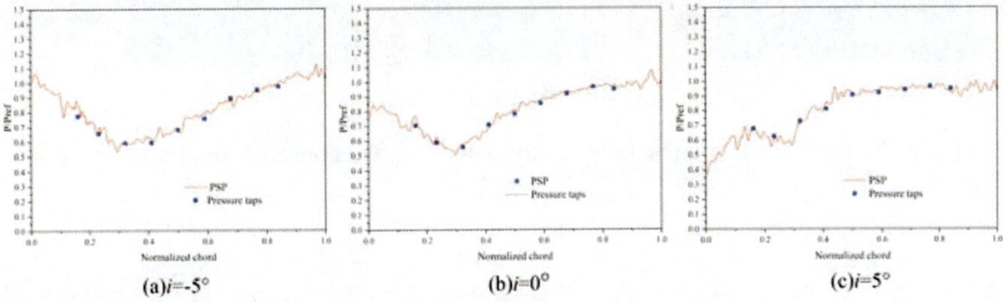

Fig. 14.15 Comparison between PSP and pressure taps measurement

measured, which makes it impossible to accurately measure the blade leading edge flow The flow in the suction surface diffuser and the peak Mach number on the surface and their positions are effectively compensated for by the PSP measurement results at these important aerodynamic positions. After quantitative comparison with the results of pressure tap, the maximum error of PSP measurement results is 1.673%, which occurs at the first pressure measuring point with $i = -5°$.

14.5.3 Impact of Incidence on Cascade Flow

Figure 14.16 gives the static pressure distribution on the suction surface under inlet Mach number of $Ma_1 = 0.7$ with different incidence angles i. All upper and lower halves of the results have good symmetry along spanwise. On the suction surface, local acceleration region represented by the static pressure valley gradually shifted upstream with the increase of incidence angle. In addition, as incidence increased from $-5°$ to $5°$, the high-pressure region around the center of trailing edge shrunk gradually, indicating a contraction of the main flow area along the cascade passage. When the incidence angle reached to $5°$, the high-pressure region reenlarged and expanded along the spanwise to the upper and lower endwall. Besides, the contour of this high-pressure region changed from concave to convex. Specifically, a sharp pressure gradient can be observed closely near the leading edge under this high incidence angle.

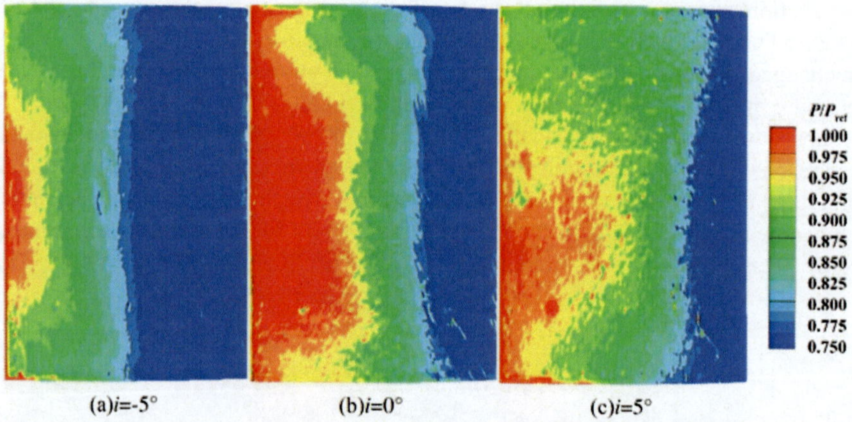

Fig. 14.16 Pressure distribution on suction surface under different incidence ($Ma_1 = 0.7$)

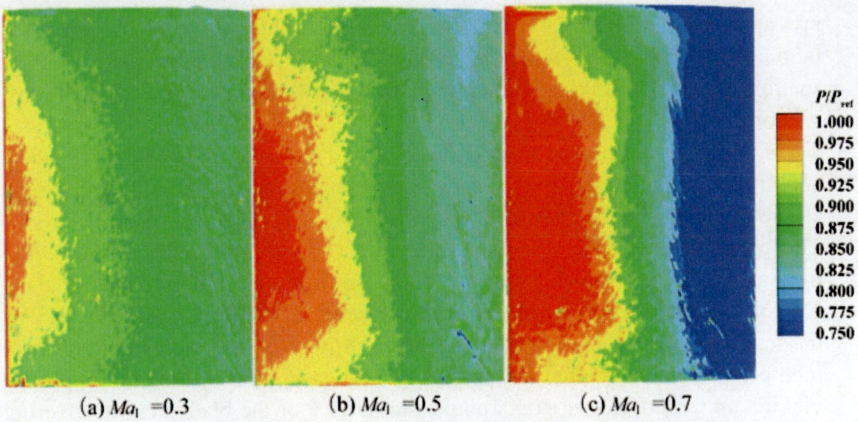

Fig. 14.17 Pressure distribution on suction surface under different inlet Mach number ($i = 0°$)

14.5.4 Impact of Mach Number on Cascade Flow

Figure 14.17 presents static pressure distribution on the suction side of the blade under an incidence of 2.5° with different inlet Mach numbers. As can be seen, it almost followed the same pattern with the increase of inlet Mach number from $Ma_1 = 0.5$ to $Ma_1 = 0.7$. The location and contour of the low static pressure region where the flow was accelerated near the leading edge, had great similarities. Yet as the inlet Mach number increased to 0.8 as shown in Fig. 14.17d, this low-pressure region was greatly expanded along spanwise and reached to the upper and lower endwalls. It could also be noticed that the shrinkage of the high static pressure region around the mid-span of the trailing edge was accompanied with the increase of inlet Mach number; that is to say, the low static pressure regions around the upper and lower corner at the trailing edge were enlarged and expanded to the midspan. This could be an indication of the deterioration of the corner flow.

14.6 Conclusion and Future Work

This paper applied the PSP measurement technology to a linear compressor cascade with a solidity of 1.53. The surface pressure distribution on the suction surface of the blade under different working conditions was obtained. A comparison was made between the PSP results and the results obtained from pressure taps. The research findings can be summarized as follows:

1. The research group developed an optical path layout method, which effectively addressed issues such as a narrow shooting field of vision and blade obstruction that arise during the internal flow measurement of the cascade using PSP. Through a three-dimensional reconstruction algorithm, the two-dimensional

pressure distribution image was transformed into a three-dimensional representation of the blade. Consequently, a comprehensive pressure distribution on the suction surface of the cascade, encompassing a wide range of working conditions, was obtained.

2. The study analyzed and compared the effects of different inflow Mach numbers and angles of incidence on the blade surface pressure distribution. The results aligned with established aerodynamic principles.
3. The research overcame the resolution limitations of traditional pressure tap measurements, providing extensive blade surface pressure distribution data. Notably, it successfully captured measurements in challenging areas such as the leading and trailing edges, which were difficult to assess using traditional methods. When compared to pressure taps measurements, the PSP approach yielded up to 200 pressure data points along 50% of the blade height, covering the entire chord length range. The deviation from the pressure hole results was within 1.673%, confirming the reliability of the PSP measurement technique.

References

1. A.E. Baron, J.D.S. Danielson, M. Gouterman et al., Submillisecond response times of oxygen-quenched luminescent coatings. Rev. Sci. Instrum. **64**(12), 3394–3402 (1993)
2. H. Sakaue, J.W. Gregory, J.P. Sullivan et al., Porous pressure-sensitive paint for characterizing unsteady flowfields. AIAA J. **40**(6), 1094–1098 (2002)
3. C. Klein, R.H. Engler, U. Henne et al., Application of pressure-sensitive paint for determination of the pressure field and calculation of the forces and moments of models in a wind tunnel. Exp. Fluids **39**(2), 475–483 (2005)
4. A.D. Gardner, C. Klein, W.E. Sachs et al., Investigation of three-dimensional dynamic stall on an airfoil using fast-response pressure-sensitive paint. Exp. Fluids **55**(9), 1–14 (2014)
5. G.E. Erickson, Wind tunnel application of a pressure-sensitive paint technique to a faceted missile model at subsonic and transonic speeds (2004)
6. J. Lepicovsky, T. Bencic, Use of pressure-sensitive paint for diagnostics in turbomachinery flows with shocks. Exp. Fluids **33**(4), 531–538 (2002)
7. L.M. Gao, J. Gao, B. Liu, Q. Zhou, Application of PSP technique to pressure measurement on cascade surface. J. Eng. Thermophys. **32**(03), 411–414 (2011)
8. L.M. Gao, J. Gao, J. Xie, B. Liu, Application of image 3D reconstruction to pressure measurement on blade surface. J. Eng. Thermophys. **33**(09), 1523–1526 (2012)
9. L.M. Gao, N. Wei, J. Gao, Y.N. Wu, Image processing of PSP technique in the internal flow. J. Exp. Fluid Mech. **01**, 93–97 (2013)
10. L.M. Gao, L.C. Tian, X.Q. Hu, R.Y. Li, A numerical method for optical layout of optical flow display technology. ShaanXi:CN105628342B, 2018-06-08
11. L. Gao, X. Hu, L. Tian et al., Pressure field measurement on compressor cascade using pressure-sensitive paint, in *Turbo Expo: Power for Land, Sea, and Air. American Society of Mechanical Engineers*, vol. 49699. V02AT37A033 (2016)
12. L. Gao, G. Yang, T. Gao et al., Experimental investigation of a linear cascade with large solidity using pressure sensitive paint and dual-camera system. J. Therm. Sci. **30**(2), 682–695 (2021)
13. Z.L. Cheng, B. Xiong, M.J. Huang, Transonic cascade pressure-field measurement based on PSP technique. Gas Turbine Exp. Res. **04**, 57–62 (2015)
14. K. Wang, H. Shao, D. Peng et al., Transonic vane film cooling with crescent-shaped craters using an endoscopic pressure-sensitive paint technique. Appl. Therm. Eng. **205**, 118081 (2022)

15. R.Y. Li, High-fidelity numerical and experimental study of corner separation in high-load compressor (Northwestern Polytechnical University, Xi'an, 2020), no. 05, pp. 100
16. M. Cai, L.M. Gao, Z. Liu, H. Cheng, H.H. Wang, Y.C. Guo, Experimental study on flow field quality of linear cascade wind tunnel under different conditions. J. Propulsion Technol. **05**, 1162–1170 (2021)
17. M. Cai, *Research on Flow Field Quality and Control Strategy of High Subsonic Planar Cascade Wind Tunnel* (Northwestern Polytechnical University, Xi'an, 2020)

Chapter 15
A Study on Underexpanded Radial Jet Impinging on Cylindrical Inner Wall

Satoshi Yamazaki, Masaki Endo, and Yoko Sakakibara

Abstract A jet radially spreading like a disc from a small gap between two circular tubes placed face to face with each other was experimentally analysed in this study. The jet is underexpanded and impinges on a cylindrical inner wall concentric with the circular tubes. Visualization by the shadowgraph method captured some ring-shaped shock waves. These shock rings indicate cell nodes of the underexpanded jet and a detached shock wave formed in front of the wall surface. The diameter of the shock ring increased as the nozzle pressure ratio increased, and the ring of the outermost node approached gradually that of the detached shock. At a certain nozzle pressure ratio, the detached shock ring jumped in the upstream cell. These rings oscillated and amplitude of the oscillation varied with nozzle pressure ratio. Dynamic strain of the inner wall generated by jet impingement was measured, and the strain histories at stations arranged in circumferential direction were analyzed. The most dominant frequency and the phase difference between neighboring stations was obtained. As a result, the cylindrical wall was found to vibrate at an integer multiple of the most dominant frequency at the nozzle pressure ratio = 2.8, 3.6 and 4.8.

S. Yamazaki (✉)
Mechanical Engineering, Graduate School of Science and Engineering, Tokyo Denki University, Ishizaka 350-0394, Hatoyama, Hiki, Saitama, Japan
e-mail: 22rmm22@ms.dendai.ac.jp

M. Endo · Y. Sakakibara
Division of Mechanical Engineering, School of Science and Engineering, Tokyo Denki University, Ishizaka 350-0394, Hatoyama, Hiki, Saitama, Japan
e-mail: mendo@mail.dendai.ac.jp

Y. Sakakibara
e-mail: sakaki@mail.dendai.ac.jp

A. Suryan et al. (eds.), *Proceedings of the 9th Asian Joint Workshop on Thermophysics and Fluid Science, 27–30 November 2022, Utsunomiya, Japan*, Springer Proceedings in Physics 396, https://doi.org/10.1007/978-981-99-9470-0_15

15.1 Introduction

An underexpanded jet, which is one of the supersonic jets, is a jet issuing from the nozzle to the atmosphere, and has a cellular structure. The expansion wave generated at the nozzle lip is reflected as the compression wave at the jet boundary, and thus the compression region and the expansion region are formed periodically in the jet. The underexpanded jet can be seen in the gas exhausted from aircraft jet engines and it radiates noise and infrared rays. Studies has been conducted in order to reduce the noise and the infrared rays [1]. In addition, the use of internal combustion engines using hydrogen fuel is being promoted as one of our efforts for a decarbonized society. Studies have been carried out on the underexpanded jet formed when hydrogen fuel is injected into the cylinder [2].

In the flow field near the intake valve of an internal combustion engine, the jet becomes radial due to the shape of the valve opening. To aim at understanding such a flow structure in the vicinity of the intake valve, studies have been conducted so far under assumption of incompressible steady [3].

Experimental methods such as visualization and acoustic measurement and analytical methods such as dynamic mode decomposition (DMD) are used in researches on the underexpanded jet structure [4–6]. The jet oscillation and the radiation noise were studied. All of these studies, however, focus on jets from circular nozzles. Therefore, there are a few study cases about radially spreading jets such as those found in intake valves of internal combustion engines, and furthermore vibration and acoustic measurements has not be carried out for the radial jets imping on the cylinder wall.

In this study, two circular pipes are placed face to face, and compression air issues from the gap between the facing pipe ends to form a radial underexpanded jet. The radial jet impinges on a cylindrical wall covering the nozzle. The strain caused by the jet impingement is measured to investigate vibration mode of the radially underexpanded jet.

15.2 Experimental Considerations

15.2.1 Piping System and Visualization

Figure 15.1a shows the piping system. The air used as the working fluid is compressed by the compressor and supplied to two high-pressure tanks in the soundproof room. The stagnation pressures p_0 at two tanks is adjusted to be equal to each other using two valves. The pressure ratio p_0/p_a was varied from 2.0 to 4.8, where p_a is the atmospheric pressure. The slit nozzle exhausting the jet consists of two circular pipes with an outer diameter of $D = 12$ mm and a wall thickness of $t = 1$ mm. In addition, a gap of $b = 2$ mm is provided at the pipe ends facing each other. The air issues from the gap of the slit nozzles and spreads radially. The inner wall shape of

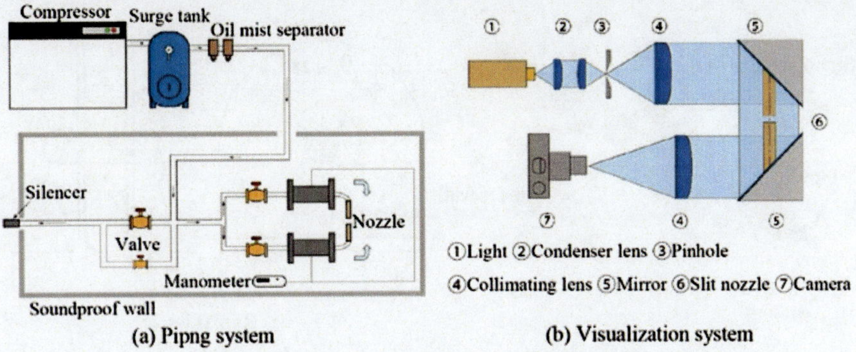

Fig. 15.1 Schematic views of piping system and visualization system

the nozzle is designed and processed so that the flow speed at the nozzle lip reaches the sonic velocity.

The flow field is visualized using the shadowgraph method. The optical system used in this study is shown in Fig. 15.1b. A nano-pulse light with a flash duration of 180 ns was used as the light source. The parallel light passing through the collimator lens is reflected by the mirror in the direction of 90 deg with respect to the optical axis and passes through the flow field. After that, the light is collected by the collimator lens and the flow field is visualized.

15.2.2 Strain Measurement and Experimental Device

A schematic view of the experimental device is shown in Fig. 15.2a and the geometry of the impinged object is shown in Fig. 15.2b. The radial jet impinges on a cylindrical wall covering the nozzle. The inner diameter, the width and the thickness of the cylinder are $D_c = 28$ mm, $w = 20$ mm and $t_c = 1$ mm, respectively. The cylinder is supported at both ends by the wall with a thickness of $t_s = 2.5$ mm. Seven strain gauges set from channel 1 to channel 7 were glued to the outer wall of the cylinder at intervals of 45 deg. The output of the strain gauge is amplified by the dynamic strainmeter and is converted by an A/D converter with a resolution of 12 bits and recorded in a computer. The sampling rate is 1 MHz.

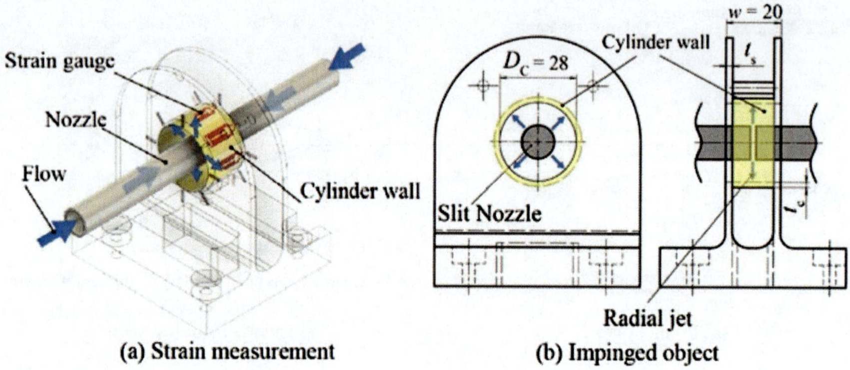

Fig. 15.2 Experimental device

15.3 Visualization

Examples of shadowgraphs taken by the visualization system in Fig. 15.2 are shown in Figs. 15.3a and b. These are grayscale images obtained by averaging 30 instant photographs with a light source flash duration of 180 ns. Since the jet is underexpanded, it has a characteristic cellular structure including oblique shocks formed at cell nodes. In addition, a detached shock is generated in front of the inner wall by the jet impingement. Thus, these shocks are photographed in a ring shape surrounding the circular pipe, i.e., the slit nozzle, by this visualization system. The shock ring corresponding to the cell nodes were observed in the visualizations from $p_0/p_a = 2.0$ to 4.8. The most upstream shock ring is called the 1st shock ring, and the next ring the 2nd shock ring. The most downstream shock ring near the inner wall is the detached shock ring. Three shock rings were visualized at $p_0/p_a = 3.0$ as shown in Fig. 15.3a, and at $p_0/p_a = 4.8$, the first shock ring and the detached shock ring appeared. Since these shock rings oscillates, the average image of the shock ring is not clear and the apparent shock thickness becomes large. Thus, the thickness corresponds to the total amplitude of the ring oscillation. The shock ring at $p_0/p_a = 3.0$ is less clear than at $p_0/p_a = 4.8$, and the former oscillates with larger amplitude. Furthermore, at $p_0/p_a = 4.8$, the detached shock ring has smaller amplitude than that of the first shock ring.

Figure 15.4 shows the relation between the diameter of each shock ring obtained by analyzing the average images and the pressure ratio. The ordinate is the value obtained by dividing the shock ring diameter d_n by the slit nozzle gap $b = 2$ mm. The abscissa is the pressure ratio p_0/p_a. At $p_0/p_a = 2.0$ to 2.6, shock ring does not appear though fragmentary shocks were observed. Hence the diameter of the shock ring at $p_0/p_a \geqq 2.8$ were analyzed. As shown in Fig. 15.4, the diameters of the first and second shock rings increase as the pressure ratio increases. The second shock ring approaches the inner wall and becomes closest to the detached shock ring at $p_0/p_a = 3.6$. After that, the shock ring jumps into the upstream cell at $p_0/p_a = 3.8$, and hence the diameter of the detached shock ring becomes smaller.

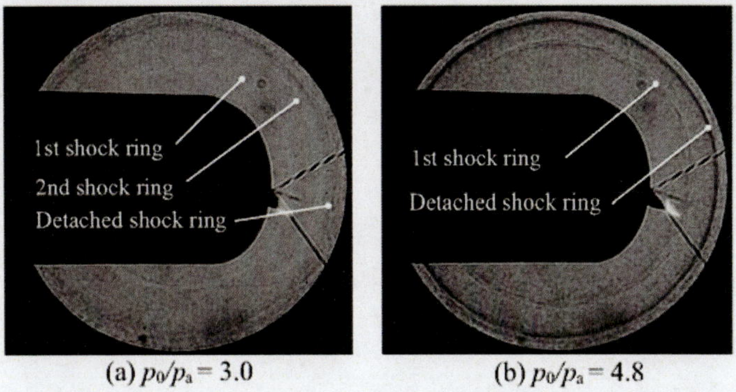

Fig. 15.3 Average image

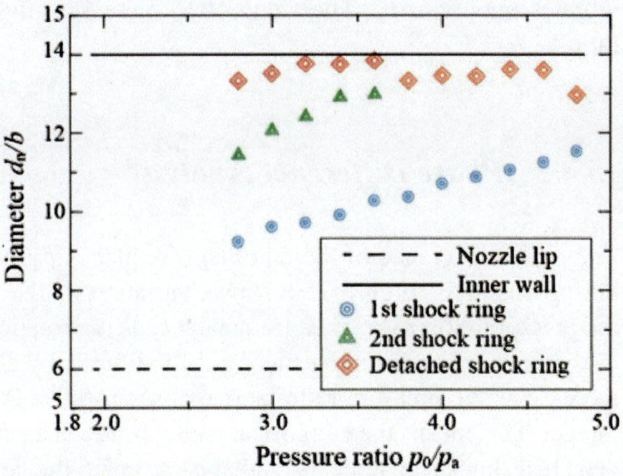

Fig. 15.4 Shock ring diameter

15.4 Strain

15.4.1 FFT Analysis

The strain measured by applying a high-pass filter to the dynamic strain, i.e., the AC component of the dynamic strain, was analyzed by FFT for each pressure ratio. The p_0/p_a was changed in steps of 0.2, three measurements were taken for each condition. Figure 15.5 shows an example of FFT analysis results for $p_0/p_a = 3.2$ and 4.2. The abscissa is frequency and the ordinate the amplitude. The strain data at seven channels were added and averaged, and the average data is analyzed. FFT results obtained by three times measurements was averaged. Some dominant frequencies appear. The dynamic strain characteristics changes along with change in the pressure ratio. The most dominant frequency f_d stands at about 3 kHz, and hence the cylindrical wall

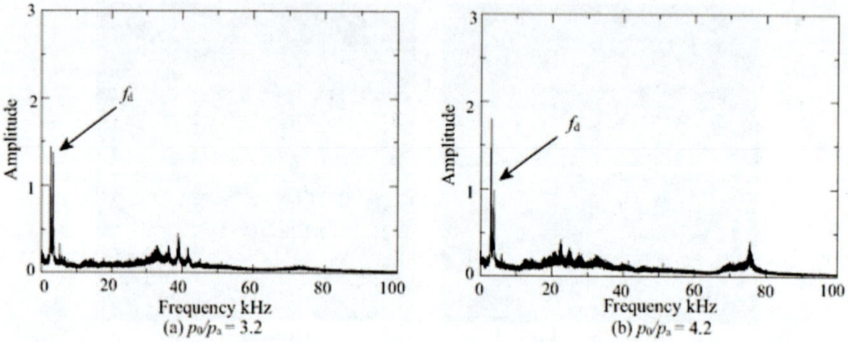

Fig. 15.5 FFT analysis

vibrates mainly at f_d. The value of f_d increases slightly with increasing pressure ratio.

15.4.2 Phase Difference Analysis

The phase difference of the dominant frequency f_d is analyzed between the neighboring channels. Figure 15.6 shows variation of the frequency ratio f_d/f_p against the pressure ratio p_0/p_a. The frequency f_p is the rotational frequency at which strain of the main vibration moves along the cylindrical wall. The value of f_d/f_p tends to increase as the pressure ratio increases. At $p_0/p_a = 2.8$, 3.6 and 4.8, f_d/f_p takes an integer. This means the cylindrical wall vibrates at a certain mode. At $p_0/p_a = 2.8$ the detached shock ring appears, and at $p_0/p_a = 3.6$ the detached shock jumps as shown in Fig. 15.4. At $p_0/p_a = 4.8$ phenomena like these is also considered to be occur.

Fig. 15.6 Phase difference analysis

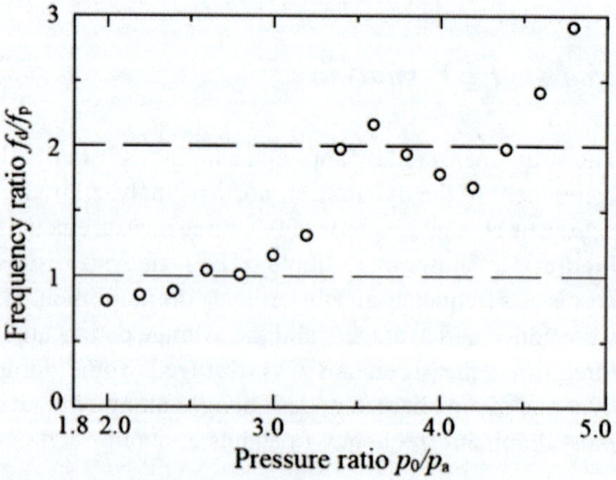

15.5 Conclusions

The underexpanded impinging jet that spreads radially was visualized by the shadowgraph method, and the visualization results were compared with the measured dynamic strain of the cylinder. The following conclusions were drawn:

1. The oscillation of the shock ring was observed by averaging the instant photographs.
2. The detached shock ring that approaches the upstream ring as the pressure ratio rises, and jumps into the inner cell at $p_0/p_a = 3.8$.
3. The most dominant frequency f_d of the cylinder vibration was around 3 kHz.
4. The rotational frequency f_p is equal to an integer multiple of the most dominant frequency f_d when the detached shock ring jumps into the inner cell.

References

1. C. Nicolas, D. Sebastien, J. Laurent, Shock patterns in a slightly underexpanded sonic jet controlled by radial injections. Phys. Fluids **19**, 1–4 (2007)
2. K. Sekine, M. Oikawa, Y. Kondo, Y. Takagi, Y. Sato, A study on hydrogen jet configuration and flame propagation by shadowgraphic visualization in high pressure direct injection hydrogen engines. Trans. Soc. Automotive Engineers of Jpn. 1209–1214 (2012)
3. G. Komatsu, T. Hosokawa, Y. Hosokawa, Induction swirl and flows from the inlet valve. Trans. Jpn. Soc. Mech. Engineers. B, 2966–2971 (1988)
4. L. Nian-Hua, L. Xiang-Ru, H. Peng-Fei, Z. Xi-Wen, H. Feng, Mode switch in tonal underexpanded impinging jets. Phys. Fluids **33**, 1–18 (2021)
5. R. Okinaka, A. Matsuo, H. Feng, Numerical investigation on the oscillation mechanism of sonic jet impinging on a flat plate. Trans. Jpn. Soc. Aeronautical Space Sci. 316–321 (2001)
6. T. Tsujimura, A. Ohta, Y. Tokunaga, J. Senda, Fujimoto: a study on underexpanded jet structure under high pressure ambient. Trans. Jpn. Soc. Mech. Engineers. B, 1348–1353 (2004)

Chapter 16
Fluctuation of Force Acting on Flat Plate by Underexpanded-Jet Impingement

Kono Tatsuya, Masaki Endo, Yoko Sakakibara, and Hiromasa Suzuki

Abstract When an underexpanded jet impinges on a flat plate, the force acts on the flat plate. The force is fluctuated due to the oscillation of the jet. A high-frequency sound called a screech tone is emitted from the underexpanded impinging jet. The screech frequency variation with l/D was examined, where l is the distance between the nozzle and the plates and D is the nozzle diameter. The screech frequency variation showed the saw-tooth characteristics. In the range from $l/D = 3.7$ to 4.2, the sound pressure of screech tone decreased. In order to clarify the relation between the force acting on the flat plate and the jet behavior, the strain of the flat plate was measured and the flow field was visualized. A thin disk was used as the flat plate. 8 strain gages were glued to the reverse side of the thin disk at equal intervals on the same circumference. The measurement of strain showed that the amplitude of the strain increased at the nozzle-plate distance where the frequency of the screech tone jumped to the high value. Visualization revealed that the jump in screech tone was closely related with the position of the plate shock and occurred when the plate shock was located near the cell node or near the boundary between the expansion region and the compression region of cell. Thus, the position of the plate shock affects the fluctuation of the force on the flat plate on which the jet impinges.

16.1 Introduction

Gas exhausted into the atmosphere through a convergent nozzle cannot fully expand at the nozzle exit when the nozzle pressure ratio is above the critical pressure ratio. So as to reduce the pressure in the jet, the expansion fan is formed at the nozzle lip, and thus the jet is underexpanded [1]. The underexpanded jet has a barrel-shaped

K. Tatsuya (✉) · M. Endo · Y. Sakakibara
Tokyo Denki University, Ishizaka 350-0394, Hatoyama, Hiki, Saitama, Japan
e-mail: 22rmm08@ms.dendai.ac.jp

H. Suzuki
Tokyo Metropolitan College of Industrial Technology, 1-10-40 Higashi-Ohi, Shinagawa 140-0011, Tokyo, Japan

A. Suryan et al. (eds.), *Proceedings of the 9th Asian Joint Workshop on Thermophysics and Fluid Science, 27–30 November 2022, Utsunomiya, Japan*, Springer Proceedings in Physics 396, https://doi.org/10.1007/978-981-99-9470-0_16

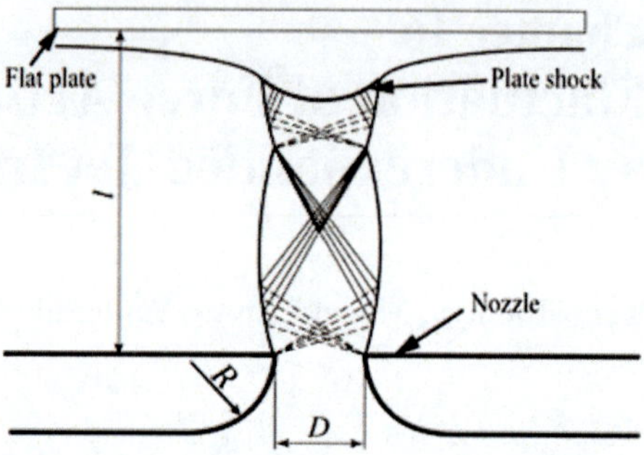

Fig. 16.1 Experimental equipment

cell structure, and the expansion and compression regions are repeated alternately in the jet. An experimental investigation was conducted in the flow field where the underexpanded jet impinges on a flat plate. Such flow fields occur in various industrial situations. A typical example is the assist gas used in laser processing [2]. Underexpanded jets are used to remove melted material produced by laser irradiation. When the underexpanded jet impinges on an object, the plate shock wave is generated and the impinging jet oscillates violently. This oscillation affects the surface pressure of the object. In addition, high-frequency noise, namely the screech tone, is emitted. This can lead to deterioration of processing accuracy and work efficiency. In this work, the relation was experimentally studied between the screech tone phenomena and the oscillation patterns of the jets.

16.2 Experimental Considerations

Figures 16.1 and 16.2 show schematic views of the experimental apparatus. The pressure ratio in the experiment was $p_0/p_a = 3.0$, where p_0 is the stagnation point pressure in the tank and p_a the atmospheric pressure. A circular type nozzle was used as a nozzle exhausting the jet. The nozzle diameter $D = 10$ mm and the curvature radius $R = 20$ mm. The experiment was carried out at different distances l between the nozzle and the flat plate. The value of l/D was varied from 3.0 to 4.5 in 0.1 increments.

In the acoustic measurements, the microphone was fixed at the position 300 mm far from the jet axis, and it pointed to the plate shock located at about 5 mm upstream of the flat plate. The output data from the microphone is recorded through an A/D converter with a sampling frequency of 1 MHz. The measurement time was 0.026 s. The number of measurements was 20 at each condition.

The underexpanded jet impinges on a thin disk perpendicularly. The thin disk is 100 mm in diameter and 0.8 mm thick. Eight strain gauges are glued to the opposite side of the impinged surface at equal intervals on the same circumference, which is 20 mm in diameter. At different values of l/D, the dynamic strain was measured to investigate the fluctuation of the force acting on the disk.

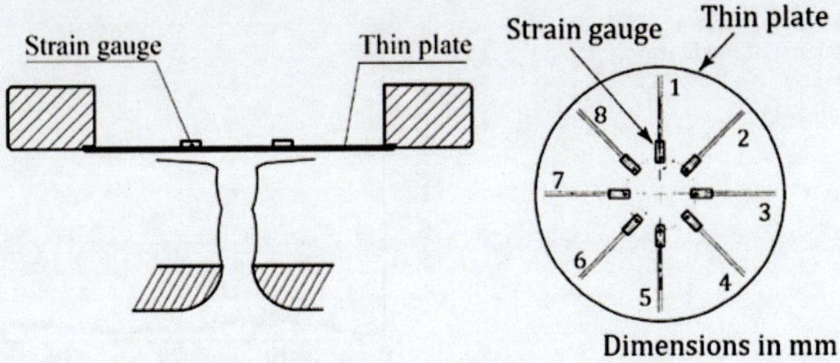

Fig. 16.2 Strain gauge layout

16.3 Experimental Results and Discussion

16.3.1 Dominant Frequency

Data obtained from acoustic and strain measurements were analyzed by FFT analysis. Figures 16.3 and 16.4 shows the results from the acoustic measurements at $l/D = 3.3$ and 3.9, respectively. These figures were obtained by averaging 20 analyses at each l/D. At $l/D = 3.3$, the dominant frequency component appears at 31.79 kHz. This component is the narrow-band noise called 'screech tone'. At $l/D = 3.9$, however, the screech tone cannot be identified. In the same way, the dominant frequency of the strain was sought at different values of l/D.

Figure 16.5 shows variation of the dominant frequency against the nozzle-plate distance. The circle represents screech tone and the triangle strain. These frequencies are in good agreement. This means that the oscillation in the pressure acting on the disk is closely related with the screech tone. In the range of l/D from 3.7 to 4.0, the screech tone was not observed, which seems to correspond to the zone of silence reported by Henderson [3].

Fig. 16.3 Frequency characteristic of sound wave ($l/D = 3.3$)

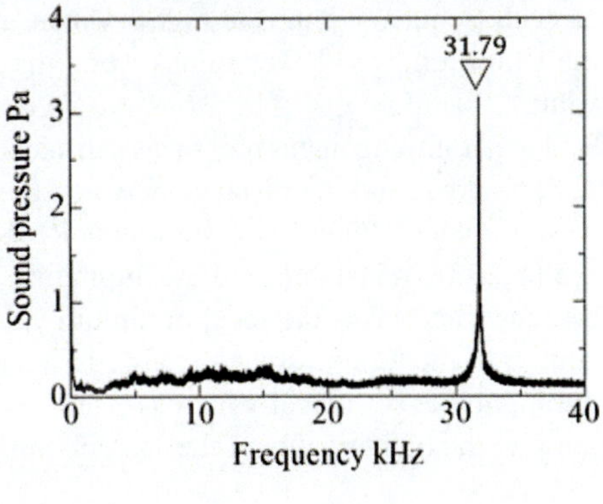

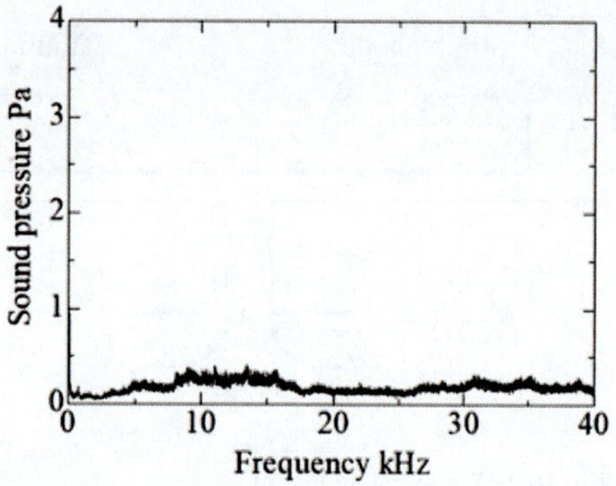

Fig. 16.4 Frequency characteristic of sound wave ($l/D = 3.9$)

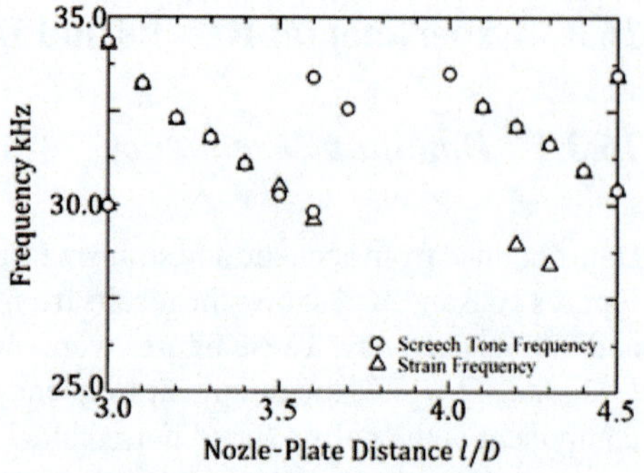

Fig. 16.5 Screech tone frequency and strain frequency

16.3.2 *Jet Oscillation Measurements*

At each value of l/D, the dynamic strains at eight stations were averaged and time-averaged value of the strain was obtained. Figure 16.6 shows variation of the time-averaged strain against the nozzle-plat distance. The strain tend to increase with increasing l/D. At $l/D = 3.5$ and 4.5, the time-averaged strain takes the local minimum values. These nozzle-plate distance l/D are close to the l/D at which the screech frequency jumps to higher values as shown in Fig. 16.6. This indicates that the strain jumps to lower values when the screech tone frequency jumps to higher values.

The jet impingement point was estimated using the dynamic strain at the stations 1, 3, 5 and 7, and the distance between the disk center and the impingement point was calculated. Figure 16.7 shows the RMS value of the distance fluctuation, which is a measure of instability of the impinging jet. The RMS periodically increases and decrease, and takes the local minimum values at $l/D = 3.0$, 3.5 and 4.4, which is similar to the time-averaged strain in Fig. 16.6. Consequently, the instability of the impinging jet periodically increases and decreases, and the jet becomes stable at the jump to the higher value of the screech tone.

Fig. 16.6 Average strain

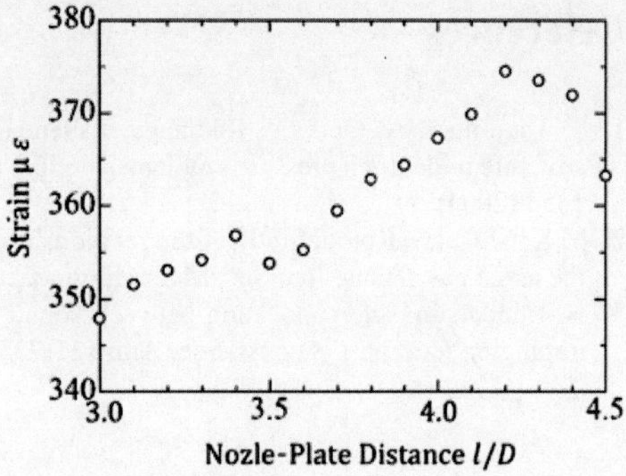

Fig. 16.7 RMS

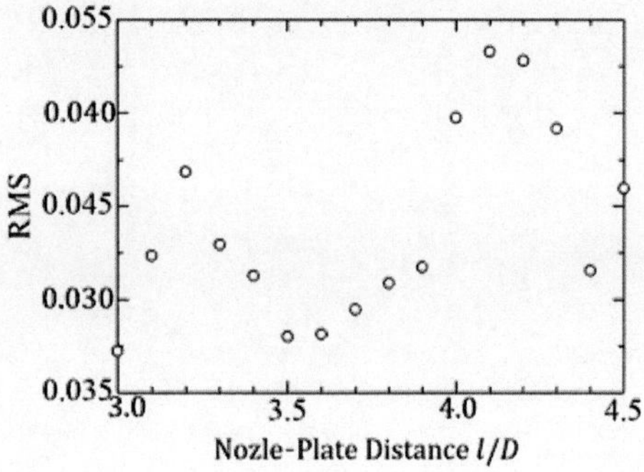

16.4 Conclusions

The strain generated in a flat plate was measured in order to investigate the relation between the screech sound from the flow field and the oscillation pattern of the jet. As a result, the following conclusions were obtained.

1. The strain frequency agrees well with the screech tone frequency, which means the oscillation in the pressure acting on the disk is closely related with the screech tone.
2. The screech frequency decreases and increases with increasing nozzle-plate distance. The screech is not observed in the range between l/D 3.7 and 4.0, and this zone is considered to correspond to the silent zone reported by B. Henderson.
3. The time-averaged strain tends to increase with increasing l/D, and takes the local minimum values at $l/D = 3.5$ and 4.5, where the screech tone frequency jumps to higher values.
4. The instability of the impinging jet periodically increases and decreases, and the jet becomes stable at the jump to the higher value of the screech tone.

References

1. T. Tsujimura, A. Ohta, Y. Tokunaga, J. Senda, H. Fujimoto, A Study on underexpanded jet structure under high pressure ambient. The Jpn. Soc. Mech. Engineers, Series B **70**(693), 1348–1353 (2004)
2. Y. Kaneoka, S. Kotou, Study of the relationship between the accuracy in CO_2 laser cutting and the assist gas. Trans. Jpn. Soc. Mech. Engineers, Series C **59**(562), 1952–1959 (1993)
3. B. Henderson, The connection between sound production and jet structure of the supersonic impinging jet. The J. Acoust. Soc. Am. **111**(735), 735–747 (2002)

Chapter 17
Pressure Pulse Wave Model for Continuous Arterial Blood Pressure Approximation

Agnes Jinu and K. S. Biju

Abstract Blood pressure (BP) is a crucial indicator of cardiovascular status. In arterial cannulation, a catheter inserted into a patient's blood vessels poses potential risks and inconveniences. Presently existing cuff based technique provide only short data without continuous BP information in an unobtrusive manner. Non-invasive type continuous BP measurements of patient is still a challenging task. The aim of this study is to use pressure pulse wave modeling to approximate continuous blood pressure. Pressure pulse wave (PPW) signal are collected from radial arteries is used for measurement. In order to estimate the systolic BP and diastolic BP values, the estimation technique adds the starting blood pressure values derived from a regression model and the pressure changes measured from the PPW. The PPW signals and its first derivative signal were used to derive eleven beneficial parameters. For estimation of initial BP, Multi-variable linear regression model using above features was constructed. The proposed regression model in estimating initial BP and the error of systolic BP is 1.33 mm of Hg and of diastolic BP is 1.14 mm of Hg respectively. The results from the proposed model indicated that PPW model has great potential for continuous blood pressure estimation, and can be used for round the clock blood pressure measurement of critical care patients.

17.1 Introduction

Absolutely, blood pressure (BP) is indeed a vital physiological parameter that plays a crucial role in the circulation of blood throughout the body. The force that the blood applies to the artery walls when the heart pumps it is referred to as blood pressure. It is represented as systolic blood pressure (SBP) and diastolic blood pressure (DBP) [1]. Medical professionals use blood pressure measurements as an important diagnostic tool to assess a person's cardiovascular health and risk factors.

A. Jinu · K. S. Biju (✉)
Department of Electronics and Communication Engineering, College of Engineering Trivandrum, Thiruvananthapuram, Kerala 695016, India
e-mail: bijukarunnya@cet.ac.in

A. Suryan et al. (eds.), *Proceedings of the 9th Asian Joint Workshop on Thermophysics and Fluid Science, 27–30 November 2022, Utsunomiya, Japan*, Springer Proceedings in Physics 396, https://doi.org/10.1007/978-981-99-9470-0_17

Normally, artery cannulation is an invasive method to monitoring continuous BP, which uses a catheter inserted into the blood vessel. The arterial catheter inserted into the arterial site is threaded over the guide wire and grip on the guide wire is always kept by the proceduralist during procedure. Later the guidewire is removed and the arterial line is set connected to pressure transducer [2]. The method provide precise measurements which help in immediate recognition of alterations, thus allowing early diagnosis of diseases. Most commonly used and known non-invasive method of BP estimation. Cuff is placed around the arm of a person for measurement in both methods [3]. The identification of Korotkoff sounds coming from the acoustic transducer signal if the bases of auscultatory method. So a non-invasive with continuous BP measuring system is essential for round the clock blood pressure measurement of critical care patients.

The concept of a piezoelectric-based continuous blood pressure measurement system is promising, the actual implementation and validation of such a technology would require rigorous testing and research to ensure its accuracy and reliability [4]. According to study, PPW obtained from a piezoelectric sensor is quite helpful for continuously monitoring blood pressure [5]. The extraction of features from PPW signal measured from radial arteries to construct a blood pressure evaluation by linear regression modeling. Additionally, cuff-free techniques based on pulse transit time and a novel indication known as the intensity ratio of the photoplethysmogram studied [6]. The estimating blood pressure using a linear multidimensional regression model based on pulse wave features is theoretically sound. The certain features measured from the pulse signal and using a regression model that takes these pulse wave features as inputs, you might be able to predict or estimate BP [7].

This study is to focus an effective model to estimate BP continuously using the pressure pulse wave. By incorporating the beat-to-beat blood pressure information from the pressure pulse wave, the regression model can enhance the accuracy of its BP predictions.

17.2 Methods

The BP is a crucial indicator of cardiovascular status. The arterial cannulation technique is frequently employed in a variety of medical situations to continuously monitor blood pressure. Presently existing cuff based technique provide only short data without continuous BP information in an unobtrusive manner. This study aims to develop a non-invasive type continuously reporting the BP recording system by wearing a single piezoelectric sensor on patients' wrist. The measurement strategy you're describing involves a combination of initial blood pressure measurements obtained through a regression model and pressure changes measured using a piezoelectric sensor. This combination aims to achieve continues monitoring of systolic BP and diastolic BP (Fig. 17.1).

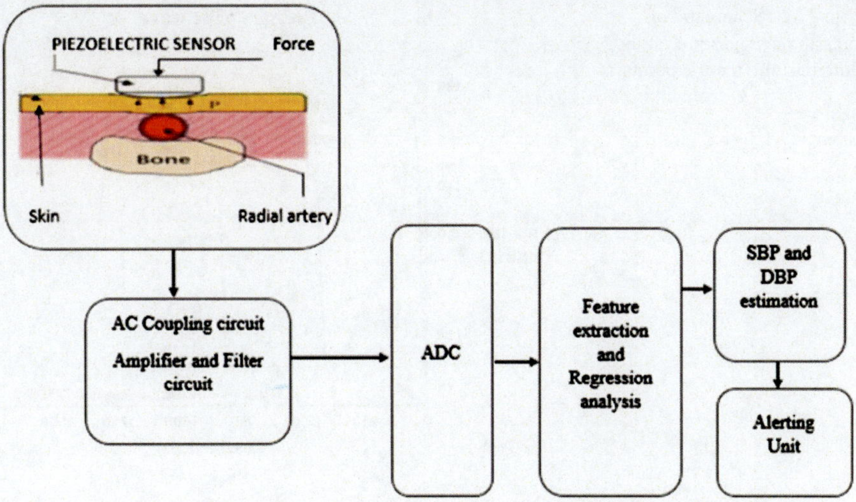

Fig. 17.1 Schematic diagram of non-invasive continuous blood pressure estimation system

Piezoelectric sensor (HK 2000B) is used to acquired pressure pulse wave from radial artery. In order to extract clear PPW signals, amplification and filtering process has to done.

In order to cover PPW signals band width (0.05–35 Hz), a combination of a Butterworth high-pass filter of 0.05 Hz and low pass filter of 35 Hz is implemented. PPW is converted to digital signal by Arduino Nano which act as ADC. Digitized signal is given to python for further feature extraction and SBP and DBP estimation. Eleven informative features were extracted from PPW and its first derivative signal. For estimation of initial BP, regression models were constructed.

This method uses a mix of an initial blood pressure prediction from a regression model with pressure changes discovered from a nearby pressure pulse wave to monitor systolic and diastolic blood pressure continuously. The systolic blood pressure and diastolic blood pressure values are calculate by the Eqs. 17.1 and 17.2. This method likely involves the use of waveform analysis to derive blood pressure values (Fig. 17.2).

$$BP_{systolic}(t) = SBP_{Regression\,(initial)} + \Delta SBP_1 + \Delta SBP_2 + \ldots \tag{17.1}$$

$$BP_{diastolic}(t) = DBP_{Regression\,(initial)} + \Delta DBP_1 + \Delta DBP_2 + \ldots \tag{17.2}$$

ΔSBP1 is the change in consecutive systolic value and ΔDBP change in consecutive diastolic value. The equation seems to involve the calculation of pressure differences between specific feature points on consecutive pressure pulse waves.

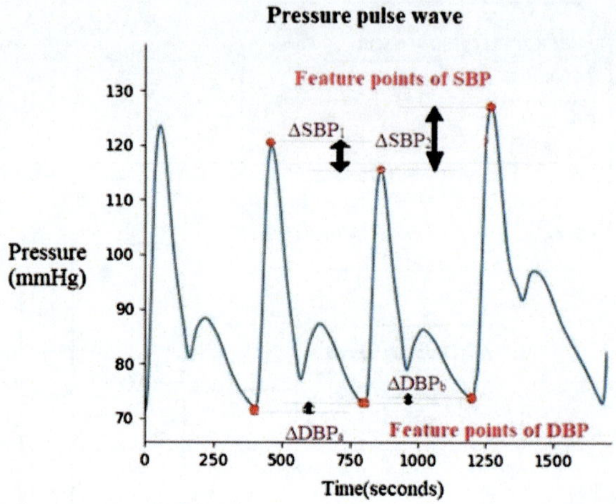

Fig. 17.2 Schematic of pressure changes in systolic and diastolic feature points

17.2.1 *Feature Extraction*

Extracting informative features from the pressure pulse wave can provide valuable insights into a person's cardiovascular health. In this study there are 11 informative features were collected from each PPW [8, 9]. These features likely help in assessing various aspects of the pulse wave and can potentially contribute to blood pressure evaluation. The first nine features were calculated from PPW and the last two features were tabulated from the first derivative of pulse wave signal [10].

17.2.2 *Multi-variable Linear Regression Model*

In a linear multivariable regression model for modeling initial blood pressure (BP) values using a set of independent variables (features). In a multivariable linear regression model, the goal is to establish a relationship between the dependent variable (in this case, BP) and multiple independent variables (features) in a linear fashion. The model can be define as per following Eq. (17.1)

$$y = \alpha_0 + \alpha_1 x_1 + \alpha_2 x_{2..........} + \alpha_n x_n + e \quad (17.3)$$

y is the SBP or DBP estimation values
$x_1, x_{2,\ldots+} x_n$ are the informative features of PPW, where n = 11.
α_0 is the intercept value of the slop.
$\alpha_1 + \alpha_2 \ldots \alpha_n$ are the coefficients of the informative features of PPW with respect to the SBP or DBP estimation values.
e is the indication of error associated with unexpected variability of measured variable which is not considered in the model.

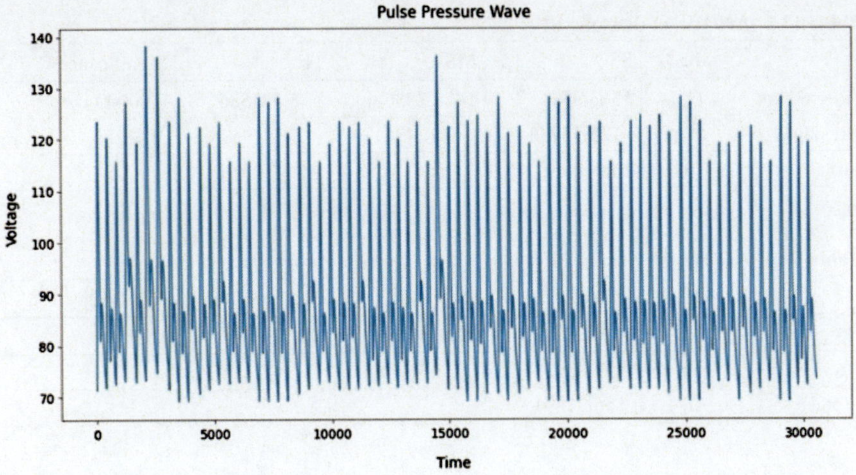

Fig. 17.3 Radial pressure pulse wave

17.3 Result

The Feature extraction, regression modeling, and beat-to-beat systolic and diastolic blood pressure estimation are important steps in this process. The sampling rate was 500 Hz is used for this study. Figure 17.3 shows a portion of radial pressure pulse wave simulated using the database collected for one minute.

In this study, developed a continuous BP prediction model, which is non-invasive and utilizes the PPW signal. A multi-variable linear regression model in estimating the initial BP using eleven features extracted from the radial artery PPW signal. Indeed, once an initial blood pressure (BP) estimation has been obtained with sufficient accuracy, the beat-to-beat continuous BP values can be calculated based on the pressure changes within the adjacent pressure pulse waveforms. This approach involves analyzing the variations in the pulse waveforms over time to track the changes in systolic and diastolic pressures from beat to beat (Tables 17.1, 17.2 and 17.3).

Table 17.1 Regression statistics of SBP and DBP

	Systolic BP	Diastolic BP
Multiple R	0.962173741	0.797611246
R square	0.925778308	0.6361837
Adjusted R square	0.896249207	0.557570767
Standard error	1.457519013	0.981480933
Observations	70	70

Table 17.2 ANOVA of diastolic BP

	df	SS	MS	F	Significance F
Regression	11	1563.3514	142.12286	73.591586	0.00000
Residual	59	125.33734	2.1243617		
Total	70	1688.6888			

Table 17.3 ANOVA of systolic BP

	df	SS	MS	F	Significance F
Regression	11	99.383922	9.034902	10.316975	0.00000
Residual	59	56.834985	0.9633048		
Total	70	156.21891			

Figure 17.4 shows the comparison of predicted DBP and actual values of DBP and the Fig. 17.5 shows the comparison of predicted SBP and actual values SBP. Mean absolute error calculated between the actual and predicted value helps in understanding the reliability of multivariable linear regression model in estimating initial BP. MAE of systolic blood pressure is 1.33 mm of Hg and of diastolic blood pressure is 1.14 mm of Hg.

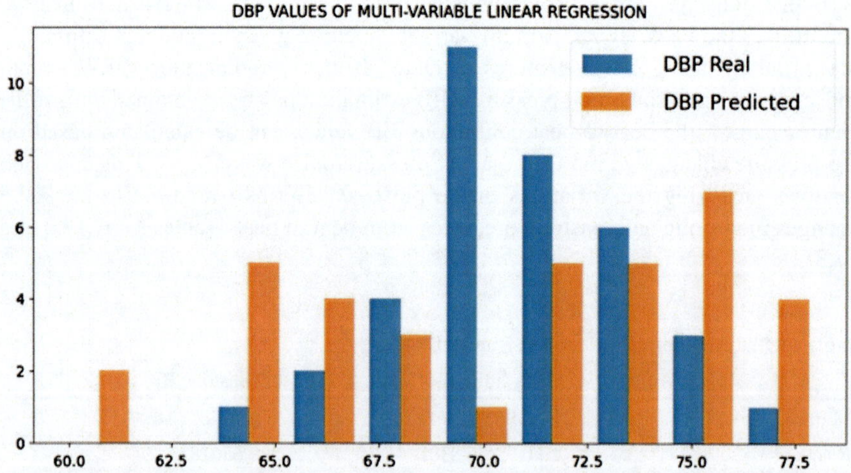

Fig. 17.4 Comparison of predicted DBP and actual values of DBP

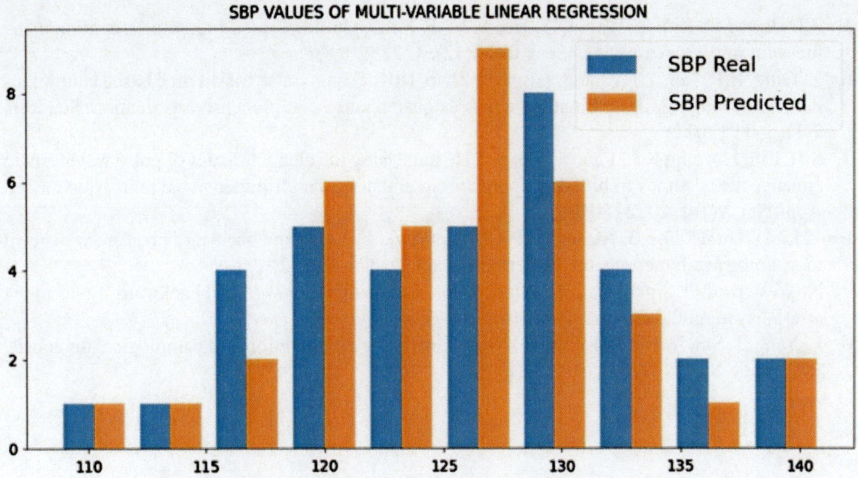

Fig. 17.5 Comparison of predicted SBP and actual values SBP

17.4 Conclusion

Here using a combination of regression modeling and analysis of the PPW signal collected from the radial artery to estimate continues systolic and diastolic blood pressure values. Proper validation and accuracy assessment are crucial for ensuring the reliability of the calculated continuous blood pressure values. Here used parametric statistical analysis for the systolic and diastolic BP. The results clearly shows the multi-variable linear regression model of both systolic and diastolic BP is good. The error between predicted and actual systolic and diastolic BP is low, with MAE were 1.33 mm of Hg for systolic BP and 1.14 mm of Hg for diastolic BP respectively. The further work can be initiated in the broad areas as, by using a flexible piezoelectric sensor, which overcomes the limitation of a solid sensor in terms of air gap for acquiring pressure pulse wave. And with the appropriate design of the analog front end continuous BP can be estimated by a wearable piezoelectric based system and it can be easily used under home health care situations with better accuracy.

References

1. L.A. Geddes, *Handbook of Blood Pressure Measurement*. (Springer Science Business Media, 2013)
2. J.A. Kitterman, R.H. Phibbs, W.H. Tooley, Catheterization of umbilical vessels in newborn infants. Pediatr. Clin. North Am. **17**, 895–912 (1970)
3. L. Peter, N. Noury, M. Cerny, A review of methods for non-invasive and continuous blood pressure monitoring: pulse transit time method is promising? Irbm. **35**(5), 271–282 (2014)
4. T.W. Wang, S.F. Lin, Wearable piezoelectric-based system for continuous beat-to-beat blood pressure measurement. Sensors **20**(3), 851 (2020)

5. Z.D. Liu, J.K. Liu, B. Wen, Q.Y. He, Y. Li, F. Miao, Cuffless blood pressure estimation using pressure pulse wave signals. Sensors **18**(12), 4227 (2018)
6. X. Ding, B.P. Yan, Y.T. Zhang, J. Liu, N. Zhao, H.K. Tsang, Pulse transit time based continuous cuffless blood pressure estimation: a new extension and a comprehensive evaluation. Sci. Rep. **7**(1), 1–11 (2017)
7. S.H. Liu, L.J. Liu, K.L. Pan, W. Chen, T.H. Tan, Using the characteristics of pulse waveform to enhance the accuracy of blood pressure measurement by a multidimensional regression model. Appl. Sci. **9**(14), 2922 (2019)
8. Y. Li, H. Gu, H. Fok, J. Alastruey, P. Chowienczyk, Forward and backward pressure waveform morphology in hypertension. Hypertension **69**(2), 375–381 (2017)
9. N. Westerhof, P. Sipkema, G.C. van den Bos, G. Elzinga, Forward and backward waves in the arterial system. Cardiovasc. Res. **6**(6), 648–656 (1972)
10. L. Yang, J. Sui, Y. Hu, Pulse wave K value averaging computation and pathological diagnosis. J. Comput. **8**(6), 1472–1479 (2013)

Chapter 18
Development of Virtual Optical Waveguide: Effect of Flatness Ratio of Open Elliptical on High-Pressure Area by 3D Numerical Calculation

Nitta Tetsuro, Ikeyama Tetsuro, Fukuoka Hiroshi, Suda Atsushi, Ishijima Ayumu, and Nakagawa Keiichi

Abstract In this paper, we propose to form a virtual optical waveguide with a focused underwater shock wave. The region where the shock wave is focused becomes high pressure and high density. In other words, a localized high-refractive index region is obtained. Increasing the pressure gradient generated by focusing shock wave, light can be confined within the refractive index region and propagate without leakage. We investigated the optimal number of divisions for 3D-Gaussian pressure distribution to model a high-pressure region generated by laser irradiation. Also, we discussed the effects of the flatness ratio of the open elliptical reflector on the focusing behavior of underwater shock wave by 3D Numerical calculation. In the calculation, the axisymmetric 3D unsteady compressible ALE (Arbitrary Lagrangian–Eulerian) and the Mie-Grüneisen equation of state were solved using the finite volume method. As a result, it was determined that three divisions Gaussian pressure distribution is optimal for modeling the high-pressure area generated by laser irradiation. Furthermore, it was confirmed that the flatness ratio decreased, the shock wave becomes more focused around the focal point.

18.1 Introduction

The high-pressure region generated by a shock wave focusing phenomena is shown interest in medical applications due to its low-invasive [1]. Furthermore, this region is expected to have applications in optogenetics.

N. Tetsuro · I. Tetsuro · F. Hiroshi (✉) · S. Atsushi
National Institute of Technology, Nara College, 22 Yata, Yamatokoriyama, Nara 639-1080, Japan
e-mail: fukuoka@mech.nara-k.ac.jp

I. Ayumu · N. Keiichi
University of Tokyo, 7-3-1Hongo, Bunkyo-Ku, Tokyo 113-8656, Japan

A. Suryan et al. (eds.), *Proceedings of the 9th Asian Joint Workshop on Thermophysics and Fluid Science, 27–30 November 2022, Utsunomiya, Japan*, Springer Proceedings in Physics 396, https://doi.org/10.1007/978-981-99-9470-0_18

Optogenetics, a technology for manipulating nerve cells with light, is widely used to study biological phenomena [2]. On the other hand, optogenetics cannot deliver light to the deep brain because the light is affected by absorption, reflection, and scattering when the light passes through cells [3]. Optogenetics needs a new system that is low-invasive and delivers light to the deep brain.

Therefore, we proposed forming an optical waveguide with a focusing shock wave using an open elliptical reflecting wall. The shock wave was generated using laser irradiation at the first focal point (f_1), and the shock wave propagated while reflecting off the reflective wall. The shock wave is then focused around the second focal point (f_2). In this way, a high reflective index region is formed at f_2. By increasing the pressure difference between the inside and outside the region, light can be confined within the high reflective index region and propagate without leakage.

In this paper, we investigated the underwater shock wave focused using the open elliptical by 3D numerical analysis. The first step is investigating the optimal number of divisions for 3D-Gaussian pressure distribution to model a high-pressure region generated by laser irradiation. Next, we investigated the effect of the open elliptical shape on the behavior of the shock wave.

18.2 Numerical Procedure and Boundary Conditions

Figure 18.1 shows the calculation area and the boundary conditions. The distance between the open end of the elliptical reflector and f_2 is 10 mm. In the numerical calculations, the axisymmetric three-dimensional unsteady compressible ALE (Arbitrary Lagrangian–Eulerian) and the Mie-Gruneisen equation of state were solved using the finite volume method. The mesh size was 0.1–0.05 mm.

This calculation expresses the pressure waveform generated by laser irradiation for initial pressure. Typically, the pressure waveform generated by laser irradiation is Gaussian distribution. However, the necessary number of grids to give 3D Gaussian pressure distribution by initial conditions is unclarified. Therefore, the 3D Gaussian pressure distribution was divided into a rectangular shape to investigate the optimal number of grids for this calculation. The parameters for this calculation are the number of divisions (N_{st} = 1, 3 and 6) for 3D Gaussian pressure distribution as shown in Fig. 18.2. Also, the open elliptical shape in this calculating is flatness ratio $F = 0.5$ ($E_L = 30$, $E_S = 15$ mm).

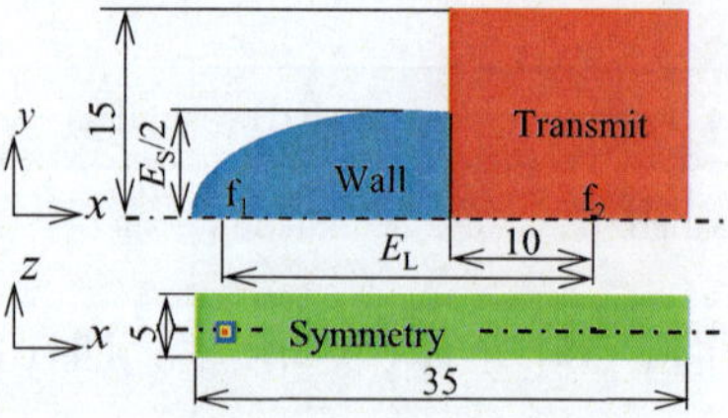

Fig. 18.1 Calculation area and boundary conditions

Fig. 18.2. 3D Gaussian pressure distribution

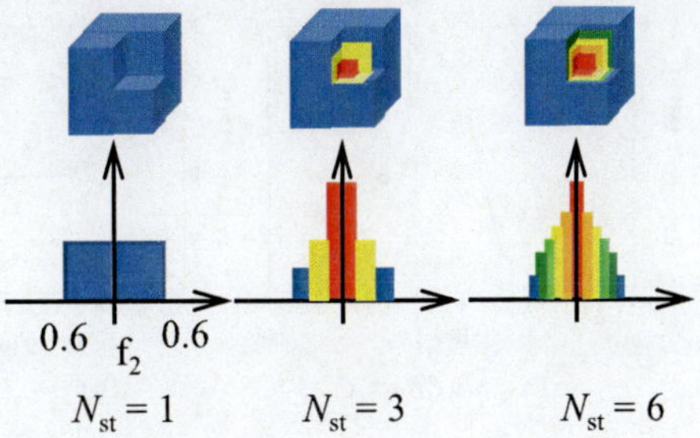

The original aim of this paper is to investigate the effect of the difference in the open elliptical shape on the shape of the high-pressure region at the time when the shock wave is focused. The parameters for the calculations are the flatness ratio $F = 0.3$ ($E_L = 30$, $E_S = 21$ mm), $F = 0.5$ ($E_L = 30$, $E_S = 15$ mm) and $F = 0.7$ ($E_L = 30$, $E_S = 9$ mm). The initial pressure is given 1000 MPa by Gaussian distribution on f_1 in Fig. 18.1.

18.3 Result and Discussion

18.3.1 Modeling Initial Pressure by 3D Gaussian Distribution

Figure 18.3a–c show the pressure contour for $N_{st} = 1$, 3 and 6 at the time when the shock wave focused. The pressure value for $N_{st} = 1$ is lower than $N_{st} = 3$ and 6 around the focal point (f_2). Also, in Fig. 18.3b and c, the pressure distribution in the case of $N_{st} = 3$ is very similar to $N_{st} = 6$.

To compare quantitatively, Fig. 18.4 shows the pressure distribution on the *x*, *y* and *z* axis around the focal point in Fig. 18.3. In Fig. 18.4a, the pressure waveforms are similar to all numbers of divisions (N_{st}). However, the peak pressure value at $x = 0.0$ mm for $N_{st} = 1$ is significantly less than $N_{st} = 3$ and 6. Also, as shown in

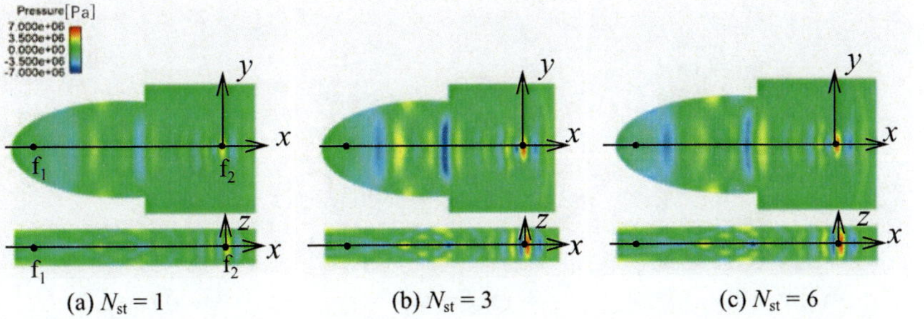

Fig. 18.3 Pressure flow field for changed number of divisions for 3D Gaussian pressure distribution

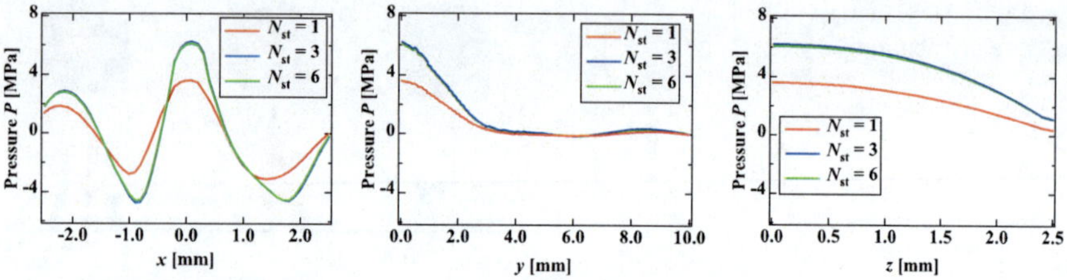

(a) x-axis direction (b) y-axis direction (c) z-axis direction

Fig. 18.4 Relation between distance for each axis direction and pressure

Fig. 18.4b, the pressure value in the range of $x = 0.0$ to 3.0 mm for $N_{st} = 1$ is very different from the $N_{st} = 3$ and 6. Figure 18.4c shows the same features as Fig. 18.4a. In Fig. 18.4a–c, the peak pressure value for $N_{st} = 3$ and 6 is higher than that for $N_{st} = 1$ on all axes. This difference is due to the peak value of the initial pressure for $N_{st} = 3$ and 6 is higher than $N_{st} = 1$. Therefore, 3 and 6 divisions are the optimal number of divisions for modeling the initial pressure by Gaussian distribution.

18.3.2 Effect of Flatness Ratio on Shock Wave Propagation

Figure 18.5 shows the pressure contour at the time when the shock wave is focused for each flatness ratio. The x–y plane of Fig. 18.5a shows that the shock wave for F = 0.3 is the most focused of all flatness ratios. On the other hand, in the x–y plane of Fig. 18.5c, the shock wave for F = 0.7 is not as focused as F = 0.3 and 0.5. Therefore, it is seen from Fig. 18.5a–c as the flatness ratio decreases, the shock wave becomes more focused around the focal point.

In the x–y plane of Fig. 18.5a and b, the pressure flow field for F = 0.3 and 0.5 are seen forming a circular high-pressure region around the focal point. Also, in the x–z plane of Fig. 18.5a and b, it is seen forming a rectangular high-pressure region around the focal point. Therefore, it was found to form the cylindrically high-pressure region

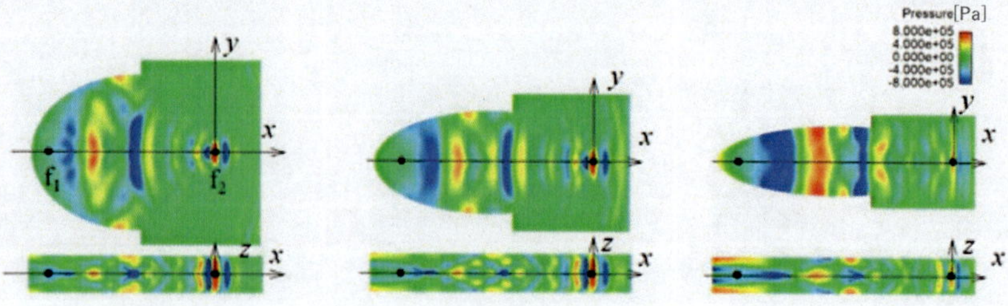

(a) $F = 0.3$ (a) $F = 0.5$ (a) $F = 0.7$

Fig. 18.5 Pressure flow field each flatness ratio at $t = 22.0\ \mu s$

in the case of $F = 0.3$ and 0.5. These results imply that we can form the cylindrically optical waveguide using the shock wave focused.

18.4 Conclusion

In this paper, the shock wave focusing phenomena using an open elliptical reflector was numerically studied by solving the axisymmetric three-dimensional unsteady compressible ALE (Arbitrary Lagrangian–Eulerian) and the Mie-Gruneisen equation of state.

First of all, we investigated the necessary number of grids to give 3D Gaussian pressure distribution. The parameters for this calculation are the number of divisions ($N_{st} = 1, 3$ and 6) for 3D Gaussian pressure distribution. We compared the pressure flow field and pressure values at the focal point for 1, 3 and 6 divisions. As a result, it was found that the 3 and 6 divisions of 3D Gaussian pressure distribution are optimal for modeling pressure waveforms generated by laser irradiation.

Second of all, we investigate the effect of the difference in the open elliptical shape on the shape of the high-pressure region at the time when the shock wave was focused. For flatness ratio $F = 0.3$, 0.5 and 0.7, as the flatness ratio decreases the shock wave was more focusing around the focal point. Also, it was found that the cylindrical high-pressure region around the focal point in the case of $F = 0.3$ and 0.5 by performing 3D Numerical Calculations. The shape of the high-pressure region implies that we can form the cylindrically optical waveguide using shock wave focused.

References

1. J. Min, L. Jae-Young, P. Eun-Joo, First report on the persist time of the free radical produced by shock wave pulses employed in clinical ESWL. Ultrasonics Sonochem. **83**, 105927 (2022)
2. L. Candice, L. Andreanne, L. Jiashu, X. Simon, B. Chen, Light up the brain: the application of optogenetics in cell-type specific dissection of mouse brain circuits. Front. Neural Circuits **14**, 1–14 (2020)
3. A. Itia, P. Daryl, A. Timo, A. Lucy, K. Ethan, R. Halina, Scattering of sculpted light in intact brain tissue, with implications for optogenetics. Sci. Rep. **5**, 11501 (2014)

Chapter 19
Experimental Investigation on Self-propulsion of Leidenfrost Droplets on Narrow Ratchet Surfaces

Praveen Arjunan, Abhinanth Jayakumar, B. S. Gireeshkumaran Thampi, P. S. Tide, K. Niju Mohammed, and Ranjith S. Kumar

Abstract Liquid drops deposited on hot solid surface at the Leidenfrost temperature form a vapour cushion, producing self-propulsion, especially on ratchet surfaces. The study investigate this phenomenon and extends it to surface-modified hot aluminum surfaces. Experiments were carried out at Leidenfrost temperatures on normal ratchet and narrow-width ratchets to explore the self-propulsion of droplet by varying the droplet diameters. Experiments reveal that narrower ratchets relative to droplet diameter limit propulsion due to lower contact area, reduced vapour layer generation, and heat loss through the sidewalls. When normal and narrow-width ratchets are compared, the contact area between droplet and ratchet surface, vapour layer, and diameter of droplet has an impact on droplet self-propulsion.

19.1 Introduction

When a liquid comes into close contact with a mass that is significantly hotter than the liquid's boiling point, an insulating vapour layer forms, preventing the liquid from rapidly boiling. When the temperature of the surface rises above 100 °C (212 °F), the water drops hiss and evaporates quickly. Later, as the temperature rises above the Leidenfrost point, the Leidenfrost effect takes place. When water droplets come into contact with the surface, they form small balls of water and skitter around for much longer than when the surface was colder. This effect continues until a significantly higher temperature causes any additional drops of water to evaporate too quickly

P. Arjunan (✉) · B. S. G. Thampi · P. S. Tide
School of Engineering, Cochin University of Science and Technology, Kochi, Kerala 682022, India
e-mail: praveenarjun@cet.ac.in

P. Arjunan · A. Jayakumar · K. N. Mohammed · R. S. Kumar
Micro/Nanofluidics Research Laboratory, Department of Mechanical Engineering, College of Engineering Trivandrum, APJ Abdul Kalam Technological University Kerala, Thiruvananthapuram 695016, India

A. Suryan et al. (eds.), *Proceedings of the 9th Asian Joint Workshop on Thermophysics and Fluid Science, 27–30 November 2022, Utsunomiya, Japan*, Springer Proceedings in Physics 396, https://doi.org/10.1007/978-981-99-9470-0_19

to produce the effect. By replacing the level ground with a surface with elevation ratchets, the direction of Leidenfrost particles can be controlled. Linke et al. [1] demonstrated this phenomenon for the first time using millimeter scale ratchets. According to Quere [2] and Mrinal et al. [3], Leidenfrost liquids directly placed on a heated material self-propel via surface ratchets, lending support to Linke et al. [1] viscous drag mechanism. The Leidenfrost effect describes droplet levitation on a vapour layer caused by a volatile liquid coming into contact with a heated solid. The transportation is carried across the entire ordered levelled area and is limited by the dissipation time of the droplet. When solids containing volatile liquids are heated to Leidenfrost temperature, vapour separates the liquid from its substrate. These droplets are extremely mobile and disperse almost immediately after being deposited. It explains more about heat transfer, evaporation, and flow within the vapour cushion.

19.2 Theory

As temperature changes above the Leidenfrost point, the bottom part of the water droplet vaporizes immediately upon contact with the hot substrate. The resulting gas suspends the rest of the water droplet just above it, preventing any further direct contact between the liquid water and the hot plate. Heat transfer between the pan and the droplet is significantly slowed because steam has much lower thermal conductivity than water. This also allows the drop to skid around the pan on the layer of gas just beneath it. The temperature at which the Leidenfrost effect begins is difficult to predict. Even if the volume of the drop of liquid remains constant, the Leidenfrost point can vary significantly depending on the surface properties. Leidenfrost propulsion applications include two phase cooling in microprocessors, drug delivery systems, high heat flux cooling equipment development, electronic devices such as high power integrated circuits and heat pumps, and low friction Leidenfrost engines[4–9]. Despite the fact that this effect is increasingly useful, studies on self-propulsion of Leidenfrost droplets on narrow ratchet surfaces are limited. Droplet diameter, substrate temperature, surface texture of the substrate (aspect ratio, wave length, etc.), surface tension and viscous mechanism of the liquid, inclination of the substrate, capillary length of the droplet and heat loss between the spaces of the droplets are found to be the influencing parameters of droplet self propulsion [10–15] (Fig. 19.1).

Aluminium alloy ratchet substrates, DI water, a high fps (5000) camera, a thermocouple, a burette, and a light apparatus were used in this experiment. Type K thermocouples are used to measure temperature in the operating range. In order to study and compare the effect of propulsion, the droplet diameters and aspect ratios were fixed. The Leidenfrost point of DI water on Aluminium alloy substrate is determined experimentally and found to be 210 °C using a heater, temperature controller, high fps camera, and lighting apparatus (Fig. 19.2).

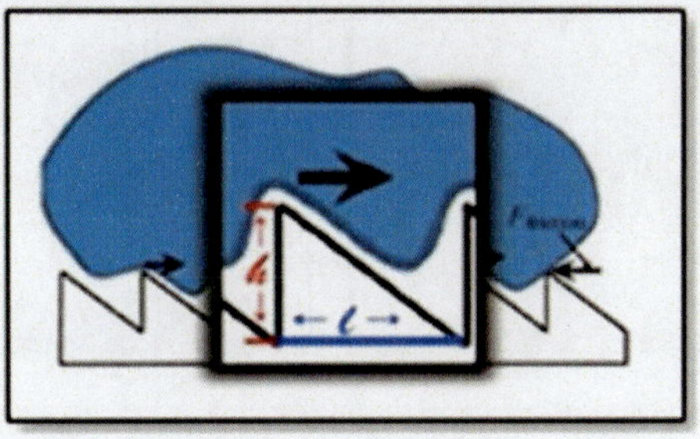

Fig. 19.1 Droplet propulsion and aspect ratio of a ratchet

Fig. 19.2 Aluminium alloy substrate prepared for testing the Leidenfrost point of liquid droplet

19.3 Methodology

The droplet diameters were calculated in order to fix the ratchet widths and to perform the analysis. The diameters of the droplets 2.65, 2.8, and 3.00 mm for the analysis are obtained using three needles of gauges 26, 23 and 18 respectively. The droplet diameters captured using camera are measured using the Image-J software. Surface textures are created on substrates for various aspect ratios using taper milling and step micromilling. After that, all of the fabricated substrates are inspected with a profilometer, optical microscope, and scanning electron microscope (SEM). Aluminium alloy was used as a substrate material in this experimental study (Fig. 19.3).

Leidenfrost temperature has been measured, as a prerequisite for conducting the experiments for finding the average velocity of Deionized water droplet. Leidenfrost temperature is an important parameter governs boiling heat transfer over hot surface. Formation of vapour layer contributes a high thermal resistance results low heat transfer characters. The Leidenfrost effect and propulsion of droplets under temperatures of 200 and 220 °C were experimentally analyzed using different diameters of needles at the substrate having aspect ratio 0.1. The droplets are of deionized water. Similarly, the Leidenfrost propulsion of deionized droplets are experimentally analyzed on the modified low width ratchet surface. The propulsion were analyzed

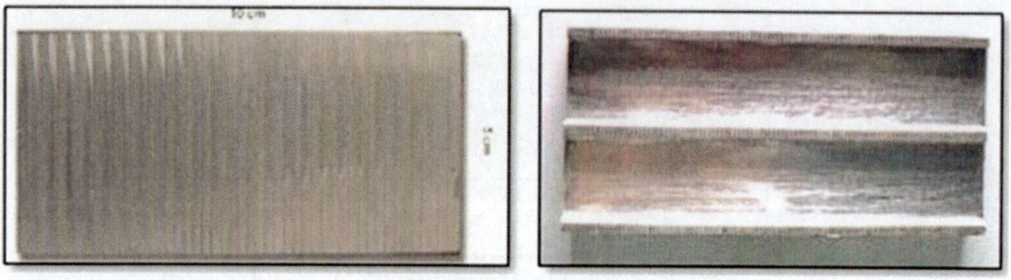

Fig. 19.3 Normal ratchet surface of aspect ratio 0.1 and three numbers of low width ratchet machined on substrate of aspect ratio 0.2 (widths 2.5, 2.7 and 3 mm)

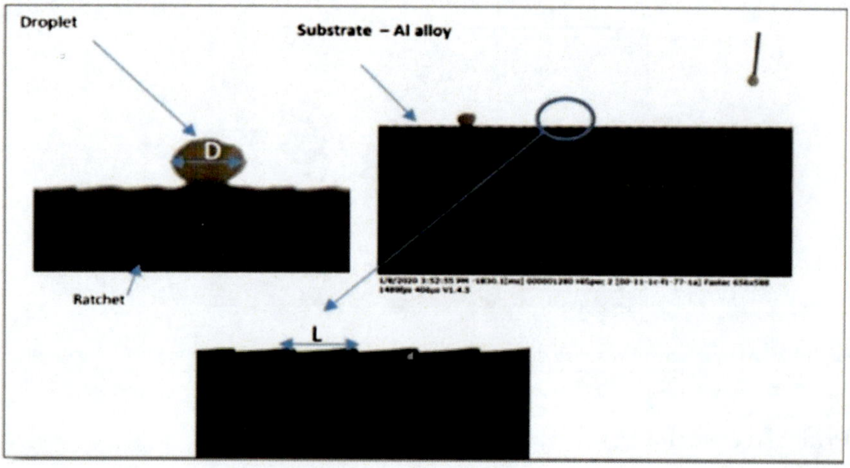

Fig. 19.4 Camera image of substrate and droplet propulsion

by scaling the images captured using the camera. The average velocity is found out by the recorded time and calculating the displacement through scaling method. The ratio of this displacement and time gives the average velocity, which resulted in plotting the curves to analyze the velocity narrow width ratchet surfaces (Fig. 19.4).

19.4 Results and Discussions

The velocity-time relations were obtained for three diameters of droplets under 200 and 220 °C on the normal ratchet surface. Time of propulsion of droplets on the ratchet surface are obtained from the Hi-Spec image capture software (Tables 19.1, 19.2, Figs. 19.5 and 19.6).

The variation in velocities with ratchet width and droplet diameter ($w = d$, $w < d$, $w > d$) is caused by a decrease in contact area, reduction in the formation of vapour

Table 19.1 Self-propulsion velocity comparison of propulsion of droplets of different diameters at 200 and 220 °C

200 °C (normal ratchet)		220 °C (normal ratchet)	
Droplet diameter (mm)	Self-propulsion velocity (cm/s)	Droplet diameter (mm)	Self-propulsion velocity (cm/s)
2.65	10.1	2.65	8.9
2.8	6.1	2.8	5.9
3.00	5.1	3.00	5.7

Table 19.2 Velocity of propulsion of droplets of different diameters on different widths at 200 °C

200 °C (low-width ratchets)	
Conditions	Self-propulsion velocities (cm/s)
Ratchet width (3 mm) = droplet diameter (3 mm)	8.9
Ratchet width (2.7 mm) > droplet diameter (2.65 mm)	9.9
Ratchet width (2.5 mm) < droplet diameter (2.8 mm)	6.5

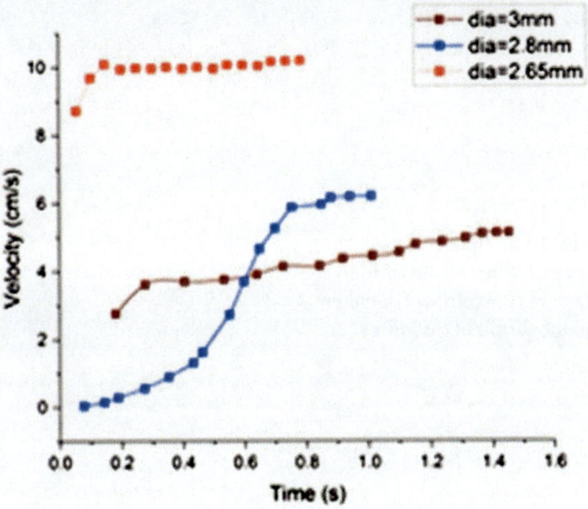

Fig. 19.5 Self-propulsion velocity–time curve of DI water droplets of temperature (200 °C) on normal ratchet surface

layer and the maintenance of Leidenfrost temperature in the contact region. The study shows with reducing the width of ratchet than droplet diameter decreases droplet propulsion over the ratchet. It has been observed that as droplet diameter increases, the contact area decreases, resulting in a reduction in vapour layer formation, which creates a barrier for self-propulsion. Furthermore, heat loss through the side-walls creates a cooling effect on the ratchet surface, reducing propulsion velocity, which is a major challenge in maintaining Leidenfrost temperature (Figs. 19.7, 19.8, 19.9, 19.10 and 19.11).

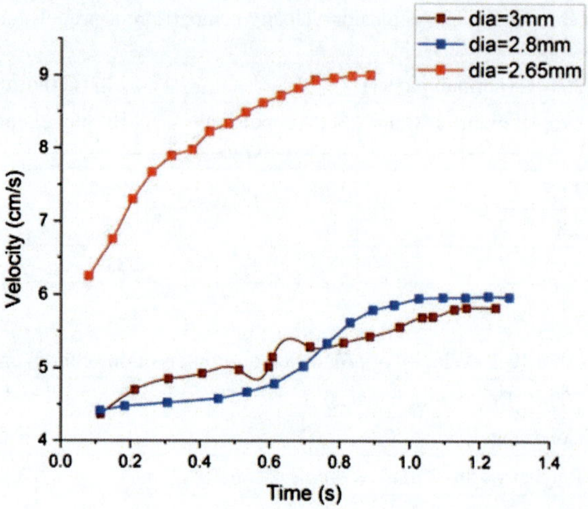

Fig. 19.6 Self-propulsion velocity -time curve of DI water droplets of temperature (220 °C) on normal ratchet surface

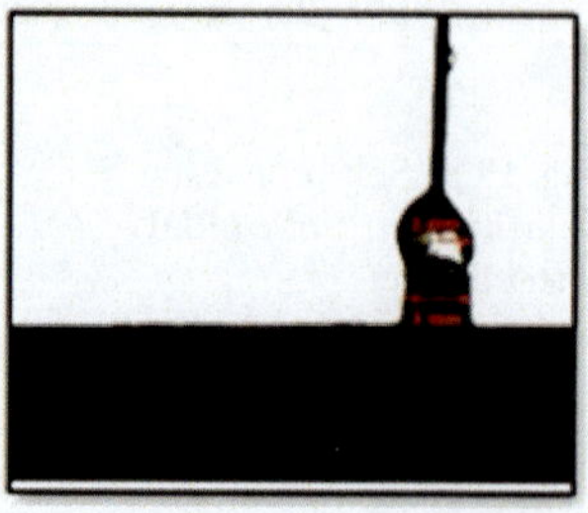

Fig. 19.7 Image of the droplet when diameter of the droplet (3 mm) = Width of the ratchet (3 mm)

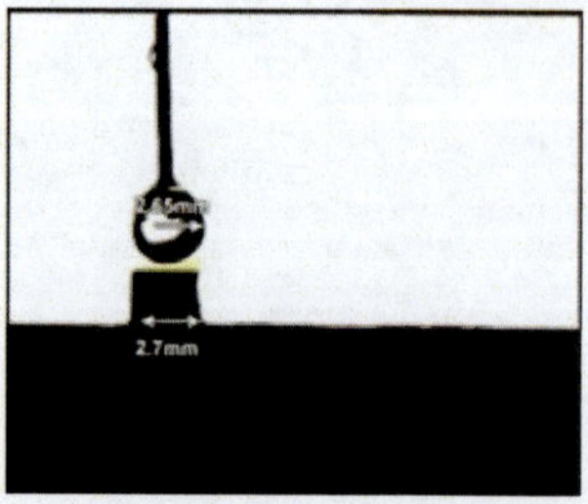

Fig. 19.8 Image of the droplet when width of the ratchet (2.7 mm) > Diameter of the droplet (2.65 mm)

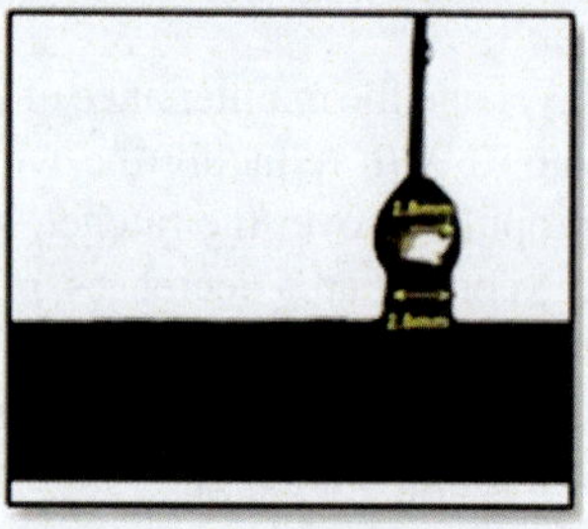

Fig. 19.9 Image of the droplet when width of the ratchet (2.5 mm) < Diameter of the droplet (2.8 mm)

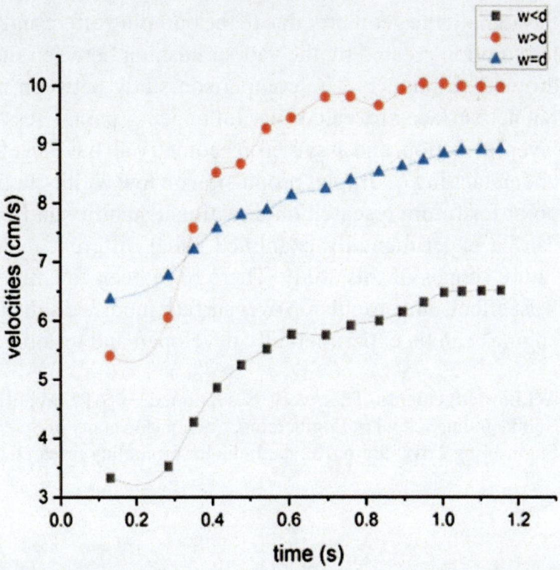

Fig. 19.10 Self-propulsion velocity–time curve of deionized water droplets of different conditions on low-width ratchet surfaces

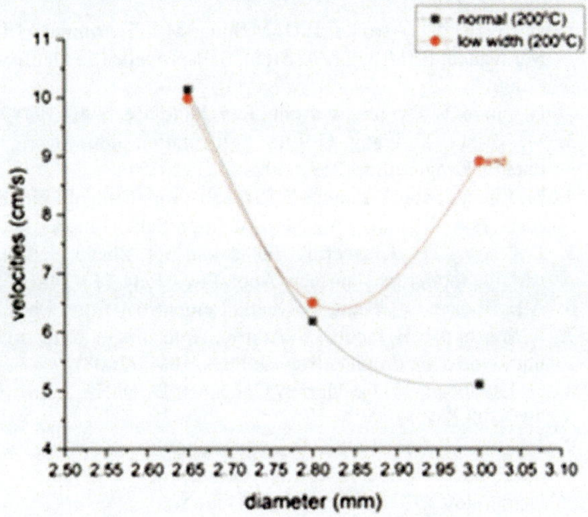

Fig. 19.11 Self propulsion velocity—diameter comparison curve of both normal and low width ratchet surfaces

19.5 Conclusions and Future Scope

According to the findings, the propulsion velocity of the droplets changes depending on the presence of vapour layer, viscous drag, and pressure-driven flow throughout the ratchet for different geometry or asymmetry. As previously stated by Linke et al. [1], for asymmetric ratchets, the solid liquid interaction creates relative motion throughout the geometry due to the pumping effect and Marangoni flow. The frictionless motion created by the vapour cushion between the ratchet teeth also allows the droplets to bounce. The comparison study between normal ratchet and low-width ratchet surfaces revealed that influencing parameters such as contact area, vapour layer formation, and assymetric geometry all have an effect on droplet propulsion. An understanding of droplet propulsion on low width ratchet surfaces provides a starting point for future research on directional stability and control. Propulsion parameters can be experimentally examined using different materials (droplets/solids) in the future scopes of this study. There have been few numerical studies on the Leidenfrost effect, and propulsion over surface modified substrates. Low friction Leidenfrost engines can be experimentally developed and tested.

Acknowledgements This work is supported by APJ Abdul Kalam Technological University, Kerala, India, School of Engineering, Cochin University of Science and Technology and College of Engineering Trivandrum, Kerala, India for promoting research in microfluidics and nano-fluidics.

References

1. H. Linke, B.J. Alemán, L.D. Melling, M.J. Taormina, M.J. Francis, C.C. Dow-Hygelund, V. Narayanan, R.P. Taylor, A. Stout, Self-propelled Leidenfrost droplets. Phys. Rev. Lett. **96**(15) (2006)
2. D. Quere, L. Dynamics, Annu. Rev. Fluid Mech. **45**, 197–215 (2013)
3. M. Mrinal, X. Wang, C. Luo, Self-rotation-induced propulsion of a Leidenfrost drop on a ratchet. Langmuir **33**(25), 6307–6313 (2017)
4. M. Chen, Z. Jia, T. Zhang, Y. Fei, Self-propulsion of Leidenfrost droplets on micropillared hot surfaces with gradient wettability. Appl. Surf. Sci. **433**, 336–340 (2018)
5. J.M. Arter, D.J. Cleaver, K. Takashina, A.T. Rhead, Self-propelling Leidenfrost droplets on a variable topography surface. Appl. Phys. Lett. **113**(24) (2018)
6. A.L. Biance, C. Clanet, D. Quéré, Leidenfrost drops. Phys. Fluids **15**(6), 1632–1637 (2003)
7. A. Snezhko, E.B. Jacob, I.S. Aranson, Pulsating–gliding transition in the dynamics of levitating liquid nitrogen droplets. New J. Phys. **10**(4) (2008)
8. G. Lagubeau, M. Le Merrer, C. Clanet, D. Quéré, Leidenfrost on a ratchet. Nat. Phys. **7**(5), 395–398 (2011)
9. J.T. Ok, E. Lopez-Ona, D.E. Nikitopoulos, H. Wong, S. Park, Propulsion of droplets on micro-and sub-micron ratchet surfaces in the Leidenfrost temperature regime. Microfluid. Nanofluidics **10**, 1045–1054 (2011)
10. G. Dupeux, M. Le Merrer, G. Lagubeau, C. Clanet, S. Hardt, D. Quéré, Viscous mechanism for Leidenfrost propulsion on a ratchet. Europhys. Lett. **96**(5)(2011)
11. Á.G. Marín, D. Arnaldo del Cerro, G.R.B.E. Römer, B. Pathiraj, D. Lohse, Capillary droplets on Leidenfrost micro-ratchets. Phys. Fluids **24**(12) (2012)
12. G. Dupeux, P. Bourrianne, Q. Magdelaine, C. Clanet, D. Quéré, Propulsion on a superhydrophobic ratchet. Sci. Rep. **4**(1) (2014)
13. T. Baier, G. Dupeux, S. Herbert, S. Hardt, D. Quéré, Propulsion mechanisms for Leidenfrost solids on ratchets. Phys. Rev. E **87**(2) (2013)
14. A. Bouillant, T. Mouterde, P. Bourrianne, A. Lagarde, C. Clanet, D. Quéré, Leidenfrost wheels. Nat. Phys. **14**(12), 1188–1192 (2018)
15. C. Luo, M. Mrinal, X. Wang, Self-propulsion of Leidenfrost drops between non-parallel structures. Sci. Rep. **7**(1) (2017)

Chapter 20
Low-Dimensional Dynamic Representation of Unsteady Flow Using Convolutional Neural Network

Yosuke Shimoda and Naoya Fukushima

Abstract To obtain low-dimensional dynamic representation and spatio-temporal mode decompositions of unsteady flow, we develop 3D-CNN Mode Decomposition Model. The model is applied to flow around a cylinder at Reynolds number, $Re_D = U_\infty D/\nu$=100, as an example of unsteady flows. The input of the model, which is a time series of flow field during one cycle of vortex shedding, are mapped into 3 modes in the latent space, and then a time series of each decomposed flow field is reconstructed from each mode. For low-dimensional dynamic representation, the models with only 3 modes can represent the flow field for one cycle of vortex shedding with high accuracy. For spatio-temporal mode decompositions, the first decomposed field, which has largest energy, shows large unsteady structures, corresponding to the Karman vortices in the wake region. The second decomposed flow fields is similar to the time-averaged flow fields of the ground truth with small and higher frequency oscillations. The third decomposed field, which has smallest energy, consists of unsteady structures of opposite phase, smaller size, and smaller magnitude in the wake. By developing 3D-CNN-MD Model, we can enhance our understanding of the flow fields with spatio-temporal variations though deep learning.

20.1 Introduction

Fluid flows, especially turbulence, have a high non-linear and of multi-scale nature. To resolve environmental and energy problems through performance improvement of various thermo-fluid devices, it is necessary to understand, predict and control flow phenomena. Therefore, many numerical and experimental studies have been conducted. In recent years, with the continuous development of computers, numerical algorithms, measuring instruments and measurement techniques, we have obtained massive numbers of flow field data from direct numerical simulations (DNS)

Y. Shimoda (✉) · N. Fukushima
Tokai University, Hiratsuka Kanagawa 259-1292, Japan
e-mail: y-shimoda@fuji.tokai-u.jp

A. Suryan et al. (eds.), *Proceedings of the 9th Asian Joint Workshop on Thermophysics and Fluid Science, 27–30 November 2022, Utsunomiya, Japan*, Springer Proceedings in Physics 396, https://doi.org/10.1007/978-981-99-9470-0_20

and laser-based flow measurements. Mode decomposition can extract important features from numerical and experimental data to enhance our understanding of flow phenomena. Proper Orthogonal Decomposition (POD) [1] and Dynamic Mode Decomposition (DMD) [2] are widely used in the area of research on fluid mechanics.

Deep Learning, which has attracted much attention in various fields in recent years, can be a nonlinear analysis method when we apply a nonlinear activation function. Therefore, deep learning is expected to have the potential for understanding flow phenomena. For example, Murata et al. have constructed a mode decomposing convolutional neural network autoencoder (MDCNN-AE) that a two-dimensional velocity field of cylinder wake are mapped into two modes in the latent space [3]. Wu et al. investigated effects of three key hyperparameters in MD-CNN-AE on the reconstruction of flow around a foil [4]. Fukami et al. proposed a hierarchical autoencoder based on CNN and Autoencoder for extracting nonlinear modes of flow fields [5]. The method is applied to two-dimensional cylinder wake and turbulent channel flow. Qu et al. applied a mode decomposing autoencoder framework using a deep variational autoencoder to cylinder flow at different Reynolds numbers [6]. Eivazi et al. proposed a deep probabilistic-neural-network architecture based on β-variational autoencoder and CNN for mode decomposition of turbulent flow through a simplified urban environment [7].

In this study, to obtain low-dimensional "dynamic" representation and spatio-temporal mode decomposition of unsteady flow, we construct 3D-CNN Mode Decomposition Model based on Convolutional Neural Network (CNN) and Autoencoder. The model extract spatio-temporal important features from a time series of flow fields and spatially and temporally decompose the flow fields to limited nonlinear modes with dimensional compression and restoration process by CNN. We apply the model to a flow around a cylinder at Reynolds number, $Re_D = U_\infty D/\nu = 100$ and evaluate characteristics of decomposed flow fields.

20.2 Method

20.2.1 Training Data

In this study, as an example of unsteady flow, a two-dimensional flow field around a cylinder is low-dimensionalized and decomposed. In order to construct the training data, a numerical simulation of a flow around a cylinder have been conducted by using OpenFoam v2006. The Reynolds number, Re_D, based on the inlet velocity of the flow, U_∞, and the cylinder diameter, D, is $Re_D = U_\infty D/\nu$=100. The pressure–velocity coupling method is PIMPLE method, which is a combination of PISO and SIMPLE methods.

As shown in Fig. 20.1, the computational domain is $24D \times 20D$ in the streamwise and spanwise directions, respectively, and the cylinder is placed $8D$ downstream

Fig. 20.1 Computational domain

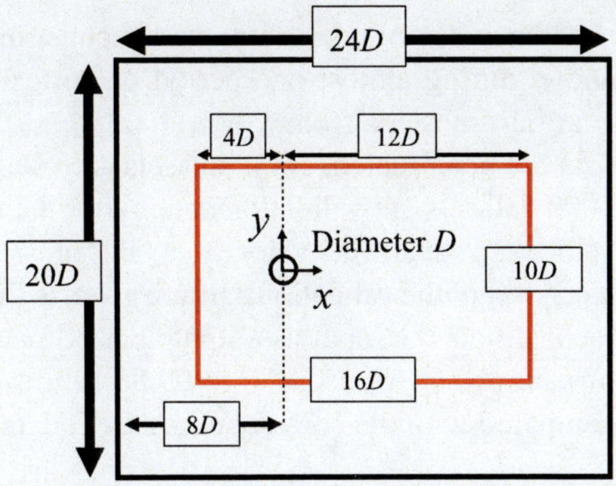

from the inlet. All the physical quantities described below in this paper are nondimensionalized by the inlet velocity U_∞ and the cylinder diameter D. The center of the cylinder is set to the origin of the coordinate system, the streamwise direction of flow is x, and the spanwise direction is y. A boundary-fitted grid system is used with the grids number of 32,768. The time step width is $\Delta t = 9.77 \times 10^{-2}$ and the total integration time is $T = 293.1$. At the Reynolds number, $Re_D = 100$, von Karman vortex street is formed in the wake of a cylinder. A dataset of the flow field in 62 timesteps corresponds to almost one period of vortex shedding. However, it is not exactly one period, and there is a slight deviation in time.

By extracting a part of the spatial data in the numerical simulation of the flow, the data for the training and validation are constructed. The velocity component in the streamwise direction, u, the velocity component in the spanwise direction, v, and pressure p in the region surrounded by the red lines in Fig. 20.1 are extracted. In addition, u, v and p on the boundary-fitting grids are interpolated to 64×64 equally spaced grids in the region. Time interval of the training data is set to $\Delta t = 9.77 \times 10^{-2}$, which is the same with the numerical simulation of the flow, and 3000 steps of the data are stored. 2000 datasets of u, v and p in 62 timesteps as the input data and the supervised data are constructed. Here, 62 timesteps is corresponding to almost one cycle of vortex shedding. 1000 datasets are used for training, and the remaining 1000 datasets are for validation.

20.2.2 Problem Setting of Training and Reconstruction

The problem setting of training and reconstruction of the two-dimensional flow field around a cylinder is set up as follows. The spatial distributions of velocities u, v and pressure p at 62 timesteps from $t_{62n+m+1}$ to $t_{62n+m+62}$ are given as input data, and the supervised data. Here, the number n denotes the number of periods for vortex shedding, $n = 0$ to 15, and the number m specifies the time step in the range of a cycle from which the reconstruction starts, $m = 0$ to 61. As described in Sect. 2.1, a dataset of the flow field at 62 timesteps corresponds to almost one period of vortex shedding. The model trains and reconstructs spatio-temporal distribution of u, v and p during almost one period of vortex shedding. 2000 datasets with u, v and p at 62 timesteps from $t_{62n+m+1}$ to $t_{62n+m+62}$ as the input data and the supervised data are constructed. 1000 datasets are used as data for training, and the remaining 1000 datasets are validation data. After the training, the distributions of streamwise velocity, u, spanwise velocity, v, and pressure, p, at 62 timesteps from $t_{62n+m+1}$ to $t_{62n+m+62}$ of the validation data are given to the trained model as input. The model then reconstructs the input as output approximately. For the validation, the distributions of u, v and p from $t = 0$ to 99.88 (992 steps) are outputted from the model and compared with the corresponding CFD data (the ground truth).

20.2.3 Deep Learning Model

To learn and extract temporal and spatial features of time series of the flow fields and to develop a compressed representation and obtain a mode decomposition of the flow fields, we have constructed 3D-CNN Mode Decomposition Model (3D-CNN-MD Model) based on Convolutional Neural Network (CNN) [8] and Autoencoder [9]. The model architecture is shown in Fig. 20.2. The model consists of $(6 + 7 \times l)$ Convolutional 3D layers: 6 layers for the encoder and 7 layers $\times$ l modes for the decoder. Here, l denotes the number of modes in the latent space.

The spatial and temporal distributions of velocity components, u, v and pressure p during almost one cycle of vortex shedding are given to the model as input data. The dimensions of the input data are $N_t \times N_x \times N_y \times N_v = 62 \times 64 \times 64 \times 3$. Here, N_t, N_x, N_y and N_v are the number of timesteps, grids in x and y directions, and variables, respectively. In the encoder part, the model compresses the dimensions of the input, through spatio-temporal convolution operations, to map the time-series flow field data with $62 \times 64 \times 64 \times 3$ dimensions into $1 \times 1 \times 1 \times l$ dimensions in the latent space. It is meant that the time-series flow field data during almost one cycle are both spatially and temporally compressed to only l modes in the latent space. In decoder part, the latent vector is decomposed into l modes with $1 \times 1 \times 1 \times 1$ array. Each mode is expanded to $62 \times 64 \times 64 \times 3$ array, which is a time series of each decomposed flow field during one cycle of shed vortices. Finally, the model approximately reconstructs the input as output. The summation of decomposed fields is the reconstructed output of the model. In this study, the model with 3 modes ($l = 3$) is applied to the flow around a cylinder.

20.3 Results

We evaluate characteristics of distributions of the streamwise velocity component, u, the spanwise velocity component, v, and pressure, p, at $t = 0$ to 99.88 (992 steps) reconstructed and decomposed by 3D-CNN Mode Decomposition Model with $l = 3$.

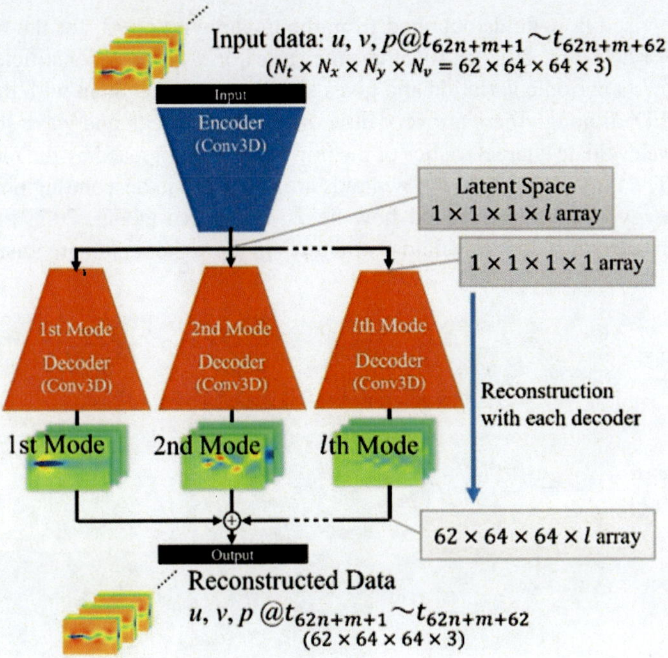

Fig. 20.2 Architecture of 3D-CNN-MD model

The distributions of velocity u, v and pressure p at 62 timesteps from $t_{62n+m+1}$ to $t_{62n+m+62}$, which corresponds to almost one period of vortex shedding, are used as the input and supervised data for training the model.

Figure 20.3 shows instantaneous distributions of u, v, and p obtained by CFD (the ground truth), reconstructed flow fields by 3D-CNN-MD Model, and three decomposed flow fields by the model, at $t = t_{62n+m+31} = 71.3$ ($n = 11$, $m = 17$). In the reconstructed flow fields, it is clearly reproduced that Karman vortices are periodically emitted from the cylinder. The model can reconstruct the flow fields quantitatively very well. In Fig. 20.4, the correlation coefficients and MSE of the flow fields obtained by CFD and reconstructed by the model with $l = 3$ are plotted against the initial time step $62n + m + 1$ of the input. The correlation coefficients and MSE are evaluated during one cycle of vortex shedding, which corresponds to each input and output of the model. The correlation coefficients for u, v, and p rise and fall repeatedly and are always higher than 0.985. The MSE of u, v, and p have similar tendences to the correlation coefficients. The flow fields reconstructed by 3D-CNN-AE Model are in qualitatively and quantitatively good agreement with the ground truth. The model gives the best correlation coefficient and MSE at $m = 17$, and the worst at $m = 57$ for over all time, when n changes form $n = 0$ to 15. Figure 20.5 shows time history and power spectrum of the streamwise velocity, u, and the spanwise velocity, v, at $(x, y) = (0.43, 0.15)$, where the Karman vortices emitted from the cylinder pass through, in the ground truth, the reconstruction, and the decomposed flow fields obtained from the model with $l = 3$, for the best cases with $m = 17$. In Fig. 20.5(a), (b), the time series of u and v reconstructed by the model shows a periodic variation and gives almost exact agreement with the ground truth of CFD, although there are very little differences in peak and wave shape.

Next, we evaluate characteristics of the flow fields decomposed by the model with 3 modes. The three decomposed flow fields are reordered in descending order of the kinetic energy of the decomposed flow fields. As shown in Fig. 20.3, in the first and the third decomposed flow fields, structures in the wake region are unsteady and

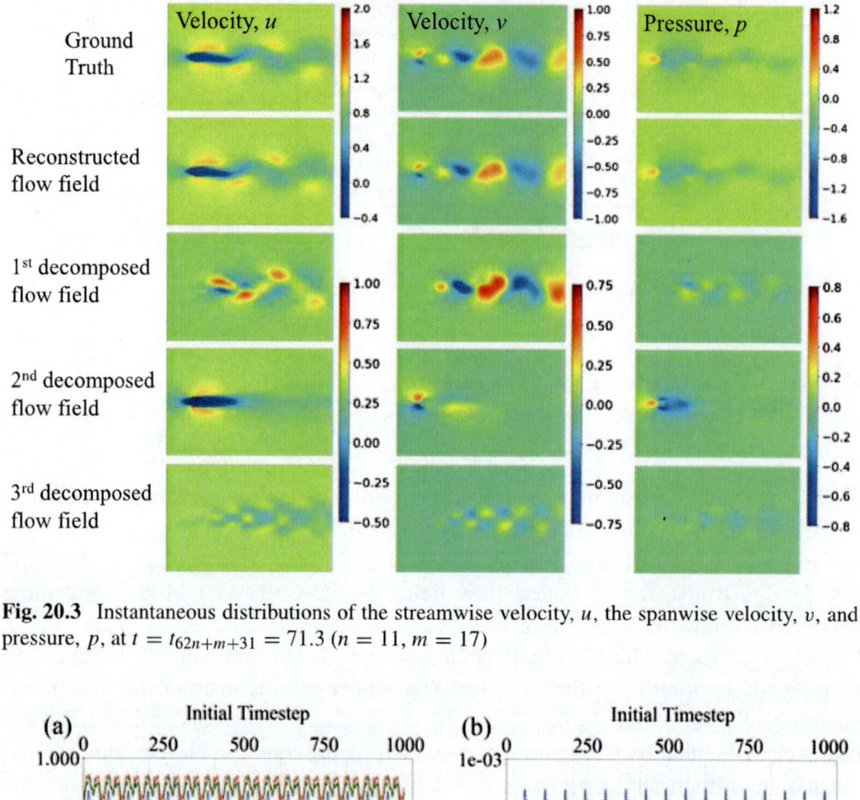

Fig. 20.3 Instantaneous distributions of the streamwise velocity, u, the spanwise velocity, v, and pressure, p, at $t = t_{62n+m+31} = 71.3$ ($n = 11$, $m = 17$)

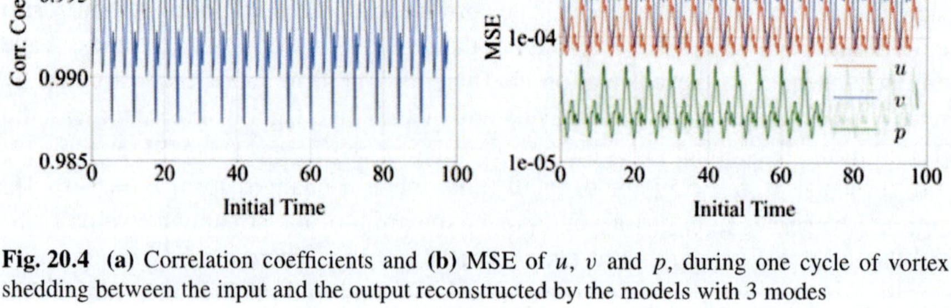

Fig. 20.4 **(a)** Correlation coefficients and **(b)** MSE of u, v and p, during one cycle of vortex shedding between the input and the output reconstructed by the models with 3 modes

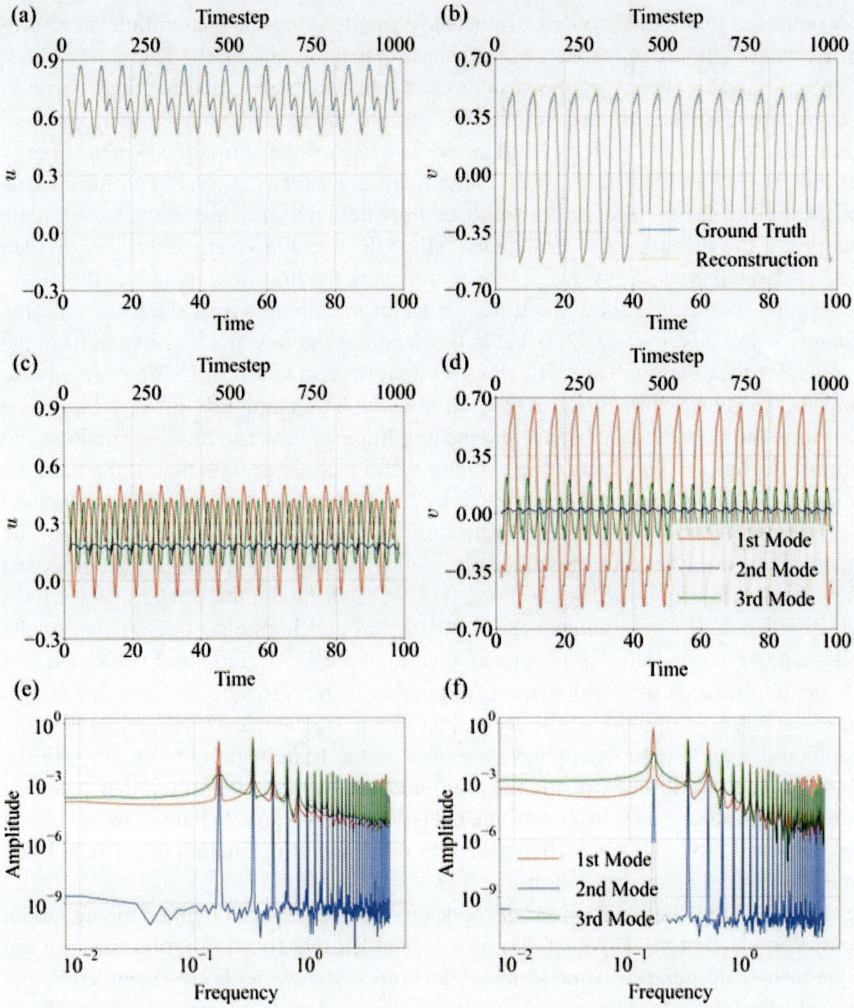

Fig. 20.5 Time history and power spectrum of the streamwise velocity, u, and the spanwise velocity, v, at $(x, y) = (0.43, 0.15)$ in the wake region, for the cases with $m = 17$ when n changes form $n = 0$ to 15. Time history of **(a)** u and **(b)** v for the ground truth and reconstruction, **(c)** u and **(d)** v of the decomposed flow fields. Power spectra of **(e)** fluctuating velocity u' and **(f)** v' of the decomposed flow fields

emitted from the cylinder periodically. In the first decomposed flow fields, large-scale unsteady structures are observed in the wake region, the size and the phase of which corresponds to the Karman vortices. The second decomposed flow fields are almost steady and similar to the time-averaged flow fields of the ground truth, with smaller-size, -magnitude, and higher-frequency oscillations. In addition, the structures in the upstream of the cylinder are reconstructed only by the second mode. The third decomposed flow fields consist of unsteady smaller-size and -magnitude structures in the wake. The structures have out-of-phase distributions with the Karman vortices. The emission frequency of the structures is twice as high as the Karman vortices. Discrepancies in the summed fields of the first and the second modes from the ground truth are compensated with the third mode. The ratio of the normalized kinetic energy of the 1st, 2nd and 3rd decomposed flow fields is 80.6 : 10.3 : 9.1. The summation of the kinetic energy of each decomposed flow field does not match the total kinetic energy of the reconstructed flow fields, since the average kinetic energy is divided into each decomposed flow fields by decomposing the flow field including the mean.

Figure 20.5(c), (d) show the time history of the streamwise velocity, u, and the spanwise velocity, v, at (0.43, 0.15) in the decomposed flow fields obtained from the model, for the cases with $m = 17$ when n changes form $n = 0$ to 15. The streamwise and the spanwise velocity changes periodically. The amplitude of u and v in 1st decomposed flow fields is largest and the amplitudes of the 2nd mode is smallest and almost negligible. The fluctuating cycle of u and v coincide with that of the Karman vortices in 1st decomposed flow fields. In 2nd and 3rd decomposed flow fields, the velocity fluctuations with half the period of the Karman vortices are observed. The power spectra of fluctuating velocity u' and v' at (0.43, 0.15) in the decomposed flow fields are shown in Fig. 20.5(e), (f). For each mode, the power spectra have the sharp peaks. They indicate the fundamental and harmonic frequencies of the Karman vortex shedding. In the power spectra of the 1st decomposed velocity u and v, the amplitude at the fundamental frequencies is largest. The 2nd and 3rd modes have the largest amplitude at the second harmonic frequencies. It corresponds to the smaller-size and higher-frequency structures in the wake region in Fig. 20.3. In the 2nd mode, the amplitude of the mean component is larger than the other peaks in fluctuating component (not shown here). It is consistent with the characteristics of 2nd decomposed flow fields, which are the almost steady with smaller-size, -magnitude and higher-frequency oscillations.

Lastly, in Fig. 20.6, variations of modes in the latent space obtained by the model with $l = 3$, are plotted against datasets. It is noted that these variations are not the time history of modes. A time series of decomposed flow fields are expanded by the decoder from a value of each mode, which correspond to each point in the graph. Each mode has a periodicity corresponding a cycle of the datasets. The variations of 1st mode have three peaks in one cycle and they are composed of three sine waves. The waveforms of 2nd mode and 3rd mode are close to single sine waves, respectively. These features indicate high expressiveness in deep learning. In addition, it should be noted that characteristics of decomposed flow fields and variations of modes can depend on training processes.

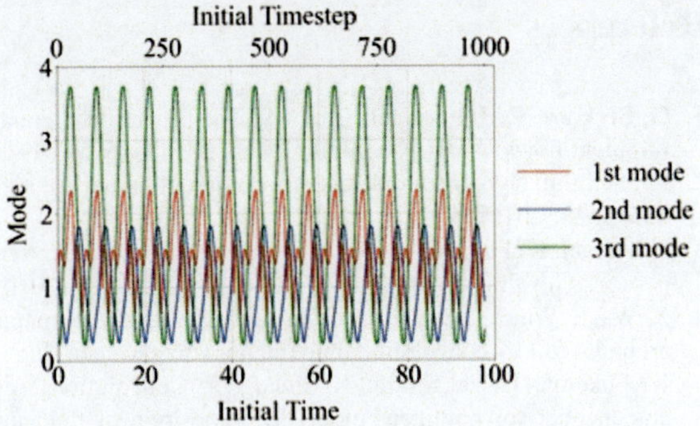

Fig. 20.6 Variations of modes in latent vectors against datasets. Each point corresponds to a time series of decomposed flow fields

20.4 Conclusion

In this study, to obtain low-dimensional dynamic representation and spatio-temporal mode decomposition of unsteady flow, 3D-CNN-MD Model has been constructed based on Convolutional Neural Network and Autoencoder. The model spatially and temporally compresses and decomposes a time series of flow fields to limited nonlinear modes, and then outputs the decomposed and reconstructed flow fields.

We apply the model to a flow around a cylinder at Reynolds number, $Re_D = U_\infty D/\nu$=100, as an example of unsteady flow, and evaluate characteristics of the decomposed flow fields. The spatial and temporal distributions of velocity components, u, v and pressure p during almost one cycle of vortex shedding are given as input data. The dimensions of the input are $N_t \times N_x \times N_y \times N_v = 62 \times 64 \times 64 \times 3$. The input is compressed into only three modes in the latent space. Each mode is expanded into a time series of each decomposed flow field. The summation of decomposed fields is the reconstructed output of the model.

A time series of the flow fields reconstructed from only three modes by the model gives quantitatively good agreements with the ground truth obtained by CFD. The correlation coefficients are very high and MSE is small enough. A time series of three decomposed flow fields have been obtained from the three modes in the latent spaces. The first decomposed flow fields are dominant in the kinetic energy and have large-scale unsteady structures in the wake region, which corresponds to the Karman vortices. The second decomposed flow fields are almost steady and similar to the time-averaged flow fields of the ground truth. The third decomposed flow fields consist of unsteady smaller-size and -magnitude, and higher-frequency structures in the wake, which have out-of-phase distributions with the Karman vortices.

These results have suggested that low-dimensional dynamic representation using deep learning can be one of effective methods for a deeper understanding of the nature of unsteady flows. By continuous development of the 3D-CNN-MD Model, we can enhance our understanding of unsteady flows, including turbulence.

Acknowledgements This work was partially supported by JSPS (Japan Society for the Promotion of Science): KAKENHI Grant Numbers 21H05007, 18H03758 and 22K03932.

References

1. G. Berkooz, P. Holmes, J.L. Lumley, The proper orthogonal decomposition in the analysis of turbulent flows. Annu. Rev. Fluid Mech. **25**, 539–575 (1993)
2. P.J. Schmid, Dynamic mode decomposition of numerical and experimental data. J. Fluid Mech. **656**, 5–28 (2016)
3. T. Murata, K. Fukami, K. Fukagata, Nonlinear mode decomposition with convolutional neural networks for fluid dynamics. J. Fluid Mech. **882**, A13 (2010)
4. X. Wu, S. Wu, X. Tian, X. Guo, X. Luo, Effects of hyperparameters on flow field reconstruction around a foil by convolutional neural networks. Ocean Eng. **247**, 110650 (2022)
5. K. Fukami, T. Nakamura, K. Fukagata, Convolutional neural network based hierarchical autoencoder for nonlinear mode decomposition of fluid field data. Phys. Fluids **32**, 095110 (2010)
6. J. Qu, W. Cai, Y. Zhao, Deep learning method for identifying the minimal representations and nonlinear mode decomposition of fluid flows. Phys. Fluids **33**, 103607 (2010)
7. H. Eivazi, S. Le Clainche, S. Hoyas, R. Vinuesa, Towards extraction of orthogonal and parsimonious non-linear modes from turbulent flows. Expert Syst. Appl. **202**, 117038 (2010)
8. A. Krizhevsky, I. Sutskever, G.E. Hinton, Imagenet classification with deep convolutional neural networks, in *NIPS'12: Proceedings of the 25th International Conference on Neural Information Processing Systems* vol. 1 (Curran Associates Inc., Lake Tahoe, 2010), pp.1097–1105
9. G.E. Hinton, R.R. Salakhutdinov, Reducing the dimensionality of data with neural networks. Science **313**(5786), 504–507 (2010)

Chapter 21
Active Wake Control of a Bluff Body Using Incompressible Numerical Simulations

J. P. Ananthu, Prasanth P. Nair, and Vinod Narayanan

Abstract The effect of active flow control using suction and blowing on the wake of a bluff body with a blunt trailing edge in a rectangular channel is investigated using URANS simulations and Dynamic Mode Decomposition (DMD). A recirculation loop is added at the rear end of the bluff body to introduce the secondary flow. Vorticity contours are plotted for different secondary flow velocities, and it is observed that the secondary flow can reduce the magnitude of vorticity in the wake of the bluff body by altering the vortex shedding and pressure distribution. The optimal secondary flow velocity is found to be 2.05 m/s. DMD is applied to analyze the flow without control and reveals two dominant modes at frequencies 29.7 and 59.48 Hz, corresponding to the horizontal and vertical motion of the vortical structures in the wake. The results demonstrate the potential of secondary flow injection as a passive control technique for wake reduction.

21.1 Introduction

Flow over a bluff body has always been the subject of interest because of its importance in engineering applications. Automobiles, Airplanes, and wind passing through structures like bridges, and lamp posts are evident examples. Vortex shedding occurs with an increased drag along the body when the Reynolds number crosses a critical value. Effective control of wake is essential in engineering to prevent unwanted effects associated with vortex shedding.

J. P. Ananthu (✉) · P. P. Nair · V. Narayanan
Mechanical Engineering Discipline, Indian Institute of Technology Gandhinagar, Palaj, Gujarat 382055, India
e-mail: ananthu.jp@iitgn.ac.in

P. P. Nair
e-mail: prasanth.n@iitgn.ac.in

V. Narayanan
e-mail: vinod@iitgn.ac.in

A. Suryan et al. (eds.), *Proceedings of the 9th Asian Joint Workshop on Thermophysics and Fluid Science, 27–30 November 2022, Utsunomiya, Japan*, Springer Proceedings in Physics 396, https://doi.org/10.1007/978-981-99-9470-0_21

Bluff bodies are those bodies that have a separate flow over a substantial part of their surface when a fluid flows over the body. The reason for this behaviour of a bluff body is explained by Prandtl in 1904. This is known as Prandtl's boundary layer theory. The boundary layer is a thin region close to the surface of a body placed in a viscous fluid flow. The velocity of a fluid at the surface of the body will be zero, and the velocity will increase as we move away from the surface. The region where this velocity variation happens due to the viscosity of a fluid is known as the boundary layer. The boundary layer corresponds to the region where the local spinning of the continuous matter is near the body surface. In continuum mechanics, this spin is known as vorticity.

A body with a non-streamlined shape placed in a streamlined flow experiences vortices of different sizes on its surface. The velocity at the center of the front side will be zero; thus, it is taken as a stagnation point. As a result of the surface vortices, a thin boundary layer is formed on both sides of the stagnation point. The thickness of this boundary layer increases as we go far downstream. At some point in the downstream direction, the vortices give rise to accelerations that are followed by strong deceleration, also called boundary layer separation. The separation leads to periodic detachment of vortices from the end of the body for a particular flow Reynold number, known as vortex shedding.

This study points toward the reduction of drag experienced by a bluff body placed in a flow field due to von-Karman vortex shedding. A bluff body with blunt edges is considered for the study since the blunt edges will make a sudden change in flow, which leads the boundary layer to form a strong wake bubble behind the trailing edge. The introduction of a secondary flow in the wake of a bluff body prevents the growth of the wake bubble. As a result, the development of periodic vortices in the wake is decreased. Most of the studies in bluff bloody with a blunt edge involve modifications in the rear surface. Park [1] modified the rear surface of the bluff body with blunt edges with uniform staggered and non-staggered arrangement of tabs. Adding tabs is a passive control method. The author suggested a non-staggered arrangement as the optimum control method. Drag reduction in a bluff body with a blunt trailing edge using an open and closed control loop was studied by Henning and King [2]. The author found that synchronizing the vortices in the wake with a feedback loop is the optimum drag reduction method.

21.2 Literature

The majority of wake experiments are aimed at reducing the drag force on bluff bodies. The drag experienced by a bluff body in a flow field is directly related to the wake region behind the body [3]. Roshko [3] was the first to calculate the shedding frequency of Karman vortex shedding for the wake of a bluff body. The dependence of shedding frequency on drag and lift forces was explained by Sumer and Fredsøe [4]. The drag force is proportional to the shedding frequency, while the lift force is proportional to twice the shedding frequency, according to the author. Buresti

[5] explains how viscous and pressure forces interact, leading to drag. According to the author, viscous forces dominate streamlined bodies, whereas pressure forces dominate in the case of bluff bodies.

Sharma et al. [6] studied the effect of front and after bodies in vortex shedding formation length, separation point, and fluctuation energy. The authors found that the lift/drag experienced by the body linearly varies with the front and after body parameters.

Controlling the wake region behind the bluff body will help to minimize drag forces. Active-open loop, Active-closed loop, and passive methods are the three types of wake control methods [7]. Passive methods control the wake by changing the surface texture. Splitter plates, surface dimples, and making rough surfaces are some of the passive approaches to wake control. External power input or feedback circuits are used in active methods [8]. Ongoren and Rockwell [9] addressed a cylinder in a flow field with active flow control. The oscillation of the cylinder here causes the formation of the wake to be disturbed. As a result, vortices with symmetric and anti-symmetric modes will be formed. Detached vortices formation will no longer be periodic due to the transition between modes.

It can be concluded from the literature survey that some unstable modes in the wake of vortex shedding are prevented by the use of passive control methods. Passive control methods are unable to adapt to the changes in the wake since they don't have a feedback loop. Active control methods can overcome this limitation with the help of external power input and a feedback loop. In general, the active control method performs better when it comes to drag reduction in the wake of a vortex shedding. The introduction of a secondary flow in the trailing edge of the bluff body is a less explored area of research in active flow control methods.

The objective of this study is to determine the effect of an active control method on the wake of a body taken. This involves the introduction of an artificial secondary vortex to the wake. Secondary vortices can be made with the help of suction and blowing at the rear end of the body. The present study focuses on the velocities in turbulent regimes. The optimum velocity ratio can be found by varying the secondary circulation velocities for a constant inlet velocity where the effect of vortices in the wake gets cancelled by the secondary flow.

21.3 Methodology

21.3.1 Flow Domain

A rectangular flow domain that resembles a 20 cm flow channel is considered for the simulations, as shown in Fig. 21.1. The left and right end of the region is taken as inlet and outlet, respectively.

The region occupied by the bluff body is taken 10 cm from the inlet and placed symmetrically to the central axis. A separate secondary recirculation region is added

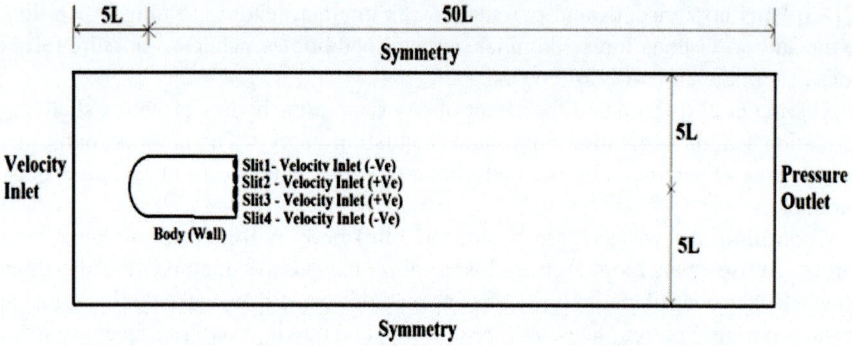

Fig. 21.1 Geometry of the domain

to the rear end of the bluff body in the active wake control study. In this case, a secondary inlet and outlet are added at the rear portion of the bluff body, as shown in Fig. 21.1.

21.3.2 Mesh

The flow domain is divided into a uniform structured grid, as shown in Fig. 21.2. Quadrilateral elements are used for the mesh. The region occupied by the bluff body is avoided from the mesh since we require the fluid region only. Boundaries of the geometry are modelled as a fixed wall for the first case without a recirculation loop. Whereas an inlet and outlet region consisting of 2 mm thickness is added at the rear portion of the bluff body in the case of a wake control study.

Mesh regions near boundary walls and bluff body walls are refined with a wall spacing ratio of 1.2 to capture boundary layer effects. The minimum and maximum size of an element in the mesh are 0.005 m and 0.02 m, respectively. The total number of cells considered for the study is 6, 31, 964.

21.3.3 Initial and Boundary Conditions

The initial inlet flow rate is fixed at 1 m/s. The outlet boundary condition is given as the atmospheric pressure condition. Water at standard conditions is taken as the medium inflow domain. The top and bottom boundaries of the channel are taken as fixed walls since the mass flow rate in the momentum equation won't penetrate through the walls. Secondary circulation flow inlet velocity is varied from 0.9 to 2.5 m/s.

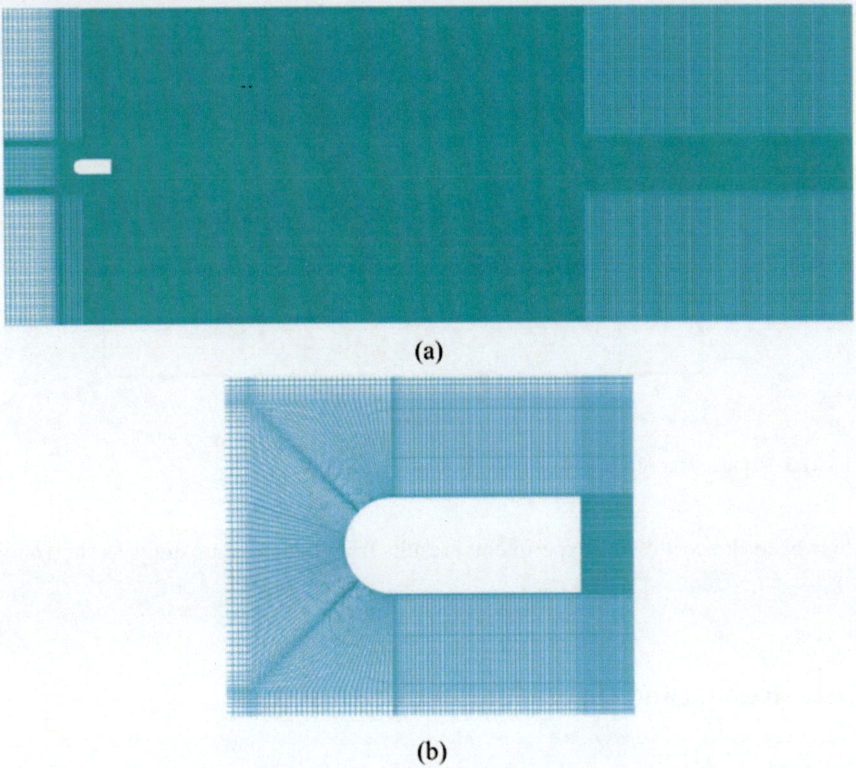

Fig. 21.2 **a** Overview of the mesh, and **b** mesh refinement near the body

21.3.4 Numerical Scheme

A pressure-based coupled solver with second-order spatial discretization is used for the numerical simulations. First-order implicit transient time formulation captures the periodic flow oscillations with a time step size of 0.001 s. A turbulence model study is conducted by taking the Strouhal number as the parameter. The Standard k-omega model gave the best results among the turbulent models. Thus, k-omega is used to model turbulence for all the numerical simulations considered.

21.4 Validation

Experimental findings of Lyn and Rodi [10] and Lyn et al. [11] from an earlier study are used to validate the developed numerical model.

The time-averaged streamwise velocity profile along y at x/D = 0.0 is plotted and compared for the numerical and experimental studies. The numerical findings

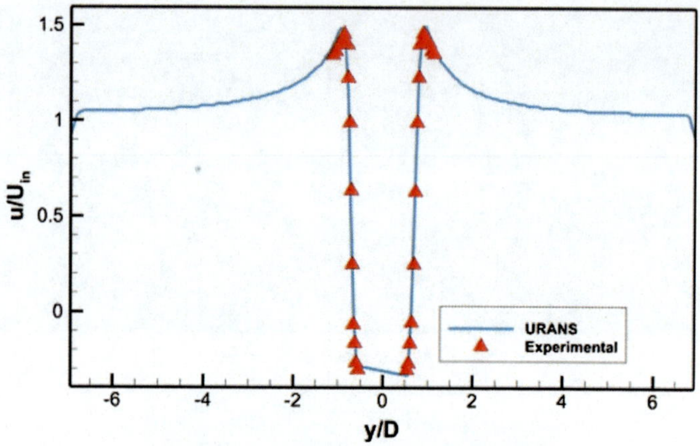

Fig. 21.3 Time-averaged streamwise velocity profile along y at x/D = 0.0

are in accordance with the experimental results from the previous study, as shown in Fig. 21.3.

21.5 Results and Discussion

The wake of a bluff body with a blunt trailing edge has been numerically simulated with a two-dimensional pressure-based transient solver. Vorticity contours are plotted for an inlet velocity of 1 m/s and secondary flow velocities of 0.9 m/s, 1 m/s, 2.05 m/s, and 2.5 m/s, respectively. A significant reduction in vorticity has been observed when a secondary flow is introduced and the trailing edge of the bluff body.

21.5.1 *Vorticity Contours*

The transportation of vortex patterns behind the blunt trailing edge is one of the main objectives of this study. Figure 21.4 shows the vorticity contours for inlet velocity 1 m/s and (a) without secondary flow, (b), (c), and (d) secondary flow velocities 0.9 m/s, 2.05 m/s and 2.5 m/s respectively. Vortex shedding is observed in all the parametric studies.

It can be observed from Fig. 21.4a that the vortices are detached from the blunt trailing edge and move towards the downstream direction. Vortices are identical in shape, having the shape of a distorted ellipse with the major axis aligned perpendicular to the flow direction till it reaches near to the halfway of the entire flow domain. Then the vortices take a 90° rotation, and the major axis gets aligned with

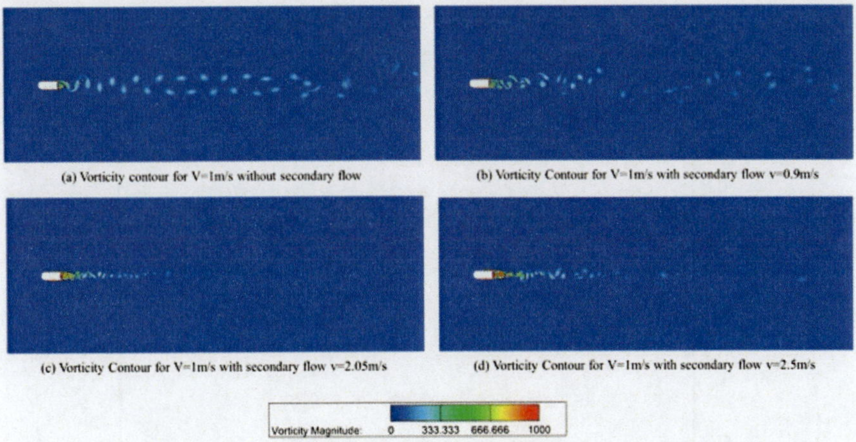

Fig. 21.4 Vorticity contour of the bluff body for different values of secondary flow with fixed inlet velocity

the flow direction, and later takes a few turns and reduces the strength of vortices while travelling towards the end of the flow domain.

A secondary flow is introduced with the help of two pairs for inlet and outlets, each at the trailing edge of the body. Initially, a flow with a velocity of 0.9 m/s is introduced at the trailing edge, while inlet velocity is kept the same (1 m/s). Now, the vortices no longer look identical. The strength of vortices decreases drastically, even before it reaches the midway of the flow domain, as shown in Fig. 21.4b. The secondary flow velocity is increased to 2.05 m/s, keeping the inlet velocity the same. Here, the vortices are arranged more near to the centerline of the flow domain compared to the previous case. The strength of the vortices are observed to be decreasing when it travels in the downstream direction, as we expected. Vortices almost disappear when they reach near the end of the flow domain, as shown in Fig. 21.4c. However, the strength of vorticity was found to be slightly increasing when we increased the magnitude of velocity to 2.5 m/s.

Overall, it can be observed that the vortex shedding has been controlled and minimized with the injection of the secondary flow, and the strength of the vortices are minimal for an injection velocity of 2.05 m/s.

21.5.2 Dynamic Mode Decomposition (DMD)

A data-driven model decomposition technique, Dynamic Mode Decomposition (DMD), has been used to analyze the results of the flow without control. The details

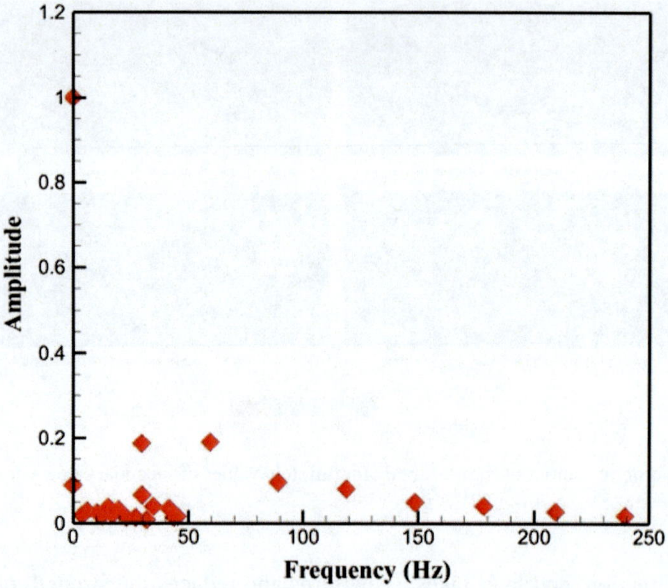

Fig. 21.5 DMD spectrum

of the formulation of DMD have been covered in previous studies [12–15]. Non-dimensionalized amplitudes corresponding to the frequencies of the DMD spectrum are shown in Fig. 21.5. Dominant modes are observed at frequencies 29.7 and 59.48 Hz.

The contour of the modes corresponding to the dominating frequency are shown in Fig. 21.6. The modes at the dominating frequency of 29.7 and 59.48 Hz correspond to the wake behind the bluff body. These modes correspond to the temporal dominating mode, which leads to both the horizontal and vertical motion of the vortical structures. It is seen that the horizontal motion of the vortical structures is of higher frequency as compared to the vertical motion of the vortical structures.

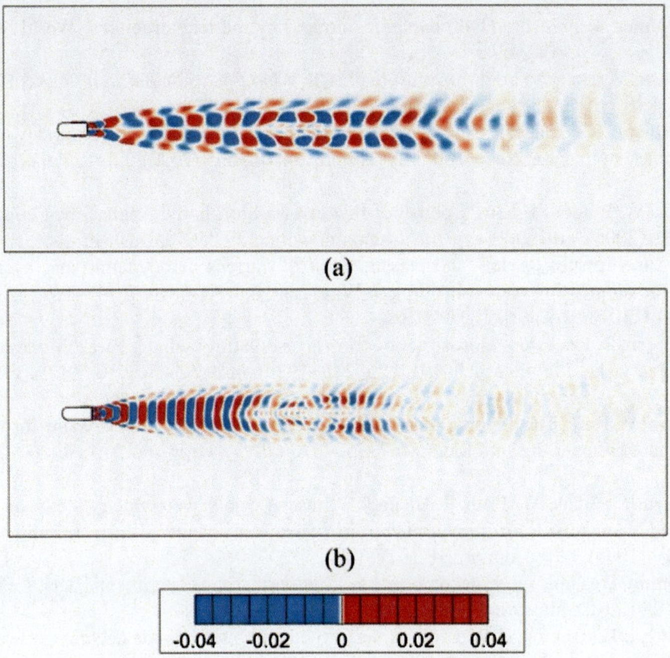

Fig. 21.6 Mode corresponding to frequency **a** 29.7 Hz and **b** 59.48 Hz

21.6 Conclusion

Numerical study on active wake control using URANS has been conducted. The inlet velocity of the flow has been kept constant. Different secondary flow has been examined to reduce the wake of the bluff body. It was observed that with higher injection velocity, the wakes behind the bluff body are reduced. DMD gives an insight into the modes responsible for the vortical structures. Two temporal dominating modes are liable for the horizontal and vertical motion of the vortical structures.

References

1. H. Park, D. Lee, W.-P. Jeon, S. Hahn, J. Kim, J. Kim, J. Choi, H. Choi, Drag reduction in flow over a two-dimensional bluff body with a blunt trailing edge using a new passive device. J. Fluid Mech. **563**, 389–414 (2006). https://doi.org/10.1017/S0022112006001364
2. L. Henning, R. King, Drag reduction by closed-loop control of a separated flow over a bluff body with a blunt trailing edge. In: Proceedings of the 44th IEEE Conference on Decision and Control. pp. 494–499. IEEE, Seville, Spain (2005). https://doi.org/10.1109/CDC.2005.1582204
3. A. Roshko, On the wake and drag of bluff bodies. J. Aeronaut. Sci. **22**, 124–132 (1955). https://doi.org/10.2514/8.3286

4. B.M. Sumer, J. Fredsøe, Hydrodynamics around cylindrical structures. World Sci. (2006). https://doi.org/10.1142/6248
5. G. Buresti, Vortex shedding from bluff bodies, in *Wind Effects on Buildings and Structures*. ed. by J.D. Riera, A.G. Davenport (Balkema, Rotterdam, 1998), pp.61–95
6. G. Sharma, A.K. Pandey, R. Bhardwaj, Effect of shape of frontbody and afterbody on flow past a stationary cylinder at Re = 100. Phys. Fluids **34**, 063605 (2022). https://doi.org/10.1063/5.0093713
7. H. Choi, W.-P. Jeon, J. Kim, Control of flow over a bluff body. Annu. Rev. Fluid Mech. **40**, 113–139 (2008). https://doi.org/10.1146/annurev.fluid.39.050905.110149
8. M.M. Zdravkovich, Review and classification of various aerodynamic and hydrodynamic means for suppressing vortex shedding. J. Wind Eng. Ind. Aerodyn. **7**, 145–189 (1981). https://doi.org/10.1016/0167-6105(81)90036-2
9. A. Ongoren, D. Rockwell, Flow structure from an oscillating cylinder Part 2. Mode competition in the near wake. J. Fluid Mech. **191**, 225–245 (1988). https://doi.org/10.1017/S0022112088001570
10. D.A. Lyn, W. Rodi, The flapping shear layer formed by flow separation from the forward corner of a square cylinder. J. Fluid Mech. **267**, 353–376 (1994). https://doi.org/10.1017/S0022112094001217
11. D.A. Lyn, S. Einav, W. Rodi, J.-H. Park, A laser-Doppler velocimetry study of ensemble-averaged characteristics of the turbulent near wake of a square cylinder. J. Fluid Mech. **304**, 285–319 (1995). https://doi.org/10.1017/S0022112095004435
12. P.J. Schmid, Dynamic mode decomposition of numerical and experimental data. J. Fluid Mech. **656**, 5–28 (2010). https://doi.org/10.1017/S0022112010001217
13. J.N. Kutz, S.L. Brunton, B.W. Brunton, J.L. Proctor, Dynamic mode decomposition. Soc. Ind. Appl. Math. (2016). https://doi.org/10.1137/1.9781611974508
14. P.P. Nair, A. Suryan, V. Narayanan, Modal analysis of mixing characteristics in scramjet combustor with passive struts. Int. J. Hydrogen Energy **47**, 34656–34675 (2022). https://doi.org/10.1016/j.ijhydene.2022.08.061
15. P.P. Nair, A. Suryan, V. Narayanan, Effect of passive strut angle on the vortical structures and mixing characteristics of scramjet combustor. Phys. Fluids **35**, 066111 (2023). https://doi.org/10.1063/5.0151676

Chapter 22
Effect of Bowed Stator Blade on Aerodynamic Performance of Low-Speed Single-Stage Axial Compressor

Masanao Kaneko and Hoshio Tsujita

Abstract The circumferential bowing of stator blade is known to possess the ability to suppress the corner separation. In this study, the influences of the positive dihedral angle of the bowed stator blade on the aerodynamic performance of low-speed single-stage axial compressors were investigated numerically. The computed results clarified the following phenomena. The bowed stator has the effects to suppress the hub corner separation and the associated loss generation. The influence of the hub corner separation on the loss generation becomes stronger according to the decrease of the flow rate. As a result, the decrease of the flow rate improves the loss reduction effect by the bowed stator, and this trend is enhanced by the increase of the dihedral angle. Meanwhile, the excessive increase of the dihedral angle induces the additional loss generation by the formation of the separation at the blade pressure surface/endwall corner at high flow coefficient.

22.1 Introduction

A decrease of flow rate through an axial compressor induces the unstable flow phenomena such as the rotating stall and the surge. In particular, the surge can generate the loud noises and the severe mechanical vibrations. Therefore, axial compressors are usually operated at larger flow rate than that at which the unstable flow phenomena occur. On the other hand, the decrease of the flow rate also increases the blade loading, and consequently raises the pressure ratio per stage. So, the enlargement of the stall margin to lower flow rate makes it possible to increase the pressure ratio under the stable operating condition.

M. Kaneko (✉)
Tokyo Denki University, Ishizaka, Hatoyama-Machi, Hiki-Gun, Saitama 350-0394, Japan
e-mail: kaneko@mail.dendai.ac.jp

H. Tsujita
Hosei University, 3-7-2, Kajino-Cho, Koganei-Shi, Tokyo 184-8584, Japan

A. Suryan et al. (eds.), *Proceedings of the 9th Asian Joint Workshop on Thermophysics and Fluid Science, 27–30 November 2022, Utsunomiya, Japan*, Springer Proceedings in Physics 396, https://doi.org/10.1007/978-981-99-9470-0_22

In order to enlarge the stall margin, a lot of studies have been devoted to clarifying the causes of the inceptions of the rotating stall and the surge. Hewkin-Smith et al. [1] reported the rotating stall inception as follows: Due to the decrease of the flow rate, the interface between the main flow and the tip leakage flow moves upstream the leading edge of the adjacent blade and then the rotating stall occurs. On the other hand, Yamada et al. [2] clarified that the hub-corner stall generated in the stator blade row significantly influences the rotating stall inception. Therefore, the suppression of the corner stall is also important to enlarge the stall margin.

At the low flow rate operating condition, the low momentum fluid on the endwall is accumulated to the corner with the blade suction surface due to the pitchwise pressure gradient. And then the large flow separation, that results in the corner stall, can be generated at the corner by the streamwise adverse pressure gradient. The control of the secondary flow to decrease the accumulation of the low momentum fluid is the one of the ways to suppress the corner stall. There is the 3-dimensional blade stacking technology known as the method to control the secondary flow [3–6]. It is apparent from the previous studies [5, 6] that the circumferential bowing of the stator blade which has the concave surface on the suction side, that is referred as the positive dihedral blade, possesses the ability to suppress the corner stall. However, at the high flow rate condition where the influence of the corner separation on the loss generation is small, the application of the bowed blade is considered to deteriorate the efficiency by the increase of the friction loss generation due to the increase of the solid blade surface area. Therefore, in order to maintain the high efficiency from the high to the low flow rate operating point, it is necessary to investigate the effectiveness of bowed blade on the compressor aerodynamic performance at the wide operating range.

In this study, the flows in the low-speed single-stage axial compressors with the typical straight stator blades and with the bowed ones were analyzed numerically. The present paper describes the effects of the bowed stator on the overall aerodynamic performance and the flow behavior in the passage of the compressor at the wide operating range.

22.2 Test Compressors

Test compressors used in this study were low-speed single-stage axial compressors. As shown in Fig. 22.1, a stator was arranged downstream a rotor. The blade shapes of the rotor and the stator were generated by defining the blade elements from the hub to the shroud at intervals of 10% span. The NACA65 airfoil was used as the blade profile. In the rotor, the outer diameter is 299.26 mm, the solidity at the midspan is 1.5, the aspect ratio is 1.63, the hub-tip ratio at the inlet is 0.5 and the number of the blades is 26. The clearance between the rotor blade tip and the shroud casing wall is 0.8% of the blade chord. Figure 22.2 shows the definition of the dihedral angle in this study and the stacking lines of the stator blades, which describe the configuration of bowed blade. The dihedral angles defined as the angle between the stacking lines

Fig. 22.1 Test compressor

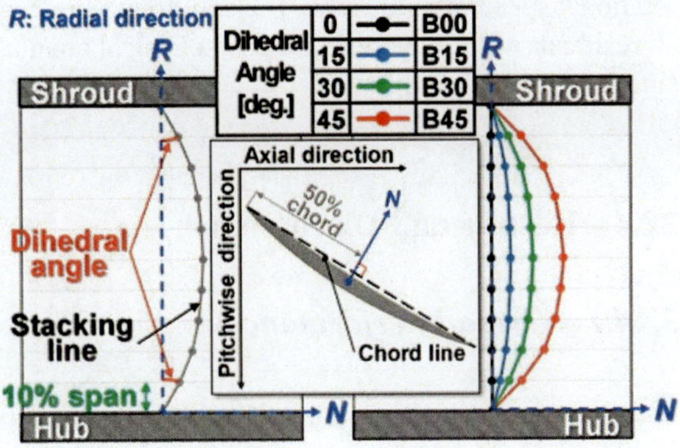

Fig. 22.2 Definition of dihedral angle and stacking lines of test stators

from the endawll to 10% span and the radial direction were varied from 0 to 45°. at intervals of 15°. The same dihedral angles were applied at the shroud side. The angles between the stacking lines and the radial direction were gradually decreased toward the midspan and these decreasing rates were fixed for all stators. All stators have the outer diameter of 300 mm, the solidity at the midspan of 1.5, the aspect ratio of 1.49, the hub-tip ratio at the inlet of 0.53 and the blade number of 30. The design rotational speed is 4,000 rpm and the design flow coefficient V_{a1}/U_t is 0.43 where the V_{a1} is the axial velocity at the rotor inlet and the U_t is the rotor tip circumferential speed. In the following, the test compressors with the dihedral angles of 0, 15, 30 and 45° are referred as B00, B15, B30 and B45, respectively.

22.3 Computational Method and Condition

In this study, the flows in the compressors were analyzed numerically by using the commercial CFD code of ANSYS CFX. The computations were conducted with the assumption that the flows through the compressors are the steady state incompressible flows. The computational domains were set to the single blade-to-blade passage in both rotor and stator regions. The computational grids were generated by using the structural grid. The total numbers of cells in the rotor and the stator domains are 7.8 million and 5.6 million, respectively. As the boundary condition, the flow rate was applied at the inlet boundary. The atmospheric pressure was set at the outlet boundary as the static pressure. At the pitchwise boundary planes, the periodic boundary conditions were imposed. The mixing-plane method was used at the interface between the rotor domain and the stator one. The k-ω based SST model was used as the turbulence model and the computations were conducted at design rotational speed of 4,000 rpm. The solutions were considered to be converged when all residuals of the equations for each physical quantity become less than 1.0×10^{-5} simultaneously.

22.4 Results and Discussion

22.4.1 Overall Performances

In this section, the overall performances will be compared at three flow coefficients V_{a1}/U_t: 0.43 (Design flow coefficient), 0.36 (The minimum flow coefficient where the computation for B00 was converged) and 0.50. Figure 22.3 shows the stage efficiency η_{14} defined as

$$\eta_{14} = \frac{Q\left(p_{04}^{*} - p_{01}\right)}{2\pi N T} \times 100$$

where, the Q, the N and the T are the volumetric flow rate, the rotational speed of the rotor and the torque, respectively. The p_{01} is the cross-sectional mass averaged total pressure at the rotor inlet. The p_{04}^{*} is the mixed-out averaged total pressure at the stator outlet. Figure 22.4 shows the stage total pressure rise coefficient C_{p14} defined by following equation.

$$C_{p14} = \frac{p_{04}^{*} - p_{01}}{0.5\rho {U_t}^2}$$

where, the ρ is the density. Figures 22.5 and 22.6 show the rotor efficiency η_{12} and total pressure rise coefficient C_{p12}, respectively. The η_{12} and the C_{p12} are defined as follows:

$$\eta_{12} = \frac{Q\left(p_{02}^{*} - p_{01}\right)}{2\pi NT} \times 100$$

$$C_{p12} = \frac{p_{02}^{*} - p_{01}}{0.5\rho {U_t}^2}$$

where, the p_{02}^{*} is the mixed-out averaged total pressure at the rotor outlet.

Figure 22.3 indicates the following characteristics: At design flow coefficient, the η_{14} for B00 is lowest. The η_{14} is improved according to the increase of the dihedral angle up to B30. However, the difference of η_{14} between B30 and B15 is decreased compared with that between B15 and B00. On the other hand, the η_{14} for B45 takes lower value than that for B30. The η_{14} for B00 at low flow coefficient is decreased compared with that at design one. In the comparison at low flow coefficient, the η_{14} for B00 is lowest. The comparisons of the η_{14} for B00 with those for B15 and B30 exhibit similar tendencies to those at design flow coefficient. However, the difference of η_{14} between B30 and B15 is increased compared with that between B15 and B00. Meanwhile, the η_{14} for B45 is higher than that for B30, but the difference of η_{14} between B30 and B45 is smaller compared with that between B30 and B15. The η_{14} for B00 at high flow coefficient is decreased compared with that at design one. At high flow coefficient, the η_{14} for B15 is improved compared with that for B00. However, the η_{14} for B30 takes similar value to that for B00. Moreover, the η_{14} for B45 is lowest.

The comparisons for the C_{p14} show similar tendencies to those for the η_{14} as shown in Figs. 22.3 and 22.4. At each operating point, the dihedral angle of bow gives no influence on the η_{12} and the C_{p12} as seen in Figs. 22.5 and 22.6. Therefore, it is obvious that the variations of the overall stage performances of compressor are caused by the flow behavior in the stators affected by the dihedral angle.

Fig. 22.3 Stage efficiency

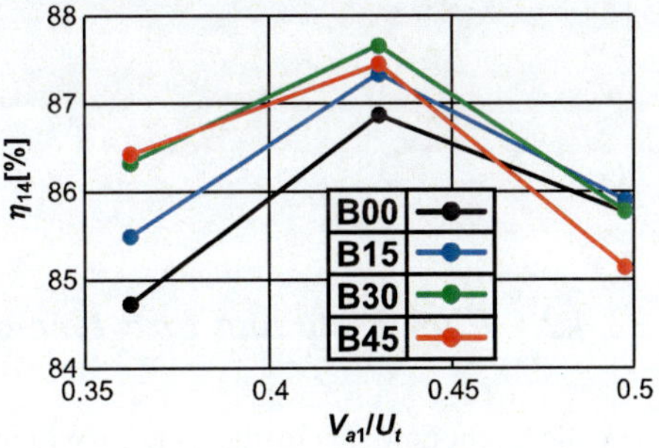

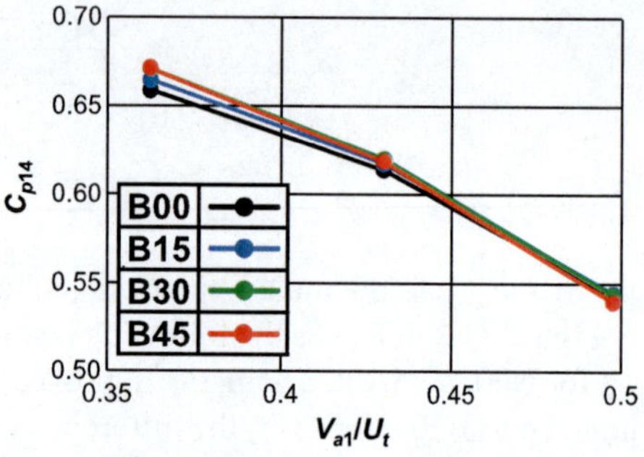

Fig. 22.4 Stage total pressure rise

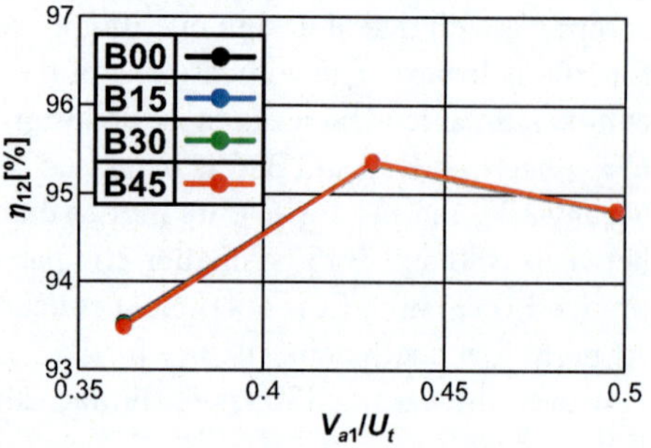

Fig. 22.5 Rotor efficiency

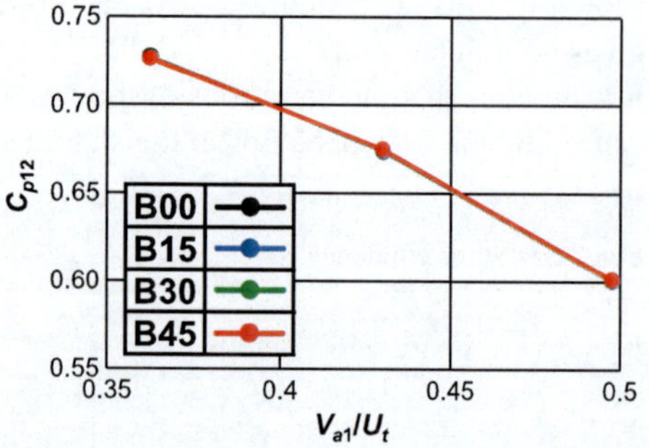

Fig. 22.6 Rotor total pressure rise

22.4.2 Flow Field and Loss Generation

The flow phenomena in the stators which cause the above-mentioned variations of overall stage performances will be examined at three operating conditions. Figure 22.7 shows the limited streamlines on the suction surfaces of the stators. In this figure, the flow directions for B00 are shown by the red arrows for the comparisons in the intensity of the secondary flows at each operating point. Figure 22.8 indicates the spanwise distribution of the circumferential mass-averaged total pressure loss coefficient at the stator outlet C_{ptc4}. The total pressure loss coefficient C_{pt} is defined by

$$C_{pt} = \frac{p_{03}^{*} - p_0}{0.5\rho U_t^{\,2}}$$

where, the p_{03}^{*} is the mixed-out averaged total pressure at the stator inlet and the p_0 is the total pressure. In this figure, the Y/H is the normalized spanwise distance. Figure 22.9 gives the limited streamlines on the pressure surface of the stator blade for B45. Figure 22.10 shows the comparison of the distribution of the absolute velocity V normalized by the U_t on the cross-sectional plane around the pressure surface/hub corner downstream the leading edge for B30 with that for B45. Figure 22.11 shows the spanwise distribution of the circumferential mass-averaged absolute flow angle at the stator inlet α_{c3}. In this figure, because the flow fields at the stator inlets are almost same for all compressors, the computed results for B00 are shown representatively. The absolute flow angle is defined as the angle between the absolute velocity vector and the axial direction.

Design Flow Coefficient. For B00, the flow separation is generated at the suction surface/hub corner as seen in Fig. 22.7a, and consequently increases the C_{ptc4} at the hub side as shown in Fig. 22.8. Similar tendency to B00 is observed for B15 as in Figs. 22.7b and 22.8. However, the separation region and the related increase of C_{ptc4} are reduced compared with those for B00 as seen in Figs. 22.7a, b and 22.8. For B15, the secondary flow from the hub toward the midspan is intensified on the blade suction surface as shown in Fig. 22.7b. This phenomenon reduces the accumulation of the low momentum fluids at the suction surface/hub corner and then suppresses the hub corner separation. Because of these flow phenomena, the overall stage performances for B15 are improved compared with those for B00 as seen in Figs. 22.3 and 22.4. Also for B30, the formation of the hub corner separation increases the C_{ptc4} as seen in Figs. 22.7c and 22.8. However, the separation region and the related increase of C_{ptc4} are smaller than those for B15 as observed in Fig. 22.7b, c and 22.8. The comparison of Fig. 22.7c with b indicates that the secondary flow from the hub toward the midspan on the blade suction surface is more intensified for B30. This phenomenon enhances the suppression effect on the hub corner separation and then improves the overall stage performances for B30 as shown in Figs. 22.3 and 22.4. The difference of η_{14} between B30 and B15 is decreased compared with that between B15 and B00 as in Fig. 22.3. This phenomenon suggests the following fact. Although the increase of the dihedral angle increases the loss by the intensification of the secondary flow and the increase of the blade surface area and simultaneously decreases the loss by the suppression of the hub corner separation, the former effect gradually becomes predominant by the increase of the dihedral angle. The hub corner separation is generated also for B45 as in Fig. 22.7d. Meanwhile, as shown in Fig. 22.7c, d, the secondary flow from the hub toward the midspan on the blade suction surface for B45 is intensified compared with that for B30. This phenomenon reduces the hub corner separation region for B45. However, in the comparison with B30, the secondary flow loss and the friction loss are considered to be more increased for B45. In addition, the increase of the dihedral angle decreases the angle between the blade pressure surface and the hub

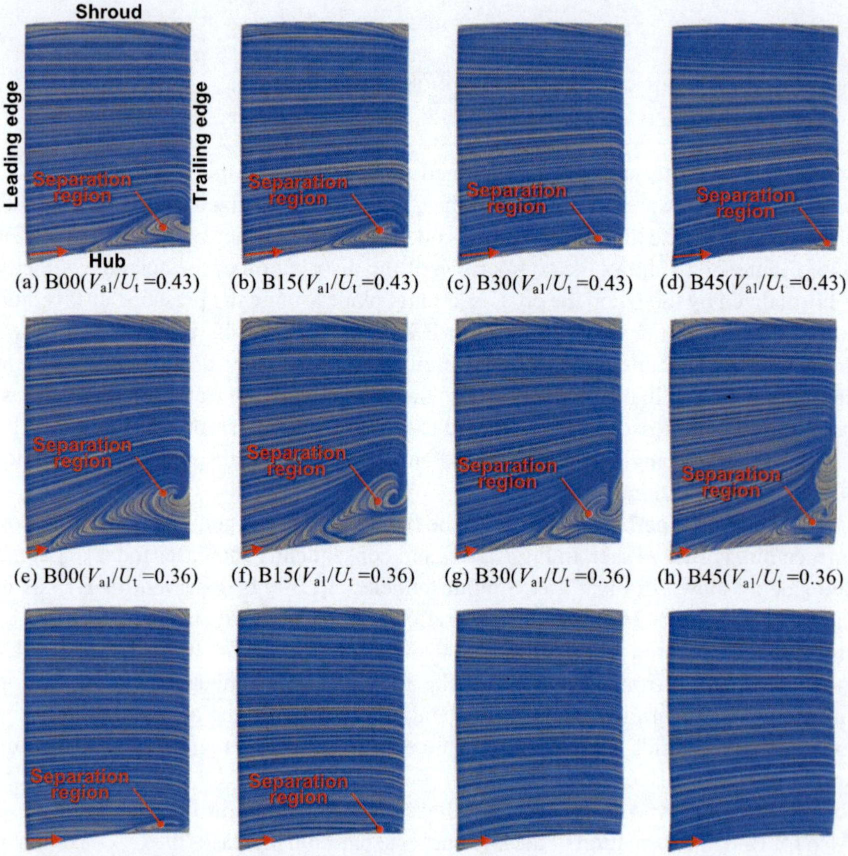

(a) B00(V_{a1}/U_t =0.43) (b) B15(V_{a1}/U_t =0.43) (c) B30(V_{a1}/U_t =0.43) (d) B45(V_{a1}/U_t =0.43)
(e) B00(V_{a1}/U_t =0.36) (f) B15(V_{a1}/U_t =0.36) (g) B30(V_{a1}/U_t =0.36) (h) B45(V_{a1}/U_t =0.36)
(i) B00(V_{a1}/U_t =0.50) (j) B15(V_{a1}/U_t =0.50) (k) B30(V_{a1}/U_t =0.50) (l) B45(V_{a1}/U_t =0.50)

Fig. 22.7 Limited streamlines on suction surface of stator blade

endwall, and consequently enlarges the low momentum fluid region in the vicinity of the pressure surface/hub corner for B45 as seen in Fig. 22.10. As a result, the flow separation is formed at this corner by the streamwise adverse pressure gradient as observed in Fig. 22.9a. These phenomena increase the C_{ptc4} as in Fig. 22.8. Thus, because these loss increase effects are activated by the increase of the dihedral angle, the overall stage performances for B45 are deteriorated compared with those for B30 as shown in Figs. 22.3 and 22.4.

The influences of the bowed stator on the flow field and the overall performances at design flow coefficient are summarized as follows: The bowed stator has the effects to suppress the hub corner separation and the associated loss generation. The increase of the dihedral angle enhances these effects, but simultaneously increases the loss generation by the intensification of the secondary flow and by the increase of the solid blade surface area. Moreover, the excessive increase of dihedral angle further increases the loss by the formation of the separation at the pressure surface/hub

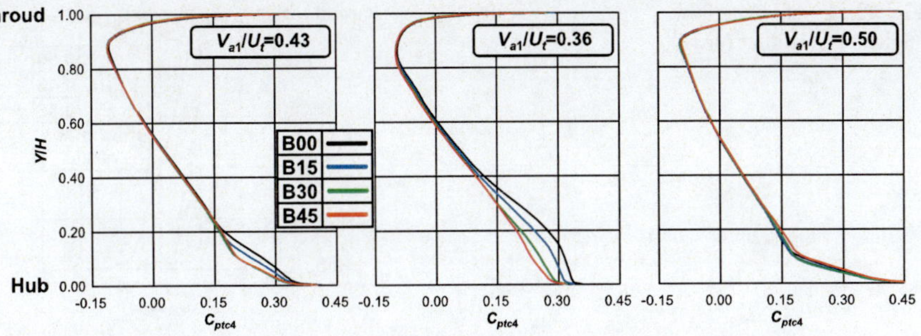

Fig. 22.8 Circumferential mass-averaged total pressure loss coefficient at stator outlet

Fig. 22.9 Limited streamlines on pressure surface of stator blade for B45

Fig. 22.10 Absolute velocity distribution (V_{a1}/U_t = 0.43)

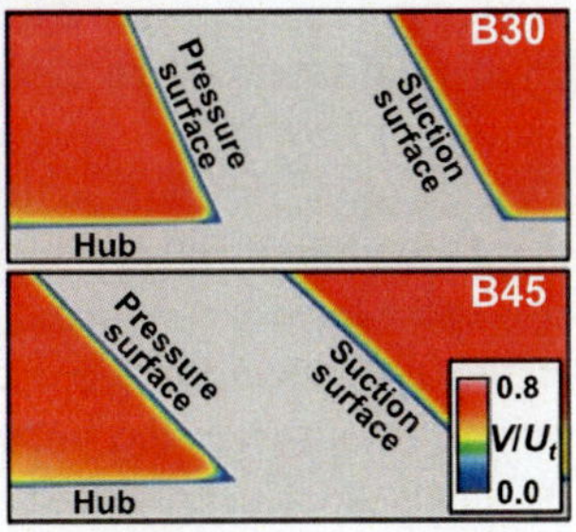

corner. As a consequence, the loss increase effects become predominant compared to the loss reduction effect and then the overall stage performances are deteriorated.

Low Flow Coefficient. For B00, the hub corner separation is formed, but its region and the associated increase of the C_{ptc4} are increased compared with those at design flow coefficient as shown in Figs. 22.7a, e and 22.8. Therefore, the influence of the hub corner separation on the loss generation becomes stronger at low flow coefficient. Also at low flow coefficient, the hub corner separations are suppressed

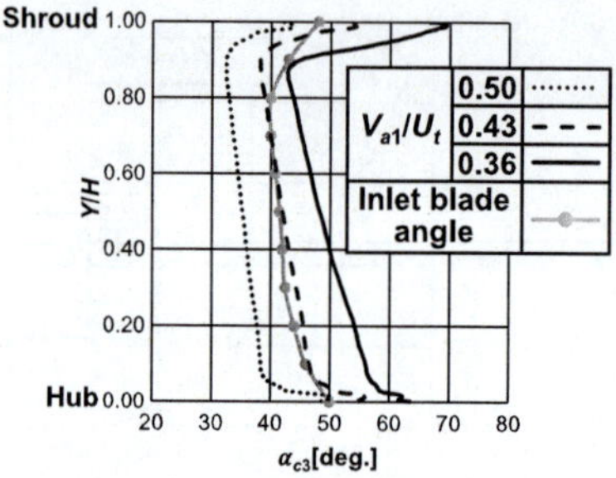

Fig. 22.11 Absolute flow angle at stator inlet for B00

for B15 and B30 by the same mechanism as mentioned at design flow coefficient as seen in Fig. 22.7e–g. As a consequence, the increases of the C_{ptc4} at the hub sides are reduced compared with that for B00 as shown in Fig. 22.8. Because the loss reduction effect due to the suppression of the hub corner separation is expected to be enhanced at low flow coefficient, the increases of overall stage performances by the increases of the dihedral angle are increased compared with those at design flow coefficient as seen in Figs. 22.3 and 22.4. For B45, the flow separation is formed at the pressure surface/hub corner, but this region is reduced compared with that at design flow coefficient as observed in Fig. 22.9a, b. This is due to the fact that the boundary layer development on the blade pressure surface is reduced by the positive incidence angle for the stator blade as in Fig. 22.11. Moreover, the hub corner separation for B45 is further suppressed compared with that for B30 as observed in Fig. 22.7g, h. Because the influence of the corner separation on the loss generation is intensified at low flow coefficient as mentioned above, the loss reduction by the increase of the dihedral angle from B30 to B45 is expected to be enhanced. Because of these facts, the overall stage performances for B45 are improved compared with those for B30 as seen in Figs. 22.3 and 22.4.

The influences of the bowed stator on the flow field and the overall stage performances at low flow coefficient are summarized as follows: The influences of the increase of the dihedral angle on the flow field and the loss generation are similar to those at design flow coefficient. However, the loss reduction caused by the suppression of the hub corner separation is intensified because the influence of this separation on the loss generation becomes predominant compared with that at design flow coefficient. As a result, the improvement effect on the overall stage performances by the increase of the dihedral angle is enhanced. Meanwhile, because of the above-mentioned facts and the suppression of the separation at the pressure surface/hub corner, the optimum dihedral angle which gives the maximum positive effect on the overall stage performances is considered to be larger than that at design flow coefficient.

High Flow Coefficient. For B00, the hub corner separation is formed but its region and the associated increase of the C_{ptc4} are reduced compared with those at design flow coefficient as shown in Figs. 22.7a, i and 22.8. Therefore, the influence of the hub corner separation on the loss generation is weakened compared with that at design flow coefficient. The influences of the increase of the dihedral angle on the flow behavior and the loss generation for B15 are similar to those at design flow coefficient as seen in Figs. 22.7a, b, i, j and 22.8. However, because these positive effects are weakened compared with those at design flow coefficient, the differences of overall stage performances between B15 and B00 are smaller than those at design condition as seen in Figs. 22.3 and 22.4. On the other hand, although the hub corner separation is disappeared for B30 as in Fig. 22.7k, the overall stage performances are similar to those for B00 as shown in Figs. 22.3 and 22.4. Therefore, the optimum dihedral angle becomes smaller than that at design flow coefficient. Moreover, because the incidence angle for the stator blade becomes negative as in Fig. 22.11, the separations are formed not only at the corner of the pressure surface with the hub endwall but also at that with the shroud one for B45 as observed in Fig. 22.9c. These phenomena further deteriorate the stage performances for B45 as seen in Figs. 22.3 and 22.4.

The influences of the bowed stator on the flow field and the overall performances at high flow coefficient are summarized as follows: The bowed stator has the effects to suppress the hub corner separation and the associated loss generation. However, because the influence of the hub corner separation on the loss generation is weakened, the optimum dihedral angle becomes smaller than that at design flow coefficient. Moreover, the excessive increase of the dihedral angle induces the separations at the corner of the pressure surface with the hub endwall as well as at that with shroud one, and consequently the deteriorations of the stage performances are promoted.

22.5 Conclusions

In this study, the influences of the bowed stator on the flow field, the loss generation and the overall performances at the low-speed single-stage axial compressor were investigated at the wide operating range. The conclusions derived from this study can be summarized as follows:

(1) The bowed stator has the effects to suppress the hub corner separation and the associated loss generation. These effects are enhanced by the increase of the dihedral angle. Meanwhile, the increase of the dihedral angle also increases the loss generation by the intensification of the secondary flow and by the increase of the solid blade surface area. Moreover, the excessive increase of the dihedral angle induces the additional loss generation by the formation of the separation at the corner of the blade pressure surface with the endwall.
(2) The influence of the hub corner separation on the loss generation becomes stronger according to the decrease of the flow rate. As a result, the decrease of the flow rate enhances the loss reduction effect attributed to the suppression of the hub corner separation caused by the increase of the dihedral angle. Moreover, the decrease of the flow rate reduces the increase of the loss generated by the separation at the pressure surface/endwall corner due to the excessive large dihedral angle. Therefore, the optimum dihedral angle which has the maximum

positive effect on the overall stage performances is increased according to the decrease of the flow rate.

References

1. M. Hewkin-Smith, G. Pullan, S. Grimshaw, E. Greitzer, Z. Panofsky, The role of tip leakage flow in spike-type rotating stall inception. ASME J. Turbomachine **141**(6):061010-1-061010-10(2019)
2. K. Yamada, M. Furukawa, Y. Tamura, S. Saito, A. Matsuoka, K. Nakayama, Large-scale detached eddy simulation analysis of stall inception process in a multistage axial flow compressor. ASME J. Turbomachine **139**(7), 071002-1–071002-11 (2017)
3. L.H. Smith Jr., H. Yeh, Sweep and dihedral effects in axial-flow turbomachinery. ASME J. Basic Eng. **85**(3), 401–416 (1963)
4. E. Shang, Z.Q. Wang, J.X. Us, The experimental investigations on the compressor cascades with leaned and curved blade, in *Proceeding of the ASME 1993 International Gas Turbine and Aeroengine Congress and Exposition* (Cincinnati, Ohio, USA, 93-GT-50, 1993)
5. H.D. Weingold, R.J. Neubert, R.F. Behlke, G.E. Potter, Reduction of compressor stator endwall losses using bowed stators, in *Proceedings of ASME 1995 International Gas Turbine and Aeroengine Congress and Exposition* (Houston, Texas, USA, 95-GT-380, 1995)
6. A. Fischer, W. Riess, J.R. Seme, Performance of strongly bowed stators in a four-stage high-speed compressor, ASME J. Turbomachine **126**(3), 333–338 (2004)

Chapter 23
Effects of Model Shape of Hemispheric-Head Cyclone Separator on Collection Performance

Shomaru Fukuzato, Kosuke Nasu, Yuki Tanaka, Kazuyoshi Yamada, Mizue Munekata, Hiroyuki Yoshikawa, Kenji Ono, and Takashi Watanabe

Abstract The hemispheric head cyclone separator has considerably higher collection efficiency (separation performance) than a conventional cyclone separator with a flat head. In our previous study, we investigated numerically the air flow in the hemispheric head cyclone separators with different shape in order to optimize the shape of the hemispheric head cyclone separator. As the numerical results of only air flow, we proposed the highest particle collection performance model of hemispheric head cyclone separator. It was predicted that the height of the inlet pipe has a significant effect on its collection efficiency. In this study, the motion of particles in the air flow field is calculated using the numerical airflow velocity field in the different shape model and then the collection efficiency of particles is estimated. However, the particle motion is calculated considering only drag and gravity without considering interaction with fluid and interference between particles. It is shown that the highest collection performance model is shown by comparing the particle motions and the particle collection efficiencies. In the model with the lowest position of the inlet pipe, it was found that the upward outflow motion of particles entering the dust holder reduced the collection efficiency.

S. Fukuzato · K. Nasu · Y. Tanaka · K. Yamada · M. Munekata (✉) · H. Yoshikawa
Kumamoto University, Kumamoto 860-8555, Japan
e-mail: munekata@kumamoto-u.ac.jp

K. Ono
Kyushu University, Fukuoka 819-0395, Japan

T. Watanabe
Shizuoka Plant, Shizuoka 421-0304, Japan

A. Suryan et al. (eds.), *Proceedings of the 9th Asian Joint Workshop on Thermophysics and Fluid Science, 27–30 November 2022, Utsunomiya, Japan*, Springer Proceedings in Physics 396, https://doi.org/10.1007/978-981-99-9470-0_23

23.1 Introduction

In our previous studies, we have shown that the hemispheric head cyclone separator developed by Shizuoka Plant Co., Ltd. forms a flow field with superior collection performance in general industrial cyclone separators [1] compared to conventional flat cyclone separators [2, 3]. We have investigated the effect of changing the dust holder size and inlet pipe height on the collection performance. It was shown that a longer dust holder increases the flow rate into the dust holder and increases the residence time, forming a flow field that improves the collection efficiency [4], and that a model with a larger dust holder size tends to cause the flow to stagnate, preventing the reentry of fine particles [5]. Lowering the height of the inlet pipe improves the centrifugal separation performance due to the increase in swirling velocity and the decrease in fine vortices around the outlet pipe, which in turn improves the collection rate [6]. In this study, the effects on the collection performance of models with a large inlet pipe diameter, models with a large outlet pipe diameter, models with horizontally and vertically tilted inlet pipe, models with shortened cone length, and a model with a lowered inlet pipe, are investigated numerically. Now we consider only the drag force and gravity without considering the interaction with the fluid or the interference between particles. Then, we create a composite model that combines shapes that contribute to the improvement of collection performance, and determine the model with the highest collection performance by performing airflow calculation and particle motion analysis.

23.2 Numerical Analysis

23.2.1 Test Model

Figure 23.1a shows the geometry of the reference default model and the values of each parameter. The representative diameter of the cyclone separator is the maximum diameter of the dome, $D = 175$ mm. The apex of the dome is set as the origin, and the x, y, and z coordinates are set as shown in the figure. Figure 23.2a shows a model with a larger inlet pipe diameter, (b) with a larger outlet pipe, (c) and (d) with the inlet pipe tilted horizontally and vertically, (e) with a shorter conical length, (f) with a lower inlet pipe height, (g)–(i) with a lower inlet pipe and shorter conical length, (j) with a larger inlet pipe and shorter conical length, and (k) with a larger outlet pipe. Figures 23.2g–i shows a model with the inlet pipe lowered and the conical length shortened, and Fig. 23.2j shows details of a model with the inlet pipe diameter increased and the conical length shortened. The inlet pipe is connected radially outward in the case of the models with the inclined and lowered inlet pipe heights in order to inscribe the domed head.

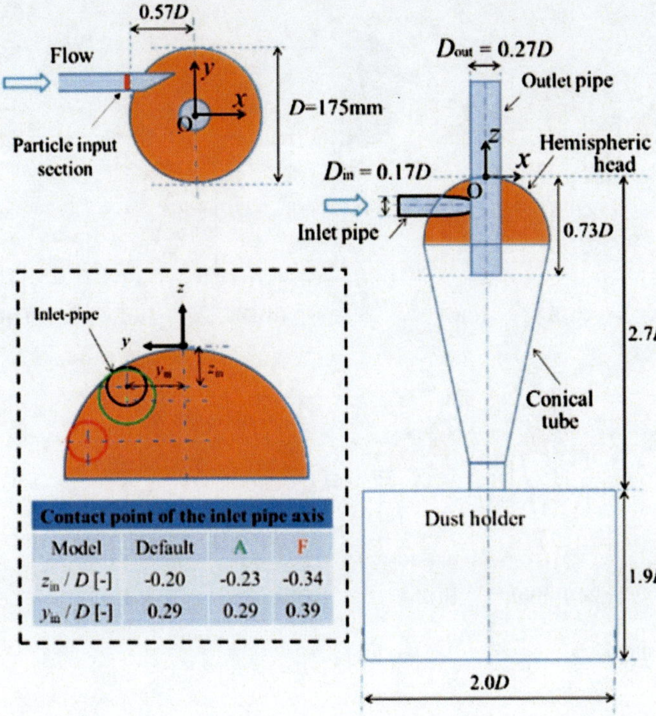

Contact point of the inlet pipe axis			
Model	Default	A	F
z_{in}/D [-]	-0.20	-0.23	-0.34
y_{in}/D [-]	0.29	0.29	0.39

Fig. 23.1 Default model

23.2.2 Numerical Procedures

The air flow field is analyzed by direct numerical simulation (DNS) using the numerical code developed by RIKEN [7] on the computer system for research use at the Research and Development Center for Information Infrastructure, Kyushu University (for general use). The governing equations are the Navier-Stokes equation and the continuity equation considering weak compressibility. The inlet pipe inflow velocity $V_{in} = 20$ m/s is set such that the design inflow flow $Q_{in} = 815$ L/min(1909 L/min at model **A** (large inlet pipe model)). The Reynolds number defined by $Re = V_{in}D/\nu_a$ is $Re = 2.21 \times 10^5$. The outlet pipe velocity V_{out} is set to satisfy the inlet pipe velocity V_{in} as a boundary condition, and the gauge pressure on the boundary surface outside the computational domain is set to $p = 0$ Pa. The size of the computational domain is $2.8D \times 2.3D \times 5.9D$ for all inlet pipe height conditions, the cell size is 1 mm, and the time interval is 5.0×10^{-6} s. Statistics are obtained by time averaging over 0.5 s after sufficient calculations.

The Lagrangian Particle Tracking method is applied to the particle behavior. A one-way coupling simulation is applied to neglect the effect of solid particles on the velocity of the gas phase, and the velocity of the airflow is obtained from the

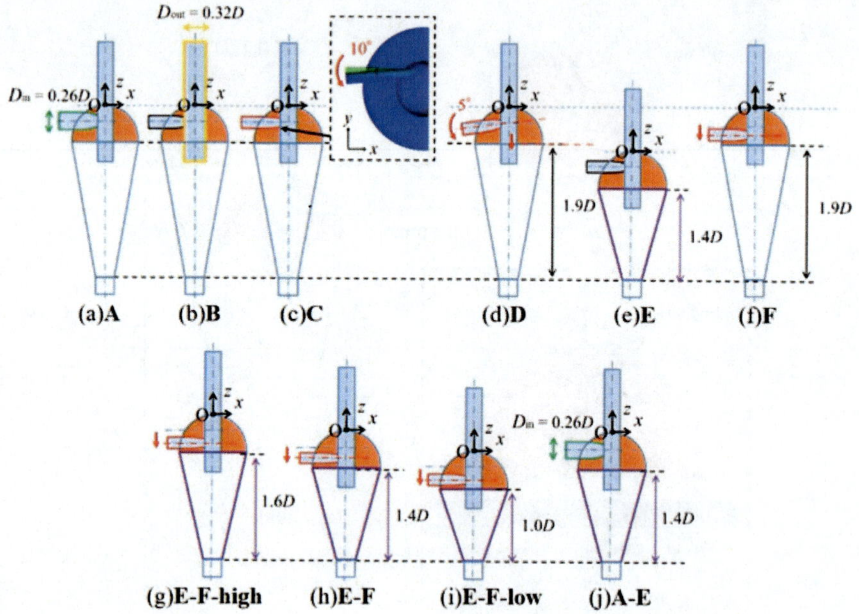

Fig. 23.2 Test models

mean flow field calculated by the airflow analysis. For external forces, only drag and gravity are considered. In the Lagrangian particle tracking method, the particle velocity and position are obtained by numerically integrating the equation of motion of the particles [6].

In this study, we assumed water and nylon particles of density $\rho_p = 1000$ kg/m^3 and put 10,000 of them equally spaced in the cross section of the inlet pipe at the position where the flow developed ($x = -0.57D$); at $D_p > 5.0$ μm, they were almost collected by all models, indicating that $D_p = 2.0$, 2.5, 3.0 and 5.0 μm, we report results for $D_p = 2.0$, 2.5, 3.0 and 5.0 μm.

23.3 Results and Discussion

Figures 23.3 and 23.4 show the results of the collection rate and the fraction of particles collected in the dust holder (**A–F**) after 10 s of particle feeding at $\rho_p = 1000$ kg/m^3, Figs. 23.5 and 23.6 show the time-averaged contour maps of the circumferential and axial velocity, Fig. 23.7 shows the time history of the collection rate at 2.5 μm and the fraction of particles collected in the dust holder, Fig. 23.8 shows the number of times particles flowed into the dust holder, binned by arbitrary number of times (total number $N = 10{,}000$), Figs. 23.9 and 23.10 show the results of the collection rate and the fraction of particles collected in the dust holder (combined model) after 10 s from particle loading at $\rho_p = 1000$ kg/m^3, and Figs. 23.9 and 23.10 show the results of the collection rate and the fraction of particles collected in the dust holder 10 s after particle loading for $\rho_p = 1000$ kg/m^3 in the combined model (**E-F-high**, **E-F**, **E-F-low**, **A-E**). Figure 23.12 shows the contour map of the time-averaged absolute velocity in the dust holder for the combined model.

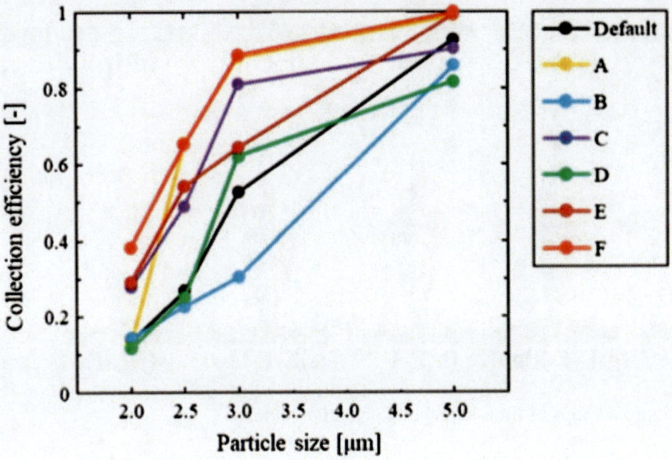

Fig. 23.3 Collection efficiency

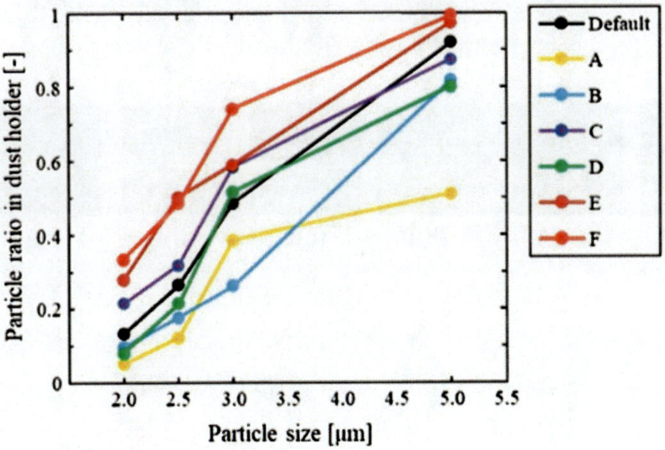

Fig. 23.4 Collected particle ratio in the dust holder

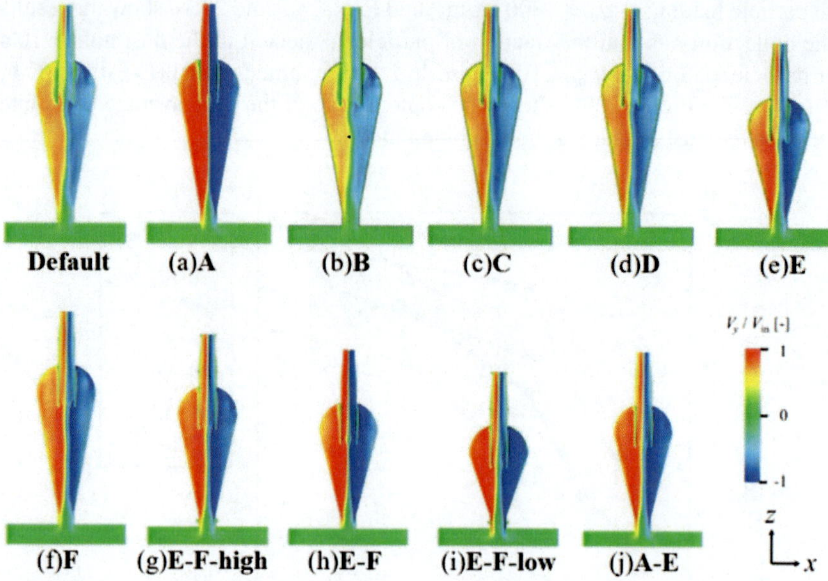

Fig. 23.5 Contour maps of time-averaged tangential velocity

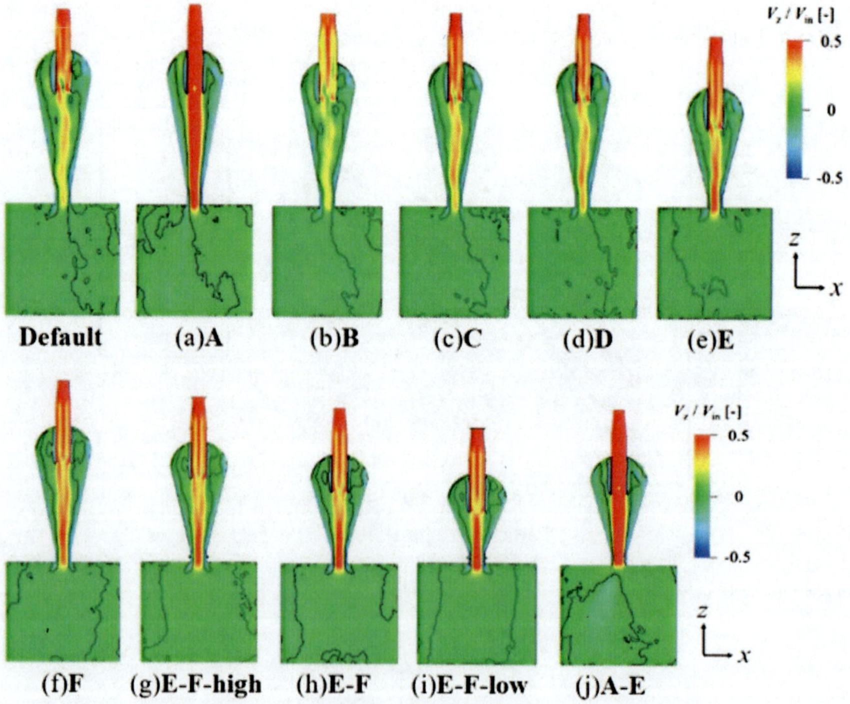

Fig. 23.6 Contour maps of time-averaged axial velocity

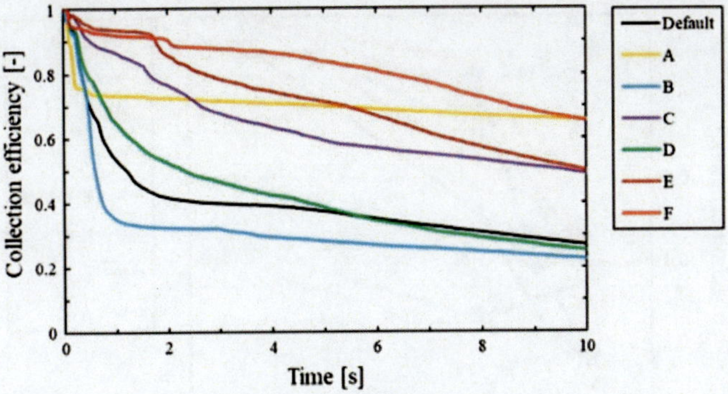

(a) Collection efficiency

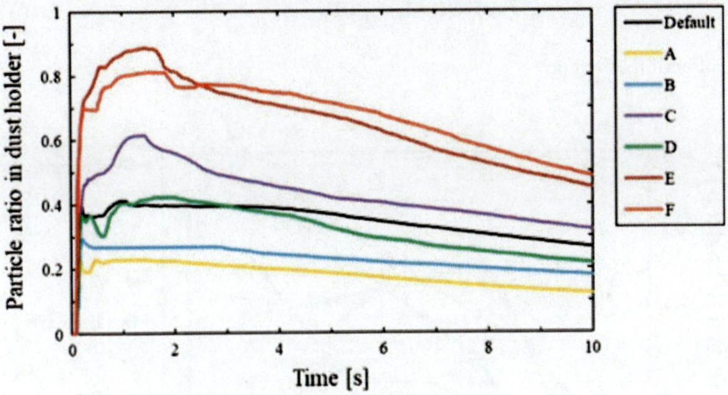

(b) Particle ratio in dust holder

Fig. 23.7 Time histories ($D_p = 2.5\ \mu m$)

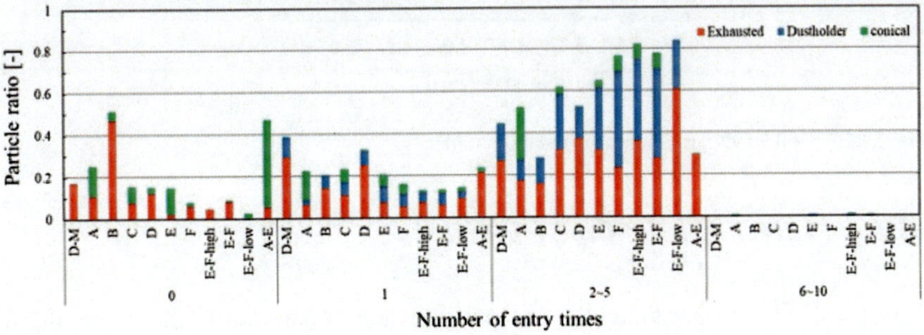

Fig. 23.8 The number of entry times into the dust holder ($D_p = 2.5\ \mu m$)

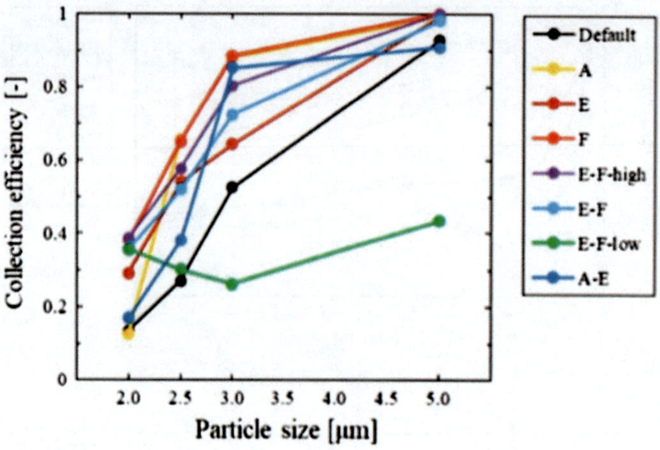

Fig. 23.9 Collection efficiency

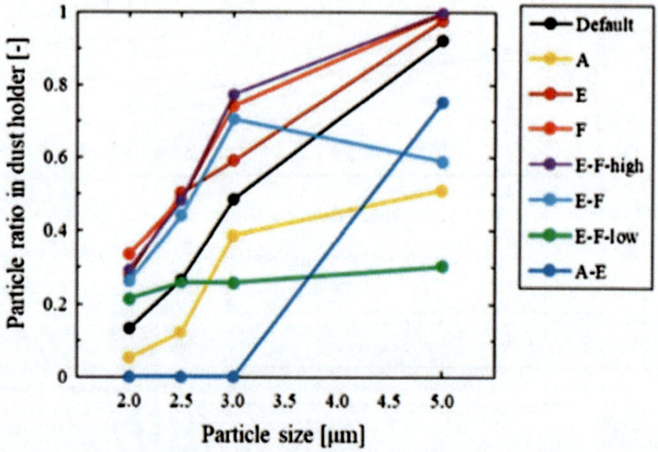

Fig. 23.10 Collected particle ratio in the dust holder

23.3.1 *Effect of Inlet Pipe Diameter*

Figure 23.3 shows that the collection performance of particles in the range of 2.5 to 5.0 µm was higher for **A** with a larger inlet pipe diameter in **Default**. The figure indicates that the increase in flow rate due to the larger inlet pipe diameter increases the circumferential velocity inside the cyclone, which facilitates the separation of particles and improves the particle collection performance. Figures 23.4, 23.6, 23.7 and 23.8 show that the increase in flow rate causes a stronger upwelling and a wider area of upwelling, which makes it easier for particles to be captured by the upwelling vortex at the bottom of the conical and harder for particles to be collected in the dust holder, and many particles continue to swirl in the conical.

23.3.2 *Effect of Inlet Pipe Diameter*

Figure 23.3 shows that the collection rate for model **B** with a larger outlet pipe diameter was the lowest among all models. Figure 23.5 shows that increasing the outlet pipe diameter decreases the swirling velocity in the free vortex region (outer circumference), which weakens the centrifugal separation action inside the conical, resulting in a decrease in particle collection performance. Figures 23.4, 23.7 and 23.8 also show that the decrease in centrifugal separation makes it difficult to transport particles to the dust holder, and many particles are discharged from the outlet pipe before entering the dust holder.

23.3.3 *Effect of Inlet Tube Height*

Figure 23.3 shows that the collection rate for all particle sizes is higher in the case with the inlet pipe inclined 10° horizontally (**C**) than in the case with the inlet pipe inclined 5° vertically (**D**), while the collection rate for all particle sizes except 3.0 μm is lower in the case with the inlet pipe inclined 5° vertically (**C**) than in the case with the inlet pipe inclined 10° vertically (**Default**). Figure 23.8 shows that in case **C**, particles continue to swirl in the conical section and are difficult to be discharged from the outlet pipe. Figure 23.5 shows that the circumferential velocity increases as the inlet pipe is tilted horizontally and vertically, resulting in an increase in the radius of gyration due to the lowering of the height of the inlet pipe. Since the centrifugal force is proportional to the square of the circumferential velocity, it can be assumed that the model with the inclined inlet pipe creates a flow field with high centrifugal separation performance, resulting in a slightly lower collection rate for **D** than for **Default**. However, the collection rate for **F**, where only the height of the inlet pipe is lowered without tilting the inlet pipe, is the highest. This may be due to the fact that the height of the inlet pipe is lower in **F** than in **C** and **D**. From Figs. 23.4, 23.5, 23.7 and 23.8, it can be seen that the particles are more easily transported to the dust holder due to the stronger centrifugal separation action.

23.3.4 *Effect of Conical Length*

Figure 23.3 shows that the **E** with a shorter conical length increases the collection rate for all particle sizes. This is because, as shown in Figs. 23.4, 23.5, and 23.7 the circumferential velocity inside the conical increases with decreasing the conical length, which improves the centrifugal separation performance and makes it easier for particles to be transported to the dust holder. Figure 23.8 shows that the number of particles discharged from the outlet pipe without ever entering the dust holder is also reduced.

23.3.5 Composite Model

23.3.5.1 Effect of Lowering the Inlet Pipe Height and Shortening the Conical Length

For the model with a low inlet pipe height (**F**) and the models with different conical lengths (**E-F**, **E-F-high**, **E-F-low**), Fig. 23.9 shows that the collection rate for **F** with the longest conical length was the highest, while the collection rate for **E-F-low** with the shortest conical length was the lowest. Figures 23.5 and 23.7 show that decreasing the conical length increases the circumferential velocity, which strengthens the centrifugal separation action and makes it easier to separate particles and transport them to the dust holder, but Fig. 23.11 shows that the shorter the conical length, the shorter the residence time in the dust holder, which causes the particles collected in the dust holder to be rolled up and the shorter the conical length, the shorter the residence time in the dust holder. Figure 23.8 shows that the longer the conical length, the more particles are swirled in the conical section and remain in the conical without being discharged from the outlet pipe.

23.3.5.2 Effect of Increasing the Inlet Pipe Diameter and Shortening the Conical Length

In the model with a larger inlet pipe diameter and a shorter conical length (**A–E**), Fig. 23.9 shows that the collection performance of **A–E** is lower than that of **A** and **E**. This is because the inlet pipe diameter is larger and the conical length is shorter than that of **A–E**. The reason for this is that although the circumferential velocity in the conical section increased as shown in Figs. 23.5, 23.6 and 23.8 the particles were more easily captured by the upward vortex due to the expansion of the region of the upward component at the bottom of the conical, making it difficult to transport particles to the dust holder, and almost no particles in the 2.0–3.0 μm range were transported to the dust holder. The increase in the air velocity in the dust holder, as shown in Fig. 23.10 shortens the residence time in the dust holder, which causes the particles collected in the dust holder to be rolled up and returned to the conical section, thereby reducing the collection rate.

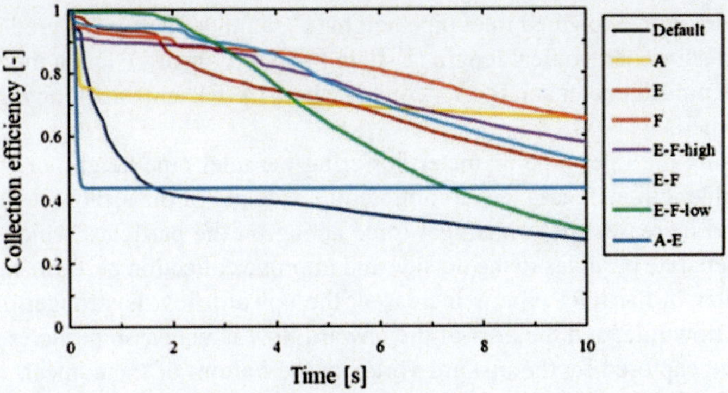

(a) Collection efficiency

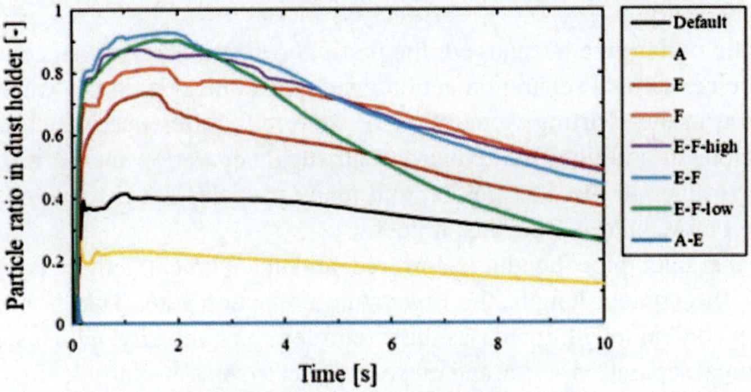

(b) Particle ratio in dust holder

Fig. 23.11 Time histories ($D_p = 2.5\ \mu m$)

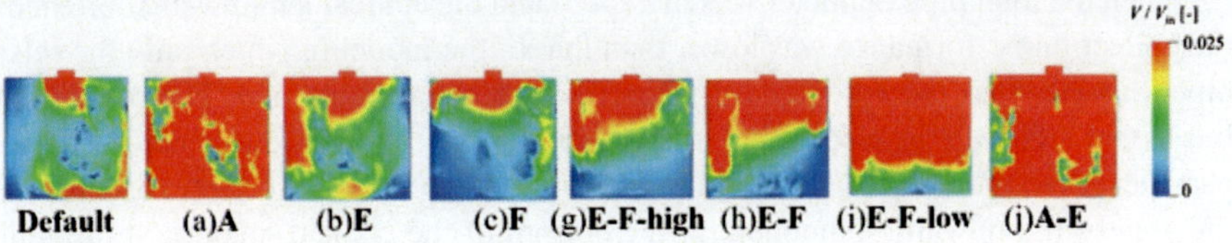

Fig. 23.12 Contour maps of time-averaged velocity in dust holder

23.4 Conclusion

In this study, the effects of the model shape of the hemispheric head cyclone separator on airflow characteristics and collection performance were investigated by performing airflow calculations and particle motion analysis on a model with a larger inlet pipe diameter (**A**), a model with a larger outlet pipe (**B**), a model with the inlet pipe tilted horizontally and vertically (**C**, **D**), a model with a shorter conical length

(**E**), a model with a lowered inlet pipe height (**F**), a model with a lowered inlet pipe height and shortened conical length (**E-F**, **E-F-high**, **E-F-low**), and a model with a larger inlet pipe diameter and shorter conical length (**A-E**), and the following findings were obtained.

Increasing the inlet pipe diameter, lowering the inlet pipe height, or shortening the conical length increases the circumferential velocity of the airflow in the conical section and increases the centrifugal force acting on the particles, which makes it easier to separate particles to the outside and improve collection performance. When the diameter of the inlet pipe is increased, the upward flow is stronger due to the increased flow rate, and the area of the upward flow is wider, so particles are more likely to be captured by the upward vortex at the bottom of the conical, making it difficult to collect particles in the dust holders. However, when the inlet pipe height is lowered and the conical length is shortened, particles are transported to the dust holder more easily.

When the outlet pipe is enlarged, the particle collection performance is reduced because the centrifugal separation action inside the conical becomes weaker due to the decrease in the swirling velocity of the outer circumference, which is the free vortex region. In addition, the reduced centrifugal separation makes it difficult to transport particles to the dust holder, and many particles are discharged from the outlet pipe before entering the dust holder.

When the inlet pipe height is lowered and the conical length is shortened, the longer the conical length, the higher the collection rate. This is because the decrease in conical length increases the circumferential velocity, which strengthens the centrifugal separation action and makes it easier to separate particles and transport them to the dust holder, but because the residence time in the dust holder is short, the particles collected in the dust holder are rolled up and returned to the conical section, which causes a decrease in the collection rate. This is a factor that reduces the collection rate.

When the inlet pipe diameter was increased and the conical length was shortened, the collection performance was lower than that of the model in which only the inlet pipe diameter was increased and the conical length was shortened. The reason for this is that although the circumferential velocity in the conical section increased, the particles were more easily captured by the upward vortex due to the expansion of the area of the upward component at the bottom of the conical, making it difficult for particles to be transported to the dust holder, and particles in the 2.0–3.0 μm range were hardly transported to the dust holder. The increased air velocity in the dust holder shortens the residence time in the dust holder, which causes the particles collected in the dust holder to be rolled up and returned to the conical section, thereby reducing the collection rate.

Model **F** is the best model for particle collection because it creates a flow field that facilitates particle collection considering the centrifugal separation performance and stagnant conditions in the dust holder.

Acknowledgements This research was supported by the Harada Memorial Foundation.

References

1. Shizuoka Plant INC., Oct. 30, 2014: International application WO2014/175083A1
2. M. Munekata, A. Narita, G. Kitamura, H. Yoshikawa, K. Ono, T. Watanabe, Flow characteristics and separation performance in a hemispheric head cycloe separator, in *Proceedings of ISAIF13*, ISAIF-S-0112 (2017)
3. M. Munekata, G. Kitamura, N. Tsuzaki, A. Tsurumaru, H. Yoshikawa, K. Ono, T. Watanabe, T.: Collecting performance in a cyclone separator with a hemispheric head, in *Proceedings of ISAIF14*, ISAIF14-066 (2019)
4. R. Furukawa, K. Terada, S. Fukuzato, M. Munekata, H. Yoshikawa, K. Ono, K. Watanabe, Numerical analysis of flow in a domed cyclone separator for fuel cell vehicles, in *Proceedings of the JSME Annual Conference 2020*, S05418 (2020)
5. S. Fukuzato, R. Furukawa, K. Nasu, M. Munekata, H. Yoshikawa, K. Ono, K. Watanabe, K. Terada, Effect of size of dust holder on the air flow characteristics in a hemispheric head cyclone separator, in *Proceedings of the 99th Lecture of the Department of Fluid Engineering*, The Japan Society of Mechanical Engineers. OS04-06, 2021.11.08-10
6. K. Nasu, R. Furukawa, S. Fukuzato, M. Munekata, H. Yoshikawa, K. Ono, Numerical study on collection performance of particulates in hemispheric head cyclone separators with different location of inlet pipe, in *Proceedings of the 75th Annual Meeting of the Kyushu Branch of the Japan Society of Mechanical Engineers*, No.121 (2022)
7. K. Ono, (ed.), Flontflow/FFV-C, https://avr-aics-riken.github.io/ffvc_package/. Accessed 17 Oct 2019

Chapter 24
Wind Tunnel Experiments on Interaction Between Three Vertical-Axis Wind Turbines Arranged in Tandem

Yoshifumi Jodai and Yutaka Hara

Abstract Wind tunnel experiments on the interaction among three closely spaced vertical-axis wind turbines arranged in tandem were conducted. The rotor spacing between an upwind turbine and a downwind turbine was kept constant at $7D$ (D is the turbine diameter) in the experiments. There are four experimental cases depending on the combinations of the rotational directions of the three turbines. Monotonic increases in the rotational speed and the rotor power of an upwind turbine and a middle turbine, with the increase in g_1, are demonstrated for all the cases tested. Here, g_1 is the rotor gap (spacing between the surfaces of these turbines. On the other hand, monotonic decreases in the rotational speed and the rotor power of a downwind turbine, with increasing g_1, are found for all the cases. There is optimal gap $g_{1,\,opt}$, in which the averaged normalized power of the three turbines $P_{norm,\,ave}$ takes maximum value, for each experimental case. The results demonstrate that a trio of turbines including an inversely-rotating pair expands the range of $g_{1,\,opt}$ in the comparison with the case only containing co-rotating turbines.

24.1 Introduction

Importance of a gap distance between two rotors [1] and overall power performance of turbines [2] have been reported on vertical-axis wind turbines (VAWTs) in a tandem arrangement. Recently, Kuang et al. [3] showed that locating downstream rotor in the gap of $5D$ in the tandem arrangement can appropriately balance the power performance and space cost by using improved delayed detached-eddy simulation (IDDES), where D is the rotor diameter of each turbine. However, changes in the

Y. Jodai (✉)
National Institute of Technology (KOSEN), Kagawa College, 355 Chokushi, Takamatsu 761-8058, Japan
e-mail: jodai@t.kagawa-nct.ac.jp

Y. Hara
Tottori University, 4-101 Koyama-Minami, Tottori 680-8552, Japan

A. Suryan et al. (eds.), *Proceedings of the 9th Asian Joint Workshop on Thermophysics and Fluid Science, 27–30 November 2022, Utsunomiya, Japan*, Springer Proceedings in Physics 396, https://doi.org/10.1007/978-981-99-9470-0_24

rotational speed of each rotor, as a result of dynamic interactions between a fluid and wind turbines, have not been considered in the above simulations assuming constant rotational speed. Unfortunately, investigations by Dessoky et al. [1], Sahebzadeh et al. [2], and Kuang et al. [3] only contain a co-rotating configuration. Moreover, as far as we know, the effect of the third turbine on overall performance in the variable-speed VAWT array arranged in tandem has not been elucidated.

Dabiri [4] conducted field experiments with three staggered counter-rotating VAWTs, in which the third turbine placed $4D$ downstream in tandem from the second turbine (i.e., the rotor spacing between the surfaces of second rotor and third rotor is $3D$). In his experiments, the performance of the third turbine was recovered to within 5% of the isolated turbine performance. Zanforlin and Nishino [5] reported the power coefficient of closely spaced VAWTs including tandem arrangement by using two-dimensional (2-D) numerical simulation assuming fixed rotational speeds. The studies by Dabiri [4] and Zanforlin and Nishino [5] only treat counter-rotating configurations.

Hara and Jodai [6, 7] reported that the rotational speed of the downwind rotor recovers 80% of the isolated rotor performance at rotor spacing of $10D$ in the case of a pair of miniature turbines arranged in tandem. This is a scaled down model [8] of a butterfly wind turbine (BWT) with a rotor diameter of 7 m developed by Hara et al. [9].

Regarding flows around two [10–16] or three [17] tandem circular cylinders, existence of critical spacing g_{crit} in wide range of Reynolds number, distinguishing reattachment regime from co-shedding regime, is well known. Alam reported [15] that the critical spacing corresponding to the inphase mode is $g_{crit} = 2.65D$ (the spacing between the surfaces of the two cylinders) and that to the antiphase mode is $g_{crit} = 4.25D$, based on 2-D numerical simulation for two tandem cylinders. In addition, Alam et al. [17] have shown that the critical spacings corresponding to the inphase mode are $(g_{1crit}, g_{2crit}) = (2.5D, 2.6D)$ and those to the antiphase mode are $(g_{1crit}, g_{2crit}) = (4.25D, 4.5D)$ for three tandem cylinders. Here, g_{1crit} is the spacing between the first and the second cylinders and g_{2crit} is that between the second and the third cylinders. It is interesting that these values of g_{1crit} are close to the abovementioned values of g_{crit} for two cylinders. The flow patterns around a circular cylinders with/without spinning might be similar to that of turbine rotors to some extent. While, Chatterjee et al. [16] numerically showed rotation induced flow suppression around two tandem circular cylinders by changing the rotor spacing as 0–$3D$ for tip speed ratio of 0–2.75. However, it is important to perceive the intrinsic difference in the manner by which vortices are shed between a solid circular cylinder and a turbine rotor comprising blades.

As for flows around rotating circular cylinders, Behara et al. [18] reported flow-induced oscillations of three tandem rotating cylinders in addition to their investigation into two tandem rotating cylinders [19]. Fluctuating fluid forces acting on two circular cylinders arranged in tandem were reported by Alam et al. [20].

Considering the studies above, this paper investigates the interaction between three closely spaced VAWTs in a tandem arrangement. This study aimed to determine the optimal rotor spacing for VAWTs, via wind tunnel experiments using miniature models assuming application to variable-speed small VAWTs seen in the field in cases of unidirectional/bidirectional wind such as accelerated flow between high-rise buildings.

24.2 Methods

24.2.1 Configuration of Flow Field

The wind tunnel experiments were executed at the National Institute of Technology (KOSEN), Kagawa College. Figure 24.1 shows the 3-D image of a miniature wind turbine used in the experiments. A BWT is a lift-type VAWT with straight-blade portions like an H-Darrieus wind turbine with high performance. Figure 24.1a shows the overview and Fig. 24.1b shows the cross section of the model at the equatorial plane. The diameter D and the height H of the miniature model are 50 mm and 43.4 mm, respectively. The chord length c is 20 mm, the blade cross section is NACA 0018 (illustrated with red hatching in Fig. 24.1b), and the blade number B is 3. The solidity σ of the turbine model defined by $Bc/(\pi D)$ was 0.382. The experiments were conducted on three turbines arranged in tandem in the open space after the wind tunnel exit at a uniform velocity of $V = 12$ m/s. An upwind turbine (Rotor 1) and a downwind turbine (Rotor 3) are fixed 100 and 500 mm downstream of the wind tunnel exit, respectively. The position of the middle turbine (Rotor 2) is changed in this study. All the three rotors are in line parallel to the main stream. The Reynolds number based on V and the rotor diameter D was $Re_D = 4.0 \times 10^4$.

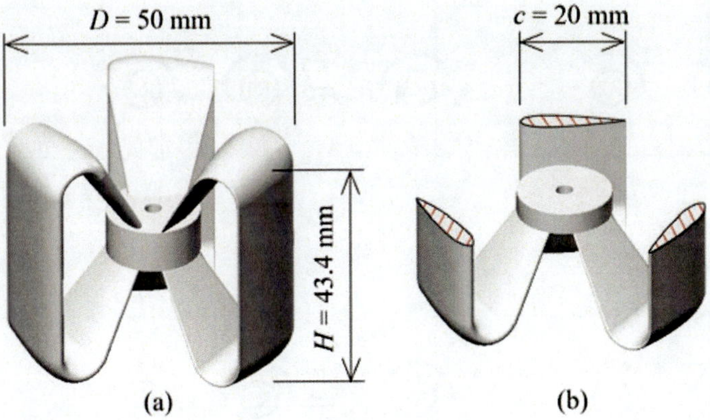

Fig. 24.1 Miniature model of a butterfly wind turbine (BWT) [9]

24.2.2 Tandem Layouts of a Trio of Vertical-Axis Wind Turbines

Figure 24.2 shows tandem layouts against a wind direction of a closely spaced VAWT trio. The rotor spacing between the upwind turbine (Rotor 1) and the downwind turbine (Rotor 3) was kept constant at $7D$ ($D = 50$ mm is the turbine diameter) through the experiments. The rotor gap between the Rotor 1 and a middle turbine (Rotor 2) is indicated by g_1 (spacing between the surfaces of Rotor 1 and Rotor 2). The rotor gap between the Rotor 2 and the Rotor 3 is indicated by g_2, which is determined by the change in g_1 since $g_2 = 7D - g_1 - D = 6D - g_1$. The VAWT trio cases with the gap ratios of $g_1/D = 0.8$–3.4 ($g_1 = 40$–170 mm) were investigated in the tandem layouts.

There are four experimental cases depending on the combinations of the rotational directions of the three turbines. In case 1 (Fig. 24.2a), three turbines rotate the same direction. In case 2 (Fig. 24.2b), Rotor 3 rotates the opposite direction to that of the Rotors 1 and 2. In case 3 (Fig. 24.2c), Rotor 2 rotates the opposite direction to that of Rotors 1 and 3. In case 4 (Fig. 24.2d), Rotor 1 rotates the opposite direction to that of Rotors 2 and 3. Rotor 1, Rotor 2, and Rotor 3 are denoted by R1, R2, and R3, respectively. The curved blue and red arrows indicate clockwise and counterclockwise directions corresponding to the rotational directions of the turbines.

Figure 24.3 shows the experimental setup of a trio of 3-D-printed VAWT models with a 150-mm gap ($g_1/D = 3$) arranged in tandem. Rotor 1, Rotor 2, and Rotor 3 are aligned with the wind direction as seen in Fig. 24.2. The rotors can easily be moved streamwise direction to obtain the required spacing.

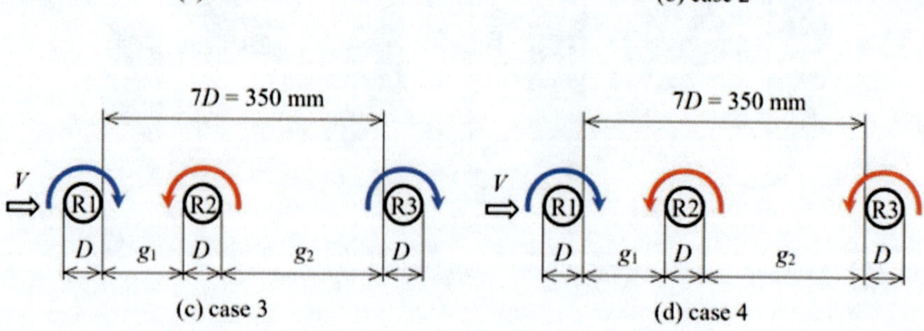

Fig. 24.2 Four cases in the tandem layout of a closely spaced VAWT trio

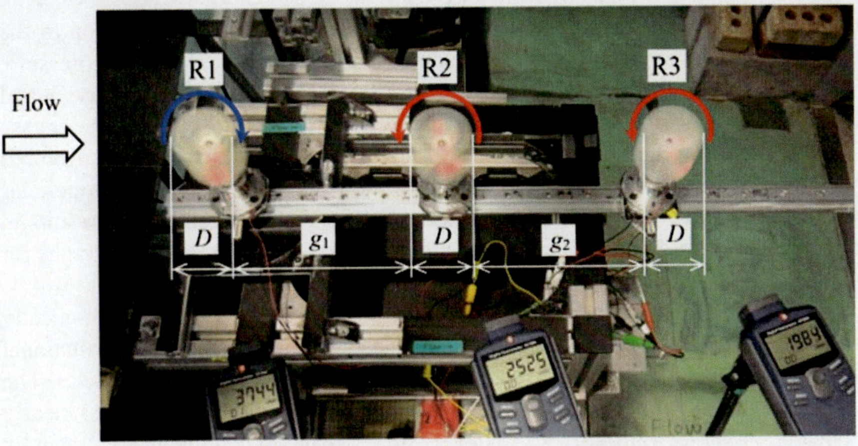

Fig. 24.3 Experimental setup of a trio of 3-D-printed VAWT models with a gap of $g_1/D = 3$ arranged in tandem under case 4

The relationship between the rotational speed N and the motor power (power P of the model turbine in equilibrium) is expressed by Eq. (24.1). The detail of the torque measurement for the power calculation has been shown in Jodai and Hara [8]. In the present experiments, a direct current motor (KIT, MT1415: Tokyo, Japan) was used only to start the rotation of a turbine in a uniform wind. Each rotor can revolve at a condition in which the rotor torque and the load torque become equal. Therefore, the present experiments are equivalent to the operation of variable-speed small VAWTs. The model's rotational speed N [rpm] was measured using a non-contact type digital tachometer (Ono Sokki, HT-5500: Yokohama, Japan) with an accuracy of ±1 rpm.

$$P\,[\mathrm{mW}] = 0.2047\left(\frac{N[\mathrm{rpm}]}{1000}\right)^3 + 0.0442\left(\frac{N[\mathrm{rpm}]}{1000}\right)^2 + 21.042\left(\frac{N[\mathrm{rpm}]}{1000}\right) - 0.0851 \tag{24.1}$$

24.3 Results and Discussion

24.3.1 *Rotational Speeds of a Trio of VAWTs in Tandem Layouts*

Figure 24.4 shows the variation in the rotational speed N [rpm] with the spacing between the surfaces of Rotor 1 and Rotor 2, g_1, at a uniform velocity $V = 12$ m/s. The values of the rotational speed N_{SI} and the power P_{SI} of an isolated turbine of each experimental case change depending on the subtle difference in alignment between the turbine rotor and the start-up direct current motor in addition to the change in the experimental condition such as temperature or atmospheric pressure. The values of N_{SI} [rpm] were: 3400 for case 1, 3500 for case 2, 3830 for case 3, and 3800 for case 4. While, the values of P_{SI} [mW] were: 80.01 for case 1, 82.88 for case 2, 92.66 for case 3, and 91.75 for case 4. In Fig. 24.4, the results of Rotor 1, Rotor 2, and Rotor 3 are illustrated in red, blue, and black, respectively. Figure 24.4 includes the averaged rotational speed of the three rotors in green. Hereafter, the results in the cases of stopping of turbine by wake interaction will not be shown in graphs for visibility.

As seen in Fig. 24.4, the rotational speeds of Rotor 1 and Rotor 2 monotonically increase with increasing g_1 for all the cases tested. On the other hand, the rotational speed of Rotor 3 monotonically decreases with increasing g_1 for all the cases. The effects of the rotational directions (the differences between the four cases) clearly appeared in the leeward turbines (Rotors 2 and 3). These phenomena are drawn by wake interference between three turbines.

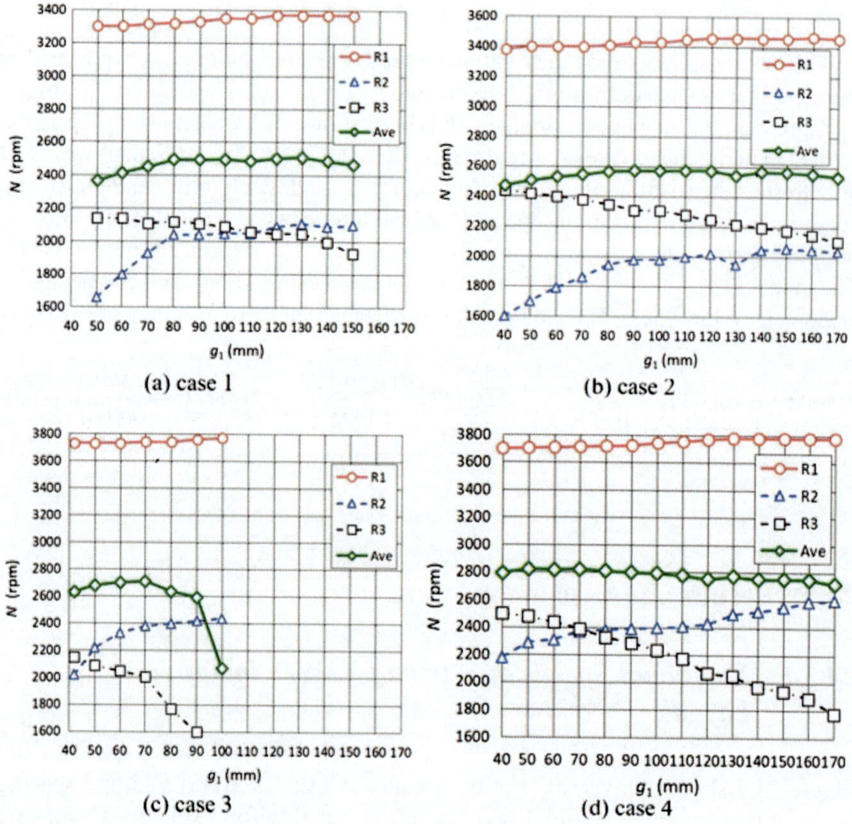

Fig. 24.4 Gap dependence of rotational speed of each rotor and averaged rotational speed

The amount of the decrease in the rotational speed N depended on a rotor gap ratio g_1/D. According to the results of our previous wind tunnel experiments on two tandem BWTs [6], the value of rotational speed of the downwind rotor is about 62% of that of an isolated rotor at $g/D = 2$ (g is the spacing between the surfaces of an upwind turbine and a downwind turbine). The values of N of Rotor 2 in this three tandem BWTs at $g_1/D = 2$ ($g_1 = 100$ mm) are 60.3%, 56.6%, 63.7%, and 63.2% for cases 1, 2, 3, and 4, respectively. These are in line with the results (~62%) of a pair of turbines [6].

Furthermore, the values of N of Rotor 3 were not significantly different from those of Rotor 2 at $g_1/D = 2$, except for case 3 (Rotor 3 stopped). This indicates that an idea adding the third turbine behind the second one can improve efficiency of usage of wind energy even in a narrow area, such as strong accelerated flow field between high-rise buildings exemplified in Jodai and Hara [6].

24.3.2 Power of a Trio of VAWTs in Tandem Layouts

Figure 24.5 plots the variation in the turbine power P [mW] with the spacing between the surfaces of Rotor 1 and Rotor 2, g_1, at a uniform velocity $V = 12$ m/s. The colors and the marks in Fig. 24.5 are the same as in Fig. 24.4. As seen in Fig. 24.5, the amount of the decrease in power also depended on a rotor gap ratio g_1/D. We define the normalized power P_{norm} as in Eq. (24.2). P_{SI} is the power of an isolated turbine.

$P_{norm} = P/P_{SI}$(24.2)

Figure 24.6 shows the normalized power P_{norm} of each turbine and averaged power of the three turbines for all experimental cases. The gap between Rotor 1 and Rotor 2, g_1, is also nondimensionalized with the diameter of each rotor D. As seen in Fig. 24.6, there exists optimal gap $g_{1,\,opt}$, in which the averaged normalized power of the three turbines $P_{norm,\,ave}$ takes maximum value, for each experimental case. Except for case 3, at a rotor gap of $g_1 \sim 2D \equiv g_{1,\,opt}$ (then, $g_2 \sim 4D$) as an optimal layout, a trio of turbines generates relatively high power of $P_{norm,\,ave} \sim 70\%$. Here, g_2 (= $6D$ - g_1) is the spacing between the surfaces of the Rotor 2 and the Rotor 3.

The approximate ranges of $g_{1,\,opt}$ for four experimental cases, in which $P_{norm,\,ave}$ takes maximum value of ~70–71%, are as follows: $g_{1,\,opt}$ = 1.6–3D for case 1; $g_{1,\,opt}$ = 1.4–3.4D for case 2; and $g_{1,\,opt}$ = 0.8–3.2D for case 4. An exception was $g_{1,\,opt}$ = 1–1.4D for case 3 having at most $P_{norm,\,ave}$ = 66–67%. These results have demonstrated that a trio of turbines including an inversely-rotating pair, as in cases 2 and 4, expands the range of $g_{1,\,opt}$ in the comparison with case 1. In contrast, a trio of turbines including two inversely-rotating pairs, as in case 3, lowers the value of $P_{norm,\,ave}$ and significantly narrows the range of $g_{1,\,opt}$.

Sahebzadeh et al. [21] reported that the averaged power of two tandem turbines is 68% of that of an isolated rotor at $g/D = 2$ (see their Fig. 24.6). The values of $P_{norm,\,ave}$ in our experiments on three tandem BWTs at $g_1 = g_{1,\,opt} = 2\,D = 100$ mm are 70.6%, 70.6%, 52.3%, and 70.2% for cases 1, 2, 3, and 4, respectively. Our results showing $P_{norm,\,ave} \sim 70\%$ on a trio of turbines at $g_1 = 2D$ (except for case

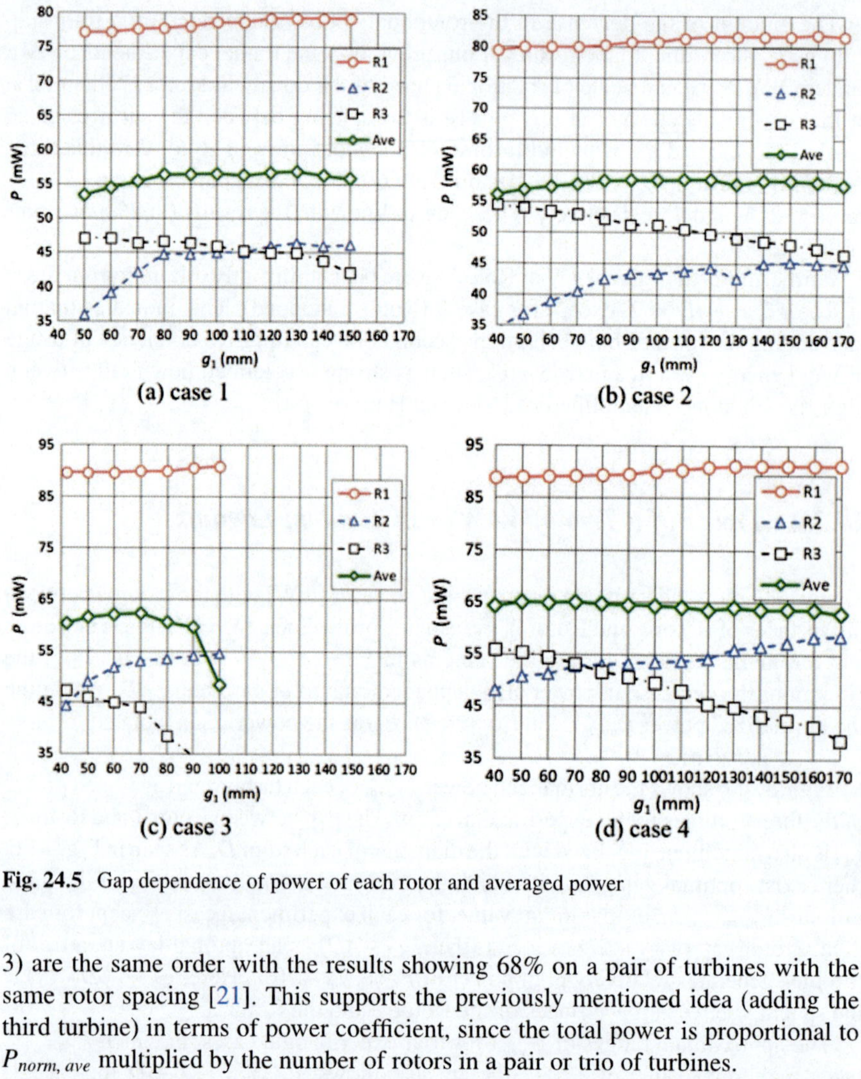

Fig. 24.5 Gap dependence of power of each rotor and averaged power

3) are the same order with the results showing 68% on a pair of turbines with the same rotor spacing [21]. This supports the previously mentioned idea (adding the third turbine) in terms of power coefficient, since the total power is proportional to $P_{norm,\ ave}$ multiplied by the number of rotors in a pair or trio of turbines.

The followings have been observed from the experiments at $g_1/D = 1$: (1) the normalized power P_{norm} of the upwind turbine (Rotor 1) is 97% of an isolated turbine regardless of the rotational directions of the three turbines; (2) P_{norm} of the middle turbine (Rotor 2) is 45% for case 1 and case 2, but 53/55% for case 3/case 4; (3) P_{norm} of the downwind turbine (Rotor 3) is 50–65% depending on the combinations of the rotational directions of the three turbines. In summary on $g_1 = 1D$, the averaged normalized power of the three turbines $P_{norm,\ ave}$ is 67–71% of an isolated turbine.

Recently, Jodai and Hara [22] reported comprehensive measurements of the wind-direction dependence on VAWTs placed equilaterally. Azadani [23] presented a review paper regarding the appropriate layouts of clusters of VAWTs.

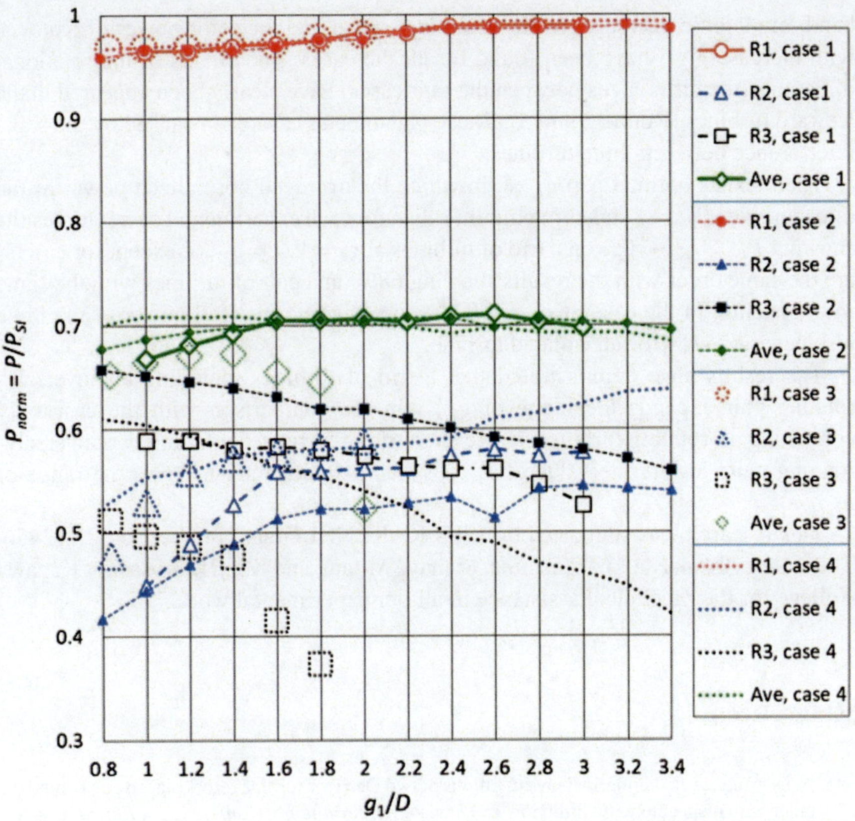

Fig. 24.6 Gap dependence of normalized power of each rotor and averaged power

24.4 Conclusions

Wind tunnel experiments on the interaction among three closely spaced vertical-axis wind turbines (VAWTs) arranged in tandem have been conducted at a uniform wind speed of 12 m/s. The rotor spacing between an upwind turbine (Rotor 1) and a downwind turbine (Rotor 3) was kept constant at 7*D* ($D = 50$ mm is the turbine diameter) in the experiments. There are four experimental cases depending on the combinations of the rotational directions of the three turbines. In case 1, three turbines rotate same direction. In case 2, Rotor 3 rotates the opposite direction to that of Rotors 1 and 2 (a middle turbine). In case 3, Rotor 2 rotates the opposite direction to that of Rotors 1 and 3. In case 4, Rotor 1 rotates the opposite direction to that of Rotors 2 and 3.

Monotonic increases in the rotational speed and the rotor power of Rotors 1 and 2, with the increase in g_1, have been demonstrated for all the cases tested. Here, g_1 is the rotor gap (spacing between the surfaces of Rotor 1 and Rotor 2). On the other hand, monotonic decreases in the rotational speed and the rotor power of Rotor 3, with increasing g_1, have been found for all the cases. The effects of the rotational directions (the differences between the four cases) have clearly been appeared in the leeward turbines (Rotors 2 and 3). These phenomena have been caused by the wake interference between three turbines.

There exists optimal gap $g_{1,\,opt}$, in which the averaged normalized power of the three turbines $P_{norm,\,ave}$ takes maximum value, for each experimental case. Our results showing $P_{norm,\,ave} \sim 70\%$ on a trio of turbines at $g_1 = 2D$ (=$g_{1,\,opt}$, except for case 3) are the same order with the results showing 68% on a pair of turbines with the same rotor spacing. In the case of $g_1 = 1D$, the averaged normalized power of the three turbines is 67–71% of an isolated turbine.

The results have demonstrated that a trio of turbines including an inversely-rotating pair expands the range of $g_{1,\,opt}$ in the comparison with the case only containing co-rotating turbines. However, a trio of turbines including two inversely-rotating pairs lowers the value of $P_{norm,\,ave}$ and significantly narrows the range of $g_{1,\,opt}$.

This research was supported by JSPS KAKENHI Grant Number JP 22K12456. The authors thank Naoki Takeshima, Manaya Mitani, and Kenji Marusasa at Kagawa College for their technical assistance in all our experimental work.

References

1. A. Dessoky, et al., Numerical investigations of two Darrieus turbine rotors placed one behind the other with respect to wind direction, in *Thirteenth International Conference of Fluid Dynamics* (ICFD13-EG-6023, Cairo, 2018), pp. 1–11
2. S. Sahebzadeh, et al., Towards optimal layout design of vertical-axis wind-turbine farms: Double rotor arrangements. Energy Convers. Manag. **226**, 113527-1–113527-17 (2020)
3. L. Kuang, et al. Characterization of wake interference between two tandem offshore floating vertical-axis wind turbines: effect of platform pitch motion. Energy Convers. Manag. **265**, 115769-1–115769-23 (2022)
4. J.O. Dabiri, Potential order-of-magnitude enhancement of wind farm power density via counter-rotating vertical-axis wind turbine arrays. J. Renew. Sustain. Energy **3**, 043104-1–043104-12 (2011)
5. S. Zanforlin, T. Nishino, Fluid dynamic mechanisms of enhanced power generation by closely spaced vertical axis wind turbines. Renew. Energy **99**, 1213–1226 (2016)
6. Y. Hara, Y. Jodai, Study on close arrangement of small vertical axis wind turbines. Turbomachinery **49**(1), 13–21 (2021)
7. Y. Hara, et al., Numerical analysis of dynamic interaction between two closely spaced vertical-axis wind turbines. Energies **14**(8), 2286-1–2286-23 (2021)
8. Y. Jodai, Y. Hara, Wind tunnel experiments on interaction between two closely spaced vertical-axis wind turbines in side-by-side arrangement. Energies **14**(23), 7874-1–7874-19 (2021)
9. Y. Hara, et al., Development of a butterfly wind turbine with mechanical over-speed control system. Designs **2**(2), 17-1–17-25 (2018)
10. S. Ishigai et al., Experimental study on structure of gas flow in tube banks with tube axes normal to flow (Part 1, Karman vortex flow from two tubes at various spacings). Bull. JSME **15**(86), 949–956 (1972)
11. M.M. Zdravkovich, Review of flow interference between two circular cylinders in various arrangements. Trans. ASME J. Fluids Eng. **99**(4), 618–633 (1977)

12. A. Okajima, Flows around two tandem circular cylinders at very high Reynolds numbers. Bull. JSME **22**(166), 504–511 (1979)
13. T. Igarashi, Characteristics of the flow around two circular cylinders arranged in tandem (1st report). Bull. JSME **24**(188), 323–331 (1981)
14. M. Zhao, L. Cheng, Two-dimensional numerical study of vortex shedding regimes of oscillatory flow past two circular cylinders in side-by-side and tandem arrangements at low Reynolds numbers. J. Fluid Mech. **751**, 1–37 (2014)
15. M.M. Alam, Lift forces induced by phase lag between the vortex sheddings from two tandem bluff bodies. J. Fluids Struct. **65**, 217–237 (2016)
16. D. Chatterjee, et al., Rotation induced flow suppression around two tandem circular cylinders at low reynolds number. Fluid Dyn. Res. **49**(4), 045503-1–045503-17 (2017)
17. M.M. Alam et al., On forces and phase lags between vortex sheddings from three tandem cylinders. Int. J. Heat Fluid Flow **69**, 117–135 (2018)
18. S. Behara, et al., Flow-induced oscillations of three tandem rotating cylinders. Phys. Fluids **30**, 113604-1–113604-24 (2018)
19. S. Behara et al., Three-dimensional transition in the wake of two tandem rotating cylinders. J. Fluid Mech. **951**(A29), 1–32 (2022)
20. M.M. Alam et al., Fluctuating fluid forces acting on two circular cylinders in a tandem arrangement at a subcritical Reynolds number. J. Wind Eng. Ind. Aerodyn. **91**(1–2), 139–154 (2003)
21. S. Sahebzadeh, et al., Impact of relative spacing of two adjacent vertical axis wind turbines on their aerodynamics. J. Phys.: Conf. Ser. **1618**, 042002-1–042002-10. TORQUE, Delft (2020)
22. Y. Jodai, Y. Hara, Wind-tunnel experiments on the interactions among a pair/trio of closely spaced vertical-axis wind turbines. Energies **16**(3), 1088-1–1088-27 (2023)
23. L.N. Azadani, Vertical axis wind turbines in cluster configurations. J. Phys.: Conf. Ser. 272, 113855-1–113855-13 (2023)

Chapter 25
Comparative Study of the Reanalysis and Instrument Data for Wind Energy Potential Assessment of Southwest Region in India

S. R. Suresh Lal, G. M. Joselin Herbert, and B. Harish Kumar

Abstract Climate change and global warming accelerated the much-needed transition from conventional energy to renewable energy globally in addition to the anxiety of depletion of fossil fuel resources. In India, the use of renewable energy is steadily growing displaying the country's commitment to sustainable development. In this study, an analysis of wind potential assessment has been made in the southwest region of India by comparing the European Center for Medium-Range Weather Forecasts Re-Analysis v5 (ERA5) and Modern-Era Retrospective analysis for Research and Applications (MERRA2) reanalysis data sets. Hourly wind speeds and directions for all days for 20 years from 2001 to 2020 are collected for a location in the region. The daily, monthly, and annual average wind speeds are determined for both data sets. The wind rose diagrams for the location are also created using the Windrose Axes library in the Python environment. These data are also compared with the instrument data for one year to find the most appropriate reanalysis data set for the region. From the analysis, it has been noted that the ERA5 data sets are closer to the instrument data. The available wind speeds are categorized as calm to moderate breeze as in the Beaufort scale in the range of 0.3–8 m/s. This points out the necessity of developing small wind turbines with small cut-in and rated speeds for maximum wind power generation. This methodology can be extended to other regions too.

S. R. Suresh Lal (✉) · G. M. Joselin Herbert
Department of Mechanical Engineering, Noorul Islam Center for Higher Education, Kanyakumari, India
e-mail: sureshlal99077@cet.ac.in

S. R. Suresh Lal · B. Harish Kumar
Department of Mechanical Engineering, College of Engineering Trivandrum, Thiruvananthapuram, India

APJ Abdul Kalam Technological University, Thiruvananthapuram, India

A. Suryan et al. (eds.), *Proceedings of the 9th Asian Joint Workshop on Thermophysics and Fluid Science, 27–30 November 2022, Utsunomiya, Japan*, Springer Proceedings in Physics 396, https://doi.org/10.1007/978-981-99-9470-0_25

25.1 Introduction

The confrontation against global warming will not be successful unless renewable energy in all possible sources is utilized effectively. At present solar, wind, biomass and hydro are the leading contributors in this sector. The unexploited or underexploited resources include solar thermal systems, offshore wind, small wind systems in low wind regions and tidal and wave energies [1, 2]. As a practical choice for sustainable and cleaner advancement, researchers are concentrating on developing renewable-based energy systems for smart cities under diverse scales [3]. In this context, the small wind energy devices working at low and moderate wind speeds can be considered along with solar energy systems. Wind energy is one of the most abundant, inexhaustible, and non-polluting resources and is broadly distributed around the globe [4, 5]. But its constancy changes with time, with the topography, at different altitudes, throughout different seasons, and year to year. Therefore, the wind power generation estimation chiefly depends on the accuracy of wind resource assessment [6, 7]. Several studies to assess the wind potential in different geographic regions have been reported like the southwestern region of Algeria [6], the East China Sea [8], the Southwestern Offshore in South Korea [9], the Mediterranean Sea coasts [10] and in India [4, 11]. The present study adopts the reanalysis data of wind speed and direction at 10 m above ground level (AGL) from two sources ERA5 and MERRA2 in a selected location in the southwestern region of India. The hourly, daily, monthly and yearly comparative results have been presented as charts, diagrams and figures to assess the wind energy potential precisely. The results will then be compared to the instrument data to assess the most suitable reanalysis data for the region.

25.2 Methods and Materials

25.2.1 ERA5

Long-term wind distribution data should be examined to make the most effective use of the wind to generate electricity via wind turbines. A projection of long-term wind characteristic data could be very helpful for investors and project managers. Variations of wind according to the season and climate can be predicted by using this data.

The fifth generation of atmospheric reanalysis of the global climate, or ERA5, was conducted by the European Center for Medium-Range Weather Forecasts (ECMWF) and covers the period from January 1950 to the present. The ECMWF's Copernicus Climate Change Service (C3S) is responsible for producing it. A large number of atmospheric, land, and oceanic climate variables are estimated hourly by ERA5. The data resolve the atmosphere utilizing 137 levels from the surface up to an altitude of 80 km, covering the Earth on a grid of 30 km.

25.2.2 MERRA-2

The second iteration of MERRA, or Modern-Era Retrospective-Analysis for Research and Applications, is known as MERRA-2. The Global Modelling and Assimilation Office of NASA's Goddard Space Flight Center is where the dataset originates. In 2009, this data set was made available. The GEOS-5 atmospheric data assimilation system version on which it is based was frozen in 2008.

The GEOS-5 Atmospheric Data Assimilation System is in charge of the MERRA review (ADAS). MERRA data were produced on a 0.5°x 0.66° grid with 72 layers, and they covered the period from 1979 to February 2016. Stand-alone reanalyses of the land surface (MERRA-Land) and atmospheric aerosols were driven by MERRA (MERRAero).

Hourly wind speeds and directions at a height of 10m AGL for all days for 20 years from 2001 to 2020 are collected for the location from both ERA5 and MERRA2 data sets.

25.2.3 Instrument Data

The instrument data acquired from June 2022 to May 2023 are used for this analysis. A cup-type anemometer, model WS1093 of Aercus instrument installed at the site at 10 m AGL is used for this purpose. Figure 25.1 shows the setup of the instrument.

Fig. 25.1 Anemometer installed for wind speed measurement

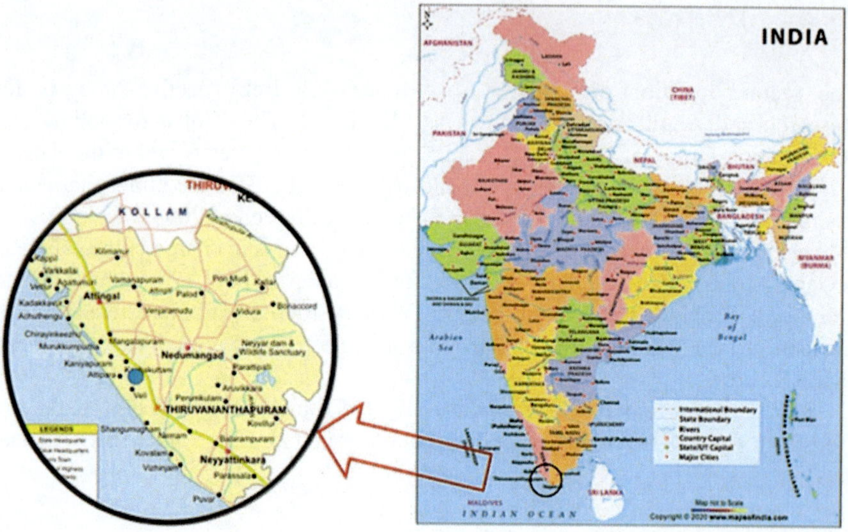

Fig. 25.2 Location map of the study

25.2.4 Location of the Study

A near coastal site at Thiruvananthapuram, Kerala, southwest of India has been considered for the study (see Fig. 25.2). The latitude and the longitude of the site are 8.54 and 76.9 respectively. The region is lying at a height of 40 m from the mean sea level.

25.3 Results and Discussion

25.3.1 Wind Data

The assessment of wind data for 20 years for the location is presented in Figs. 25.3, 25.4 and 25.5 and Tables 1 and 2. The southwest monsoon from the month of June to September [12] has a significant effect on the wind potential of this region. Both analyses show that the wind speeds during this period are exceptionally high compared to other periods of the year and repeat the pattern from year to year. The mean wind speed for the period is determined as 3.16 m/s for ERA5 and 4.1 m/s for MERRA2 data sets. There is a significant upper hand of about 1 m/s average wind speed of MERRA2 over ERA5. In all sets, around 80% of the wind speeds are between 1.6 and 8 m/s.

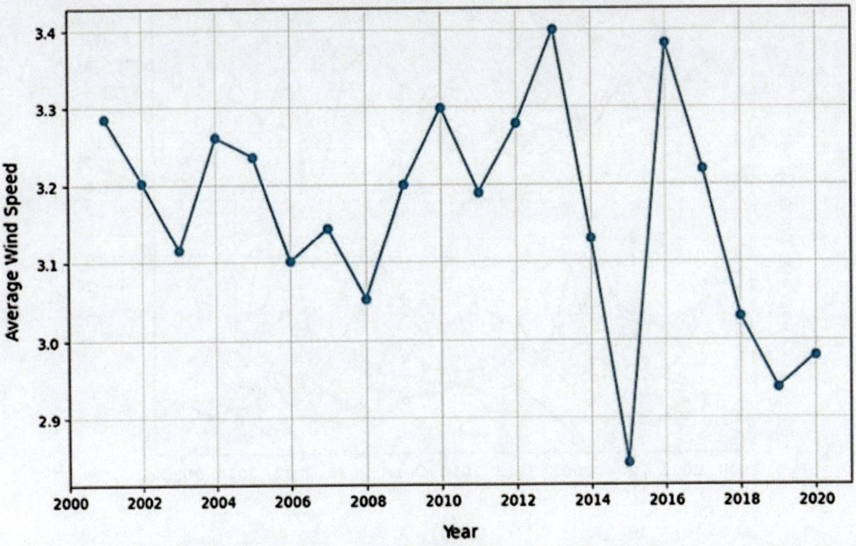

(a)

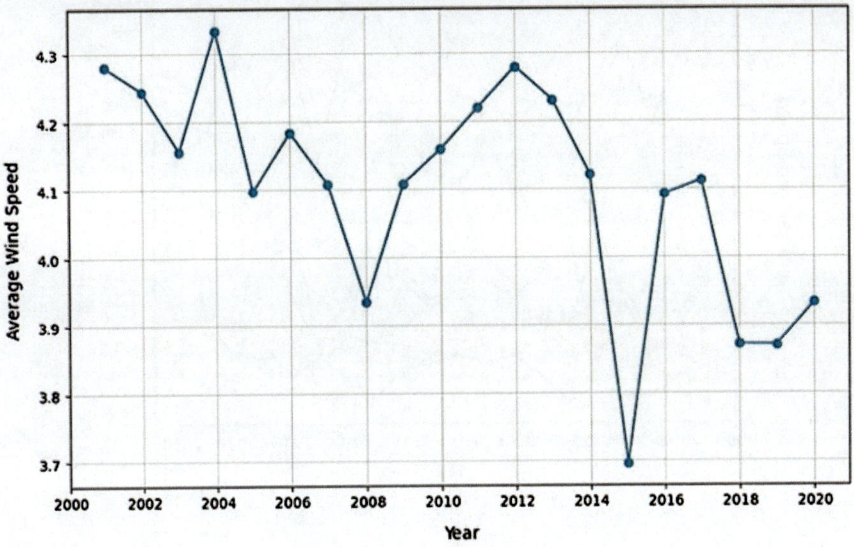

(b)

Fig. 25.3 Average annual wind speed for 20 years **a** ERA5 **b** MEERA2

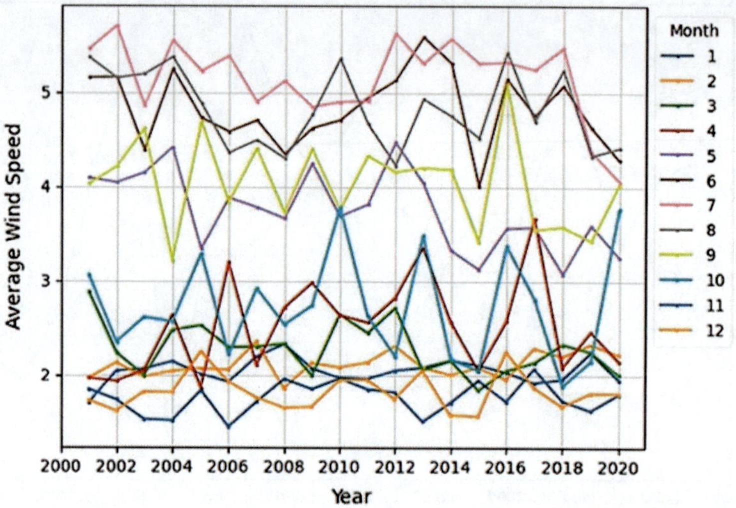

(a)

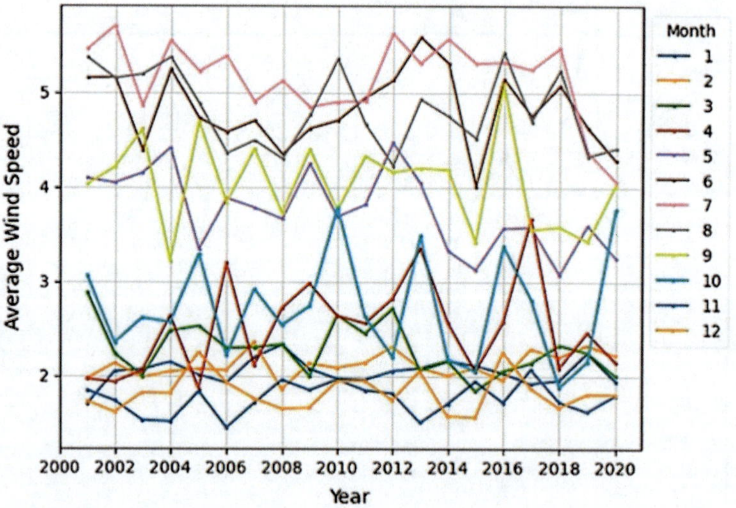

(b)

Fig. 25.4 Average monthly wind speed for 20 years **a** ERA5 **b** MEERA2

25.3.2 Wind Rose Diagram

A wind rose diagram is the typical visual representation of wind direction and speed [13]. Concentric circles on it stand in for the Beaufort scale-classified wind speed and direction. In this study, the WindroseAxes package in the Python environment is used to create the Windrose diagrams. Figure 25.6 presents the wind rose plots for

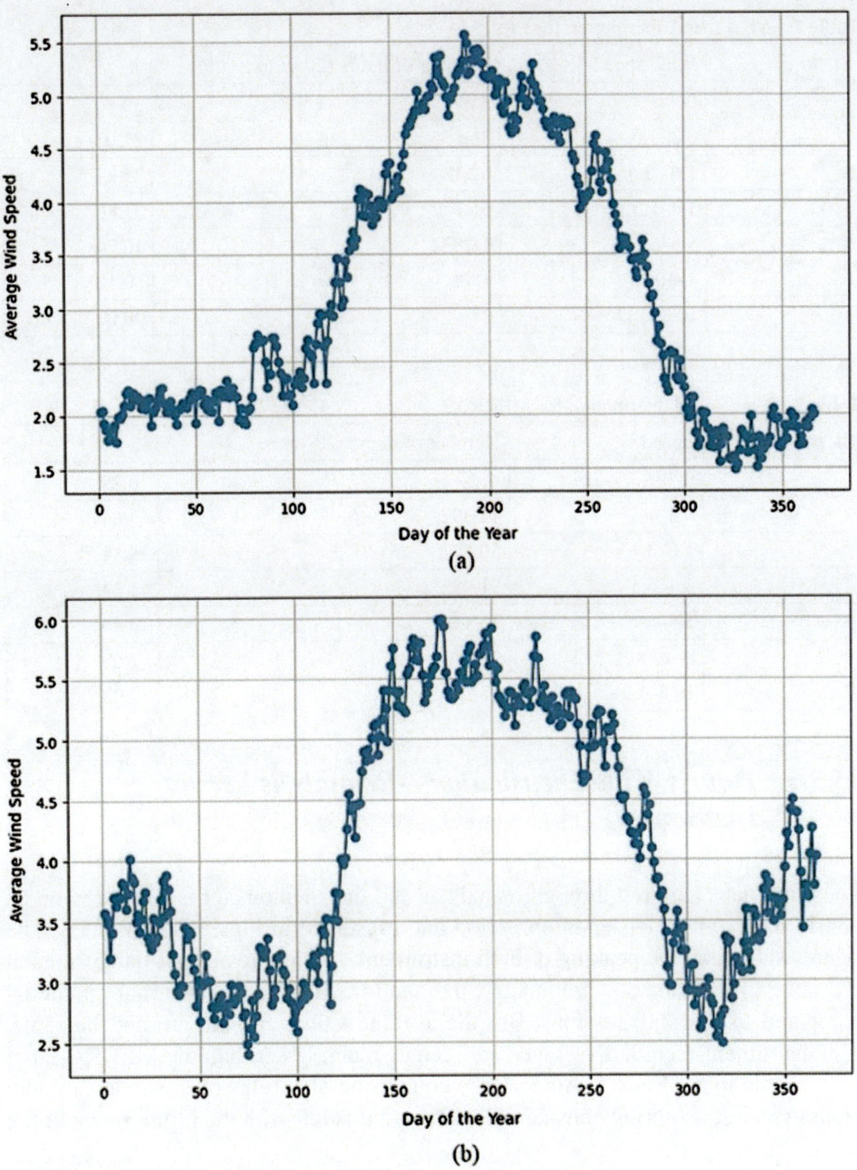

Fig. 25.5 Average daily wind speed for 20 years **a** ERA5 **b** MEERA2

the two locations in both data sets. The diagrams generated with ERA5 show that the major wind direction is northwest (NW) but for MERRA it is north of northwest (NNW). The validation of the exact direction is the future scope of this study.

Table 1 Wind speed distribution for ERA5

Sl. no	Wind speed	Number of hours (20 years)	%
1	0–0.3	1504	0.86
2	0.3–1.6	36,866	21.03
3	1.6– 3.4	62,184	35.47
4	3.4–5.5	55,752	31.80
5	5.5–8	18,832	10.74
6	8–10.8	180	0.10
7	10.8–13.8	2	0.00

Table 2 Wind speed distribution for MERRA2

Sl. no	Speed	Number of hours (20 years)	%
1	0–0.3	432	0.25
2	0.3–1.6	13,392	7.64
3	1.6–3.4	50,571	28.84
4	3.4–5.5	71,952	41.04
5	5.5–8	36,116	20.60
6	8–10.8	2812	1.60
7	10.8–13.8	45	0.03

25.3.3 Hourly Wind Distribution—Reanalysis Versus Instrument

The wind data acquired through reanalysis and the anemometer is compared on an hourly basis for the observation period and presented in Figs. 25.7 and 25.8. The figures show that the peaking of both instrument data and reanalysis data occurs at the same time of the day around 1–2 p.m. and the afternoons have more potential compared to morning sessions. But the average values in both are not the same. The instrument-recorded averages are seen as much lesser than the MERRA-2 but comparable to the ERA-5. An accurate conclusion about this observation can only be drawn after acquiring considerable data and it is left with the future scope of the work.

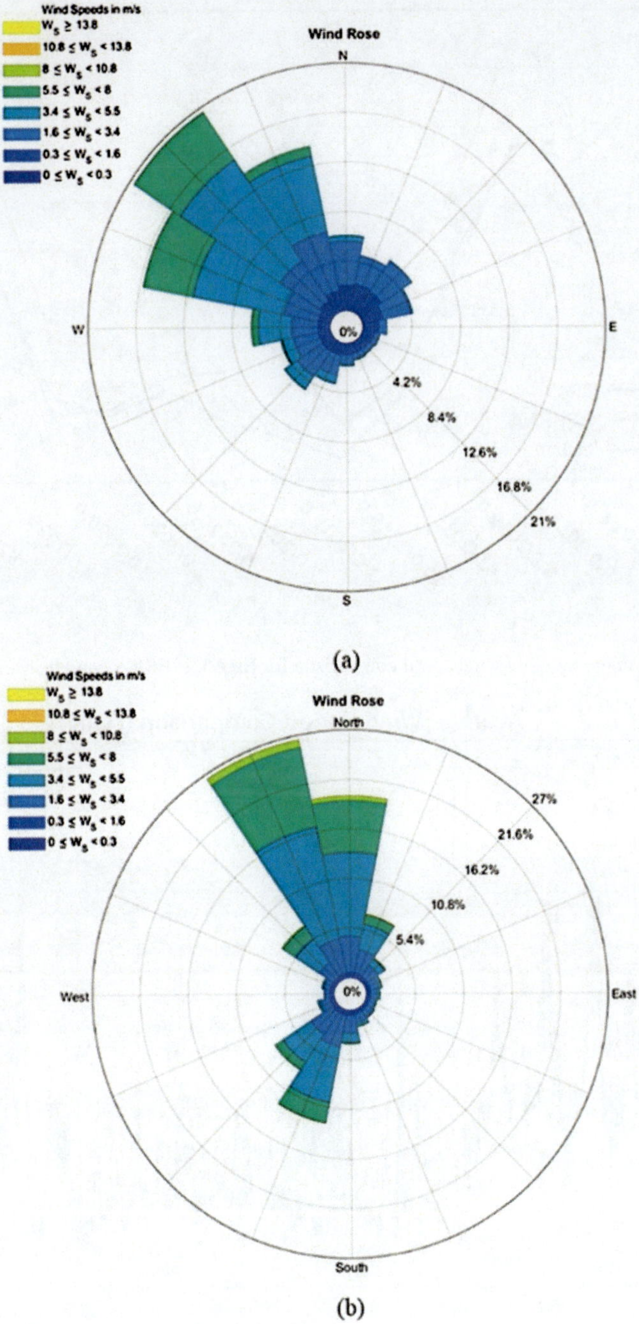

Fig. 25.6 Wind rose diagram **a** ERA5 **b** MERRA2

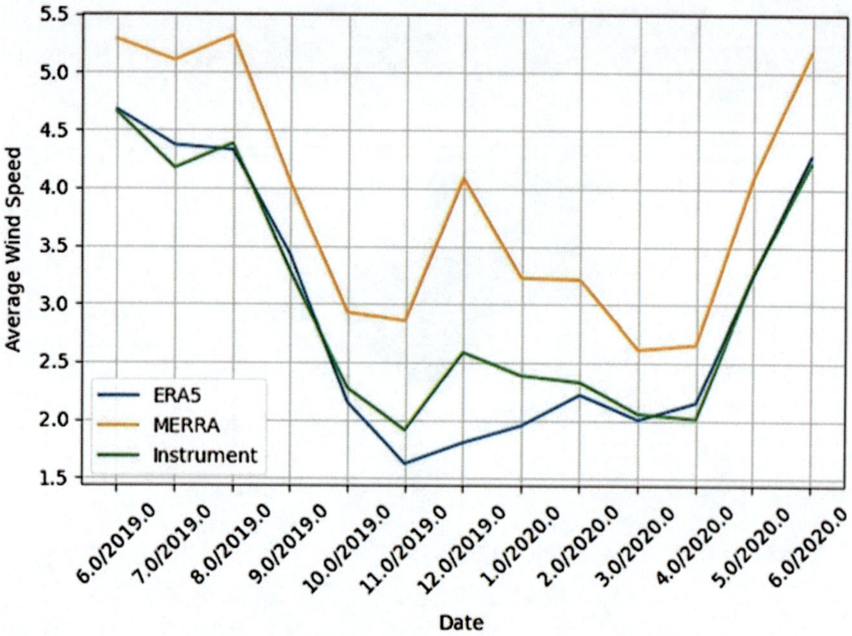

Fig. 25.7 Average monthly wind speed comparison for ERA5, MERRA2 and instrument data

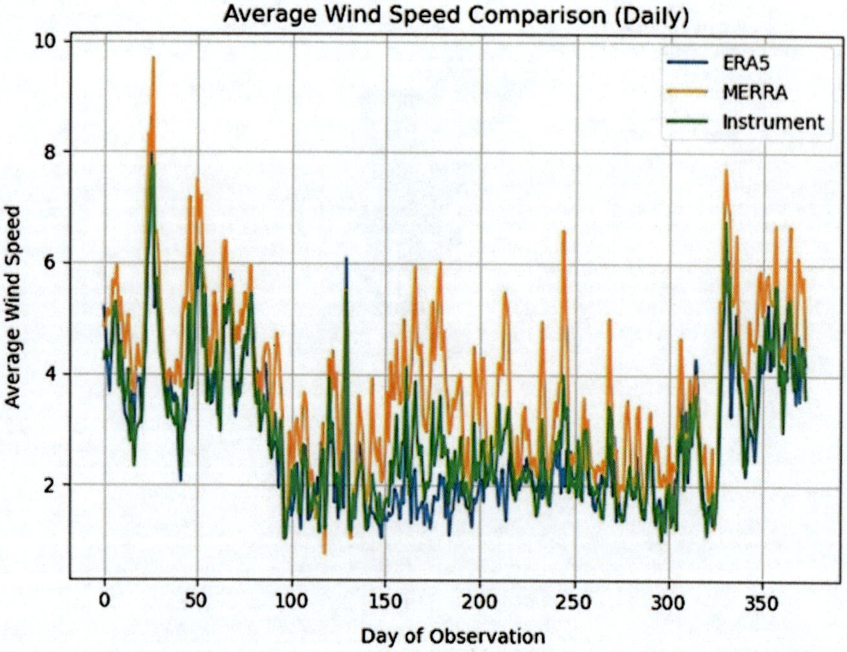

Fig. 25.8 Average daily wind speed comparison for ERA5, MERRA2 and instrument data

25.4 Conclusion and Future Work

Wind data for 20 years is extracted from the ERA5 and MERRA2 reanalysis data sets and analyzed for mean wind speed. This study shall provide crucial wind data for wind turbine developers to design suitable small and microturbines for small and moderate wind sites across the country.

The key findings could be summarized as follows:

- The major portion of the wind speeds obtained in the study ranged from 1.6 to 8 m/s. Therefore, a small or micro wind turbine with a 1.5 m/s cut-in speed and 8 m/s rated speed shall be designed or selected for the location.
- The anemometer data provides the same distribution of wind over days as in reanalysis data.
- The mean wind speed is determined as 3.16 m/s for ERA5 and 4.1 m/s for MERRA2 during the 20 years from the year 2001 to 2020.
- ERA-5 data sets are comparatively closer to the actual measured values than the MERRA-2 data sets.
- This opens a future scope of the research to continue the recording of instrument data for the location, to find more suitable reanalysis data for the region.

References

1. S.S.R. Lal, J.G.M. Herbert, P. Arjunan, A. Suryan, Advancements in renewable energy transition in India: a review. Energy Sources Part Recov. Util. Environ. Eff. 1–31 (2022). https://doi.org/10.1080/15567036.2021.2024921
2. P. Arjunan, J.H. Gnana Muthu, S.L. Somanasari Radha, A. Suryan, Selection of working fluids for solar organic rankine cycle—a review. Int. J. Energy Res. **46**(14), 20573–20599 (2022). https://doi.org/10.1002/er.7723
3. A.T. Hoang, V.V. Pham, X.P. Nguyen, Integrating renewable sources into energy system for smart city as a sagacious strategy towards clean and sustainable process. J. Clean. Prod. **305**, 127161 (2021). https://doi.org/10.1016/j.jclepro.2021.127161
4. A. Sharma, J. Srivastava, S.K. Kar, A. Kumar, Wind energy status in India: a short review. Renew. Sustain. Energy Rev. **16**(2), 1157–1164 (2012). https://doi.org/10.1016/j.rser.2011.11.018
5. P.K. Chaurasiya, V. Warudkar, S. Ahmed, Wind energy development and policy in India: a review. Energy Strategy Rev. **24**, 342–357 (2019). https://doi.org/10.1016/j.esr.2019.04.010
6. Y. Himri, S. Rehman, B. Draoui, S. Himri, Wind power potential assessment for three locations in Algeria. Renew. Sustain. Energy Rev. **12**(9), 2495–2504 (2008). https://doi.org/10.1016/j.rser.2007.06.007
7. Y. Himri, S. Himri, A.B. Stambouli, Wind power resource in the south-western region of Algeria. Renew. Sustain. Energy Rev. **14**(1), 554–556 (2010). https://doi.org/10.1016/j.rser.2009.07.018
8. L. Song, Z. Liu, F. Wang, Comparison of wind data from ERA-Interim and buoys in the Yellow and East China Seas. Chin. J. Oceanol. Limnol. **33**(1), 282–288 (2015). https://doi.org/10.1007/s00343-015-3326-4

9. H.-G. Kim, J.-Y. Kim, Y.-H. Kang, Comparative evaluation of the third-generation reanalysis data for wind resource assessment of the southwestern offshore in South Korea. Atmosphere **9**(2), 73 (2018). https://doi.org/10.3390/atmos9020073
10. F. Onea, L. Deleanu, L. Rusu, C. Georgescu, Evaluation of the wind energy potential along the Mediterranean Sea coasts. Energy Explor. Exploit. **34**(5), 766–792 (2016). https://doi.org/10.1177/0144598716659592
11. R. Narayana Sarma, V. Kumar, S.S.R. Lal, M. Reghunath, A. Jayan, A. Suryan, Wind power resource assessment and wind-hydrogen generation potential: a case study. Energy Sources Part Recov. Util. Environ. Eff. 1–17 (2022). https://doi.org/10.1080/15567036.2022.2026537
12. H. Varikoden, J.V. Revadekar, J. Kuttippurath, C.A. Babu, Contrasting trends in southwest monsoon rainfall over the Western Ghats region of India. Clim. Dyn. **52**(7–8), 4557–4566 (2019). https://doi.org/10.1007/s00382-018-4397-7
13. S.H. Pishgar-Komleh, A. Keyhani, P. Sefeedpari, Wind speed and power density analysis based on Weibull and Rayleigh distributions (a case study: Firouzkooh county of Iran). Renew. Sustain. Energy Rev. **42**, 313–322 (2015). https://doi.org/10.1016/j.rser.2014.10.028

Chapter 26
Proposal of Design Method for Large Tidal Current Turbine Incorporating Cavitation Occurrence

Wakana Tsuru, Masaki Sakaguchi, Tengen Murakami, Norimasa Shiomi, and Yoichi Kinoue

Abstract Among renewable energy sources, tidal current power generation systems, which are predictable over long time periods, are attracting attention. One way to achieve high power output from tidal current turbines is to increase the turbine diameter. However, the turbine diameter is limited by cavitation. In this study, a simple large turbine design method was proposed to obtain high power output while suppressing cavitation. The turbine design was based on blade element momentum (BEM) theory, which is widely used in tidal current turbines. A process for the modification of the angle of attack was incorporated into the design method based on BEM theory to suppress cavitation. The blade was NACA 63–415. Lift and drag characteristics of the blade and cavitation inception pressure were calculated numerically. The turbines were designed using a conventional design meth-od and the design method incorporating the angle of attack modification process. The power output of the turbine designed by the proposed method was increased by more than 50%. These results indicate that the proposed design method has the capability to design turbines with high output. In addition, the proposed design method is simple and applicable to a various blade. Therefore, the performance of tidal current turbines could be further improved by using more appropriate blade and the pro-posed design method.

W. Tsuru (✉) · T. Murakami · N. Shiomi · Y. Kinoue
Saga University, 1 Honjo, Saga 840-8502, Japan
e-mail: tsuru@ioes.saga-u.ac.jp

T. Murakami
e-mail: murakami@ioes.saga-u.ac.jp

N. Shiomi
e-mail: siomi@me.saga-u.ac.jp

Y. Kinoue
e-mail: kinoue@me.saga-u.ac.jp

M. Sakaguchi
Graduate School of Saga University, 1 Honjo, Saga 840-8502, Japan
e-mail: sakaguchi-masaki@buntoku-h.ed.jp

A. Suryan et al. (eds.), *Proceedings of the 9th Asian Joint Workshop on Thermophysics and Fluid Science, 27–30 November 2022, Utsunomiya, Japan*, Springer Proceedings in Physics 396, https://doi.org/10.1007/978-981-99-9470-0_26

26.1 Introduction

Renewable energies are attracting attention in order to achieve a decarbonized society. Power generation using ocean energy, which is a form of renewable energy, includes offshore wind, ocean thermal, wave, salinity difference, ocean currents, and tidal power generation. Tidal power generation uses regularly fluctuating tidal currents with long-term predictability. In addition, wind and photovoltaic power generations are greatly affected by weather conditions, while tidal current power generation is less affected by the weather.

Many studies have been conducted on horizontal-shaft propeller type turbines for tidal current power generation [1–3]. Since the operating principle is the same as that of wind turbines, design based on blade elementary momentum (BEM) theory [4] is widely used [5–7], as is the design of wind turbines. In BEM theory, the blade shape is determined and designed at an angle of attack that provides the maximum lift-drag ratio. In wind turbines, the turbine diameter has been enlarged to increase the output power. On the other hand, in tidal current power generation, the turbine diameter is limited to avoid cavitation. Investigations on cavitation limits [8, 9] and studies on diameters and chord lengths to avoid cavitation [5] have been conducted by incorporating cavitation limits into BEM theory. The blade geometry has also been designed to suppress cavitation [10]. Therefore, to avoid cavitation, it is required to reduce the diameter or use special blades. However, for the practical application of tidal current power generation, it is important to obtain large output power while keeping low costs for design and manufacturing, and a design method that enables larger size with simpler geometry is desired.

In this study, a simple design method for a larger turbine than the conventional method of BEM theory is proposed in order to obtain high power output without cavitation. Incorporate cavitation limits based on BEM theory. In addition, the angles of attack for each blade element were adjusted to avoid cavitation and to achieve a larger size than that of the conventional BEM theory design. Cavitation limits and lift and drag coefficients were obtained by computational fluid dynamics using a single blade with a two-dimensional geometry. The performance of turbines with diameters up to the cavitation limit based on conventional BEM theory and larger turbines with adjusted angle of attack were compared to investigate the effectiveness of this design method.

26.2 Turbine Design Method

The operating principle of tidal current turbines is the same as that of wind turbines. Therefore, the BEM theory, which is commonly used in the design of wind turbines, is also widely used in the design of tidal current turbines. In this design, based on the BEM theory, cavitation limitation and angle of attack modification processes were incorporated to design a large tidal turbine.

26.2.1 BEM Theory

The BEM theory consists of blade element and momentum theories. Figure 26.1 shows a conceptual diagram of BEM theory.

In the blade element theory, a single blade is divided into multiple blade elements, to obtain the hydrodynamic forces acting on each blade element as a two-dimensional blade. In the momentum theory, the cross-section of a turbine is divided into circular elements, and the hydrodynamic force acting on each element is obtained from the change in momentum. Based on the equivalence of the blade element theory and the momentum theory hydrodynamic forces, the induced velocities in the axial and tangential directions are obtained by iterative calculations.

Using the axial and tangential induced velocity coefficients a and a', the axial component w_a and the tangential component w_t of the relative velocity W at radius r are obtained as follows:

$$w_{\mathrm{a}} = (1 - a)V_{\mathrm{m}}, \tag{26.1}$$

$$w_{\mathrm{t}} = (1 + a')\Omega r, \tag{26.2}$$

where V_m is the mainstream velocity, Ω is the rotational velocity of the turbine, and ϕ is the inflow angle of W into the turbine. Using the local tip speed ratio λr $(= r\Omega/V_m)$, ϕ is expressed as follows:

$$\tan\phi = \frac{(1 - a)V_{\mathrm{m}}}{(1 + a')\Omega r} = \frac{1 - a}{(1 + a')\lambda_r}. \tag{26.3}$$

From the blade element theory, the lift dL and drag dD of a blade element can be given as follows:

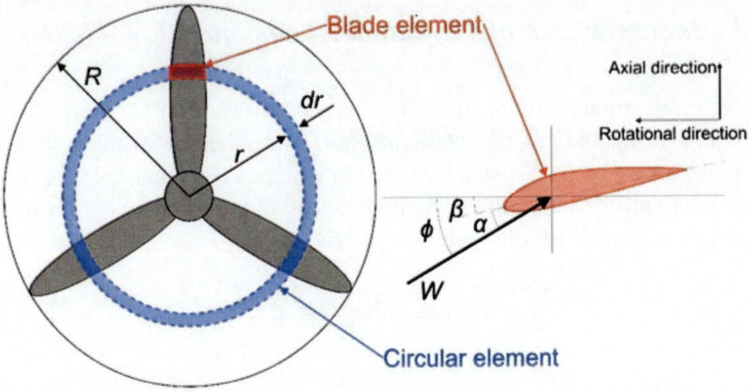

Fig. 26.1 Blade element and circular element

$$dL = \frac{1}{2}\rho C_L W^2 C dr, \tag{26.4}$$

$$dD = \frac{1}{2}\rho C_D W^2 C dr, \tag{26.5}$$

where ρ is the density of the fluid, C_L and C_D are the coefficients of lift and drag, and C is the chord length. From Eqs. (26.4) and (26.5) and the number of blades B, the thrust dT and the torque dQ are expressed as follows:

$$dT = \frac{1}{2}B\rho W^2(C_L \cos\phi + C_D \sin\phi)C dr, \tag{26.6}$$

$$dQ = \frac{1}{2}B\rho W^2(C_L \sin\phi - C_D \cos\phi)C r dr. \tag{26.7}$$

The momentum theory also gives dT and dQ for a circular element from the following equations:

$$dT = 4a(1-a)\rho V_m^2 \pi r dr, \tag{26.8}$$

$$dQ = 4a'(1-a)\rho V_m \Omega \pi r^3 dr. \tag{26.9}$$

From the equivalence of Eqs. (26.6) and (26.8), and Eqs. (26.7) and (26.9), respectively, a and a' are obtained. In addition, Prandtl's tip loss correction [4] was applied to account for the effects of blade tip vortices. In the general BEM theory, the angle of attack α, and C_L and C_D at the maximum lift-drag ratio are used for the design. In this design, the blade shape is that of NACA63-415, which is often used in tidal power generation.

26.2.2 *Incorporation of Cavitation Limits into BEM Theory*

Incorporate cavitation limits into the design method based on BEM theory. Cavitation limits were investigated by Computational Fluid Dynamics (CFD) analysis for a two-dimensional single blade. When the cavitation volume is less than 0.1% of the blade volume, it is considered as the cavitation inception, and the cavitation inception number σ_i is obtained from the following equation:

$$\sigma_i = \frac{p_\infty - p_v}{\frac{1}{2}\rho U_\infty^2}, \tag{26.10}$$

where p_∞ is the pressure at infinity, p_v is the saturation pressure, and U_∞ is the mainstream velocity, and cavitation occurs when it is less than σ_i. Since σ_i varies with the angle of attack, it was investigated from 0 to 10 deg. in 1 deg. increments. In a turbine, the lowest pressure occurs at the top of the turbine. Therefore, the cavitation number at the top of the turbine σ_{top} is defined as:

$$\sigma_{\mathrm{top}} = \frac{p_{\mathrm{top}} - p_{\mathrm{v}}}{\frac{1}{2}\rho W_{\mathrm{top}}^2}, \tag{26.11}$$

then,

$$\sigma_{\mathrm{top}} > \sigma_{\mathrm{i}} \tag{26.12}$$

is the condition for without cavitation. The calculation starts from the hub side, and W is obtained from a and a' obtained by BEM theory, confirming that $\sigma_{\mathrm{top}} > \sigma_{\mathrm{i}}$. The turbine radius is increased until $\sigma_{\mathrm{top}} \leq \sigma_{\mathrm{i}}$.

The larger the turbine diameter, the greater the output power, so further expansion is considered. As the blade angle of attack decreases, the peak negative pressure on the blade surface decreases, hence σ_{i} becomes smaller. Therefore, when $\sigma_{\mathrm{top}} \leq \sigma_{\mathrm{i}}$ at initial α, the turbine diameter was further increased by decreasing α until $\mathrm{d}Q \leq 0$ to avoid cavitation. Because the tip loss correction uses the turbine diameter, the design was iterated until the turbine diameter converged.

26.2.3 *CFD Method for Cavitation Limits*

Cavitation limits were investigated by CFD. The three-dimensional steady-state uncompressed Navier–Stokes equations were solved using Ansys CFX 2021 R1. The blade is a NACA63-415 with a chord length of 100 mm and a span of 316 mm, and has a two-dimensional blade geometry. For the turbulence model, SST was applied, and for the cavitation model, the Rayleigh-Plesset model of the Ansys CFX 2021 R1 standard was applied. The calculation was performed from an angle of attack of 0 to 10 deg. The computational domain and the grid around the blade are shown in Fig. 26.2.

The height of the domain and its length in the flow direction were set to 1000 mm. The depth is 316 mm. The left side shown in red is the inflow boundary and the green side is the opening boundary where inflow and outflow can occur. The surfaces on both sides of the blade were assumed to be slip-wall walls, while the blade surface was assumed to be a non-slip wall. The number of grid points was 7.31 million. The cavitation inception number, where cavitation occurs in less than 0.1% of the blade volume, and the lift and drag coefficients, C_{L} and C_{D}, were obtained. In the non-cavitation state, the experimental data [11] and the present CFD results were compared for C_{L} and C_{D}. Figure 26.3 shows the C_{L} and C_{D}.

Figure 26.3 shows that C_{L} is slightly smaller and C_{D} is larger than the experimental data. However, in the range of α from 0 to 4 deg., the CFD and experimental data

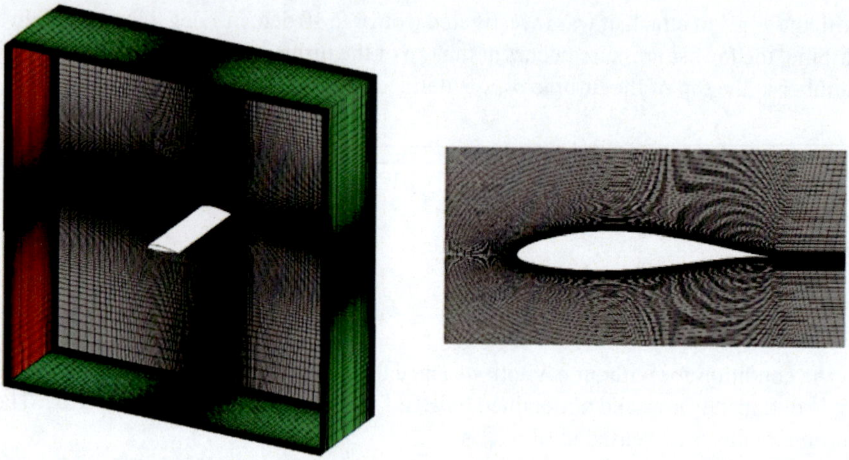

Fig. 26.2 Calculation region and grid around blade

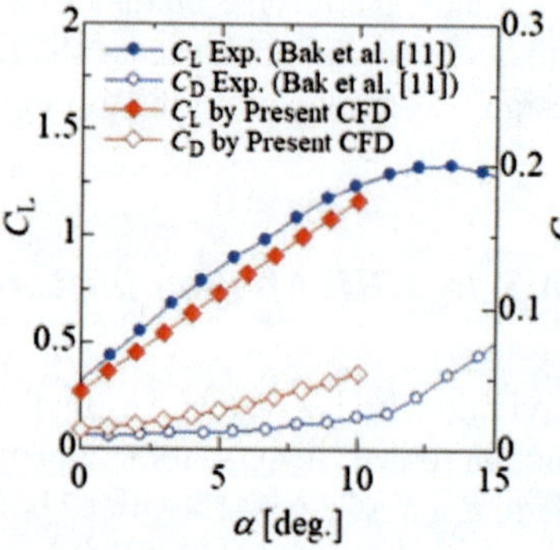

Fig. 26.3 Comparison of CL and CD between experimental data [11] and present CFD

generally agree well. The results for cavitation inception need to be verified in the future.

26.3 Designed Turbines and Comparison Performances

In this design, the design was conducted on a model scale for experiments to be conducted in the future. Main velocity $V_m = 4$ m/s, rotation speed $\Omega = 59.7$ rad/s, number of blades $B = 3$, hub radius $R_h = 0.08$ m, water depth at the turbine rotation axis $H = 0.5$ m, and saturation pressure $p_v = 4500$ Pa. The initial angle of attack was set to 4 deg. at the maximum lift-drag ratio. The designed turbine is shown in Fig. 26.4.

Fig. 26.4 Designed turbines

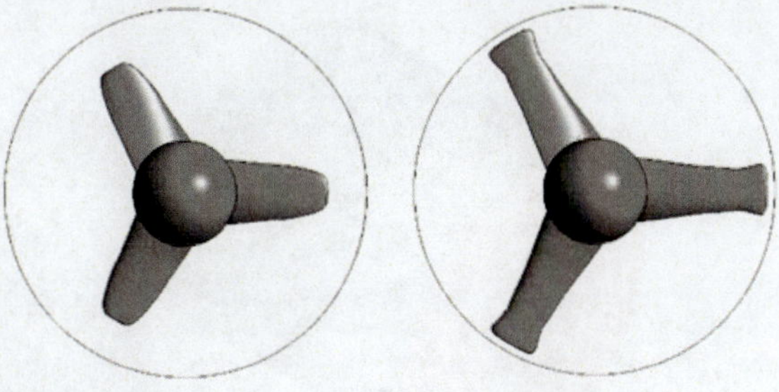

Figure 26.4a shows the turbine enlarged to the cavitation limit with constant α. Figure 26.4b shows the turbine further enlarged by varying α. Figure 26.5 shows α and C for varying α against radius.

R_0 on the horizontal axis is the maximum radius at which α remains 4 deg. The design performance is shown in Table 1.

Figure 26.4 and Table 1 show that the turbine radius for varying α is approximately 1.2 times larger than for constant α. From Fig. 26.4a, the chord length for constant α

Fig. 26.5 Distribution of α and C

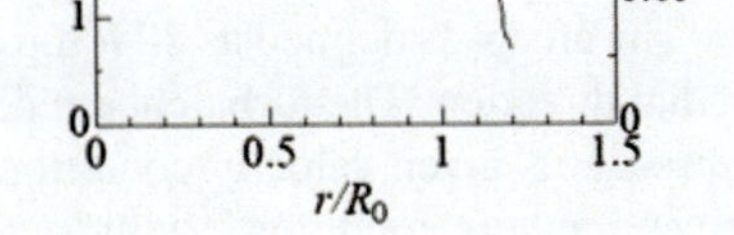

Table 1 Design performance

	Constant α	Varying α
Turbine radius, R_t [m]	0.220	0.268
Tip speed ratio, $\lambda = \Omega R_t/V_m$	3.28	4.0
Thrust, T [N]	273	415
(Thrust coefficient, $C_T = T/(0.5\rho V_m{}^2\pi R_t{}^2)$)	0.224	0.230
Torque, Q [Nm]	26.9	42.1
(Torque coefficient, $C_Q = (Q/0.5\rho V_m{}^2\pi R_t{}^3)$)	0.10	0.0870
Power, P [kW]	1.60	2.51
(Power coefficient, $C_P = P/(0.5\rho V_m{}^3\pi R_t{}^2)$)	0.330	0.348

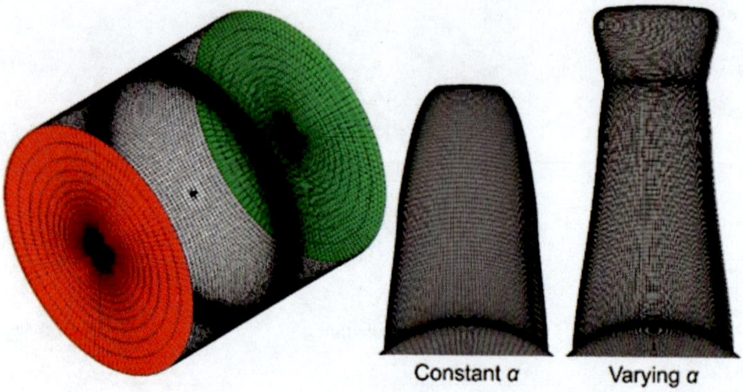

Fig. 26.6 Turbine calculation region and grid around blade

decreases as the radius increases. On the other hand, when α is varied in Fig. 26.4b, Fig. 26.5 shows that the chord length becomes longer when α becomes less than 4 deg.

Table 1 shows that the tip speed ratio λ was 1.2 times lager due to the larger radius. The thrust T was 1.5 times lager (thrust coefficient C_T was 1.03 times lager), and torque Q and power P were 1.6 times lager (torque coefficient C_Q was 0.87 times lager and power coefficient C_P was 1.06 times lager) when α was varied compared to the constant α case. It was found that although T increased with turbine diameter expansion, the power per turbine cross section also increased.

To confirm the design performance, CFD calculations of steady-state incompressible in non-cavitation conditions were performed. As with the cavitation limit calculations, Ansys CFX 2021 R1 was used. The computational domain and the grid around the blade are shown in Fig. 26.6.

The area was defined as 20 R_t upstream and downstream of the turbine and in the radial direction. The turbulence model was applied k-ε. The inlet boundary, shown in red, was given velocity condition and the outlet boundary, shown in green, was given pressure condition. Cylindrical surfaces are assumed to be slip wall surfaces. The number of grid points is 14 million when α is constant and 15 million when α is varied. Table 2 shows the CFD results as well as the design performance.

Table 2 shows that the T and C_T of the design and CFD results differ significantly. Since the BEM does not include the thrust acting on the hub, the difference due to the hub is considered to be significant. Torque and power output are close to CFD and design values with constant α. For varying α, the design values and CFD results are significantly different. This is considered to be due to the secondary flow, which is not treated by the BEM theory. On the other hand, both the design performance and CFD results show that increasing the turbine size by varying α increases the power output. In addition, the cavitation limit is expected to different due to blade tip vortices, so further experimental validation and cavitation studies are needed. This design method can be applied to any blade geometry, and it may become a useful design method by improving the accuracy of the cavitation limit.

Table 2 Performance of design and CFD results

	Constant α		Varying α	
	Design	CFD	Design	CFD
Tip speed ratio, $\lambda = \Omega R_t/V_m$	3.28	3.28	4.0	4.0
Thrust, T [kN]	1.09	1.31	1.66	1.97
(Thrust coefficient, $C_T = T/(0.5\rho V_m{}^2\pi R_t{}^2)$)	0.897	1.08	0.920	1.09
Torque, Q [Nm]	26.9	27.6	42.1	32.4
(Torque coefficient, $C_Q = (Q/0.5\rho V_m{}^2\pi R_t{}^3)$)	0.10	0.103	0.0870	0.0670
Power, P [kW]	1.60	1.65	2.51	1.94
(Power coefficient, $C_P = P/(0.5\rho V_m{}^3\pi R_t{}^2)$)	0.330	0.338	0.348	0.268

26.4 Conclusion and Future Work

In this study, a design method for a large tidal turbine was proposed based on BEM theory. By varying the angle of attack to avoid cavitation, a larger turbine than a general BEM was designed. The designed turbine produced more power per turbine area by adjusting the angle of attack and enlarging the turbine diameter.

However, in addition to pressure conditions, cavitation is related to various other factors such as the cavitation core, non-condensable gas, and turbulence. In addition, the cavitation limit is expected to differ from the actual limit due to blade tip vortices. In the future, experimental validation and cavitation investigations are needed, and the accuracy of the design method should be improved by incorporating rigorous cavitation limits through more advanced cavitation prediction.

References

1. A. Roberts, B. Thomas, P. Sewell, Z. Khan, S. Balmain, J. Gillman, Current tidal power technologies and their suitability for applications in coastal and marine areas. J. Ocean Eng. Mar. Energy **2**, 227–245 (2016)
2. C.H. Frost, P.S. Evans, M.J. Harrold, A. Mason-Jones, T. O'Doherty, D.M. O'Doherty, The impact of axial flow misalignment on a tidal turbine. Renewable Energy **113**, 1333–1344 (2017)
3. M.A. Hannan, Y.A. Ahmed, N.D.B. Zaid, I.Z.M. Kamal, Performance characteristics of horizontal axis tidal turbine with tidal current interaction. IOP Conf. Ser. Mater. Sci. Eng. **1051**, 1–10 (2021)
4. O.L.M. Hansen, *Aerodynamics of Wind Turbines*, 2nd edn. (Routledge, 2015)
5. Z.C. Sun, D. Li, Y.F. Mao, L. Feng, Y. Zhang, C. Liu, Anti-cavitation optimal design and experimental research on tidal turbines based on improved inverse BEM. Energy **239**, 1–20 (2022)
6. A.J. Williams, T.N. Croft, I. Masters, M.R. Willis, M. Cross, Combined BEM-CFD modelling of tidal stream turbine, in *International Conference on Renewable Energies and Power Quality* (Granada, 2010), pp. 1–6
7. K. Ai, M. Wang, D. Wang, H. Wang, R. Cao, E. Avital, Numerical study of a generic tidal turbine using BEM optimization methods. China Ocean Eng. **35**, 344–351 (2021)

8. C.E. Douvi, P.D. Margaris, Computational investigation of cavitation on horizontal axis tidal turbines blades. Int. J. New Technol. Res. **3**(12), 11–18 (2017)
9. W. Aidan, V. Christopher, W. Richard, Cavitation limits on tidal turbine performance. Ocean Eng. **152**, 223–233 (2018)
10. W. Shi, R. Rosli, M. Atlar, R. Norman, D. Wang, W. Yang, Hydrodynamic performance evaluation of a tidal turbine with leading-edge tubercles. Ocean Eng. **117**, 246–253 (2016)
11. C. Bak, P. Fuglsang, J. Johansen, I. Antoniou, *Wind Tunnel Tests of the NACA 63–415 and a Modified NACA 63–415 Airfoil*. (Denmark, Forskningscenter Risoe, Risoe-R-1193, 2020), pp. 6–28

Chapter 27
Analysis of Parabolic Trough Collector Performance Enhancement Using Nanofluids

Pradeep Arjunan, G. M. Joselin Herbert, and Praveen Arjunan

Abstract One of the most significant sources of renewable energy is solar power. Solar energy may be captured in two different ways: photovoltaic systems convert solar energy into electricity, while solar concentrating systems convert solar energy into heat energy. Using concentrated solar collectors, usable heat may be generated at medium and high temperatures. The most effective solar collector among concentrating collectors is the parabolic trough collectors (PTC). As a result, they may be employed in a wide range of applications, including the production of electricity, industrial heating, space cooling, and chemical reactions. In solar collectors, water is usually used as the working fluid. However, it experiences several challenges that reduce the thermal efficiency of the solar collector. The use of nanofluids as the working fluid is one of several potential strategies to enhance the performance of the concentrating collectors. The focus of this study is on determining how the use of nanofluids would improve the thermal performance of the parabolic solar trough collector. In order to do this, numerical analysis is conducted using both water and nanofluid as the working fluids. Additionally, the results of applying various volume fractions of nanofluids were assessed. Al_2O_3 is the chosen nanofluid for the investigation, with water as the base fluid. The numerical analysis was performed using the k-ε model and a uniform heat flux distribution. The results showed that the thermal performance and heat transfer of parabolic solar collectors were greatly enhanced when nanofluid was used as the working fluid. It was also discovered that increasing the volume fraction of nanoparticles increases effective thermal conductivity, which improves heat transfer coefficients and collector thermal efficiency.

P. Arjunan (✉) · G. M. Joselin Herbert
Noorul Islam Center for Higher Education, Kanyakumari, Tamil Nadu 629180, India
e-mail: pradeeparch@cet.ac.in

P. Arjunan · P. Arjunan
College of Engineering Trivandrum, Kerala 695016, India
e-mail: praveenarjun@cet.ac.in

P. Arjunan
School of Engineering, Cochin University of Science and Technology, Kerala 682022, India

A. Suryan et al. (eds.), *Proceedings of the 9th Asian Joint Workshop on Thermophysics and Fluid Science, 27–30 November 2022, Utsunomiya, Japan*, Springer Proceedings in Physics 396, https://doi.org/10.1007/978-981-99-9470-0_27

27.1 Introduction

Humans are consuming more energy on a daily basis, which increases the consumption of fossil fuels, which has an adverse effect on the environment in the form of acid rain, global warming, air pollution, and fossil fuel depletion. The advancement of renewable energy conversion technologies, such as solar, biomass, geothermal, etc., is the solution to this issue. Solar energy is the most prevalent in nature when compared to other renewable energy sources. Photovoltaic cells and solar thermal collectors are the two technologies used to harvest solar energy. The two types of solar thermal collectors that are the subject of this article are concentrating and non-concentrating. Parabolic trough collectors and parabolic dish collectors are solar concentrating collectors that can be utilized to produce very high-temperature ranges. The heat transfer coefficient in the flow can be raised in a number of ways to increase the concentrating collector's thermal efficiency. Utilizing nanofluids as a working fluid is one method for increasing the heat transfer coefficient. This article uses water and nanofluid (Al2O3) as working fluids, and their performances are compared.

27.2 Numerical/Experimental Studies

The nanofluids used in these investigations are based on both thermal oil and wáter. Ghasemi and Ranjbar [1] examined the Al_2O_3 and CuO in water for different concentrations up to 1.5%. They discovered that Al_2O_3 had a 35% increase in heat transfer coefficient, compared to a 28% increase with Cu. Mwesigye et al. [2] examined the Parabolic Trough Collector with Syltherm 800/Al_2O_3 and found a 7.6% thermal efficiency enhancement with 8% nanoparticle concentration. Additionally, Mwesigye et al. [3] discovered that adding 6% CuO to Syltherm 800 results in up to a 15% increase in thermal efficiency and a 35% increase in heat transfer coefficient. Khakrah et al. [4] investigated the usage of synthetic oil/Al2O3 nanofluid with concentrations of 1, 3, and 5% nanoparticles. They discovered a mean thermal efficiency improvement of approximately 7% and a maximum of nearly 12.5%.

For experimental experiments, the majority of researchers favoured using water as the basic fluid. De Los Rios et al. [7] experimental testing of the water/Al2O3 nanofluid in a PTC revealed improvements in thermal efficiency of up to 24%. Rehan et al. [5] investigated that Al2O3 and Fe2O3 nanoparticles dispersed in water at different concentrations (0.20, 0.25, and 0.30%). They investigated at several flow rates for both nanofluids (1.0, 1.5, and 2.0 l/min). According to their results, the maximum enhancement with Al2O3 was 17%, while the enhancement with Fe2O3 was 13%.

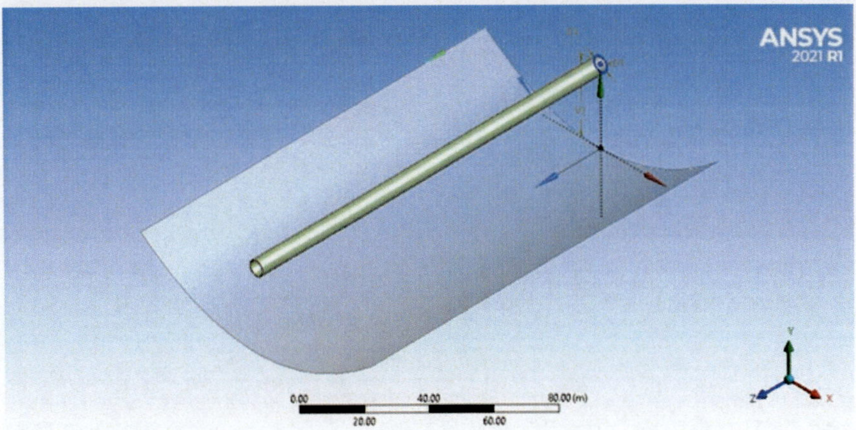

Fig. 27.1 PTC geometrical model

Table 1 Description of PTC geometrical model

Parameter	Value
Width of PTC	1 m
Length of PTC	1.5 m
Focal distance of PTC	31.25
Aperture area of PTC	1.5 m^2
Concentration ratio of PTC	26.5
Receiver outer diameter	1.2 cm
Receiver inner diameter	1 cm
Receiver outer surface area	0.0565 m^2

27.3 Model Description

The geometrical model of the solar Parabolic Trough Collector system is shown in Fig. 27.1. The line circular receiver, which is positioned on the focus of the parabolic collector, receives the majority of the solar radiation that falls on the collector. Convection heat transfer subsequently transfers the concentrated radiation in the receiver to the working fluid (Table 1).

27.4 Governing Equations

The continuity, momentum, and conservation of equations for fluid flow in a turbulent domain that correspond to a steady-state condition are given [6].

$$\frac{\partial(\rho_m \overline{u_i})}{\partial x} = 0 \tag{27.1}$$

$$\frac{\partial}{\partial x_j}\left(\rho_m \overline{u_i u_j}\right) = -\frac{\partial \overline{P}}{\delta x_i} + \frac{\partial}{\partial x_j}\left[\mu_m\left(\frac{\partial \overline{u_i}}{\delta \overline{x_j}} + \frac{\partial \overline{u_j}}{\partial x_i}\right) - \frac{2}{3}\mu_m \frac{\partial \overline{u_i}}{\delta \overline{x_j}}\partial_{ij} - \rho_m \overline{u_i\prime u_j\prime}\right] + (\rho\beta)_m (T - T_0)\,\overrightarrow{g} \tag{27.2}$$

$$\frac{\partial}{\partial x_j}\left(\rho_m \overline{u_j} C_{p,m} \overline{T}\right) = \frac{\partial}{\partial x_j}\left(\lambda_{eff,m}\frac{\partial \overline{T}}{\partial x_j} + \frac{\mu_t}{\sigma_{h,t}}\frac{\partial\left(C_{p,m}\overline{T}\right)}{\partial x_j}\right) + \overline{u_j}\frac{\partial \overline{P}}{\partial x_j} + \frac{\partial \overline{u_i}}{\partial x_j}\left[\mu_m\left(\frac{\partial \overline{u_i}}{\partial x_j} + \frac{\partial \overline{u_j}}{\partial x_i}\right) - \frac{2}{3}\mu_m \frac{\partial \overline{u_i}}{\partial x_i}\partial_{ij} - \rho_m \overline{u_i\prime u_j\prime}\right] \tag{27.3}$$

27.5 Results

The numerical analysis is done with the use of commercial software called ANSYS FLUENT. For the simulation of forced convection in the receiver tube, the k- ε model with conventional wall functions is used. Separate numerical analyses are performed for each of the three operating fluids. The fluid used are wáter, Al2O3 nano fluid with concentration 0.5 and 1% (Figs. 27.2 and 27.3 and 27.4).

When water is used as the working fluid the temperature rise at the outlet is 2.06 °C. The temperature rise at the exit was 3.7 and 4.04 °C, respectively, when

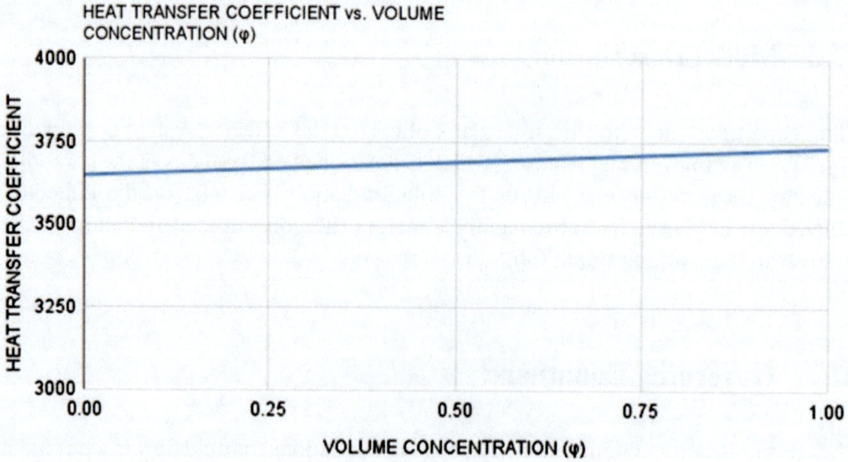

Fig. 27.2 Variation of heat transfer coefficient with volume concentration

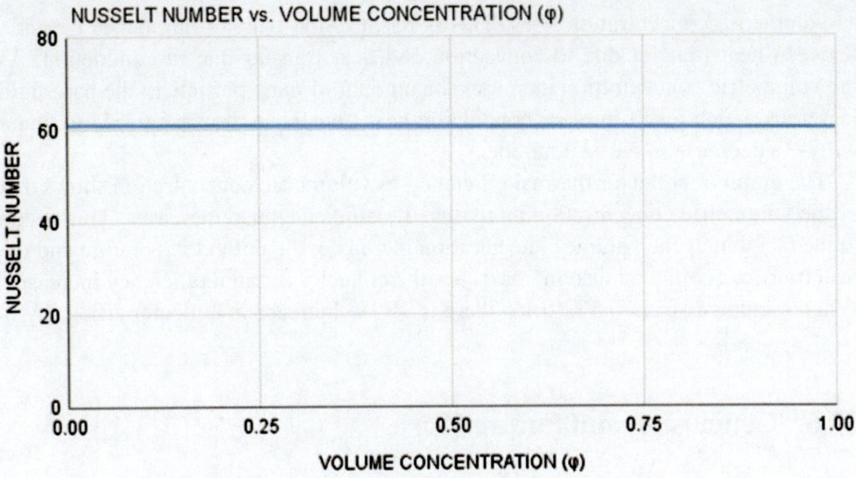

Fig. 27.3 Variation of Nusselt number with volume concentration

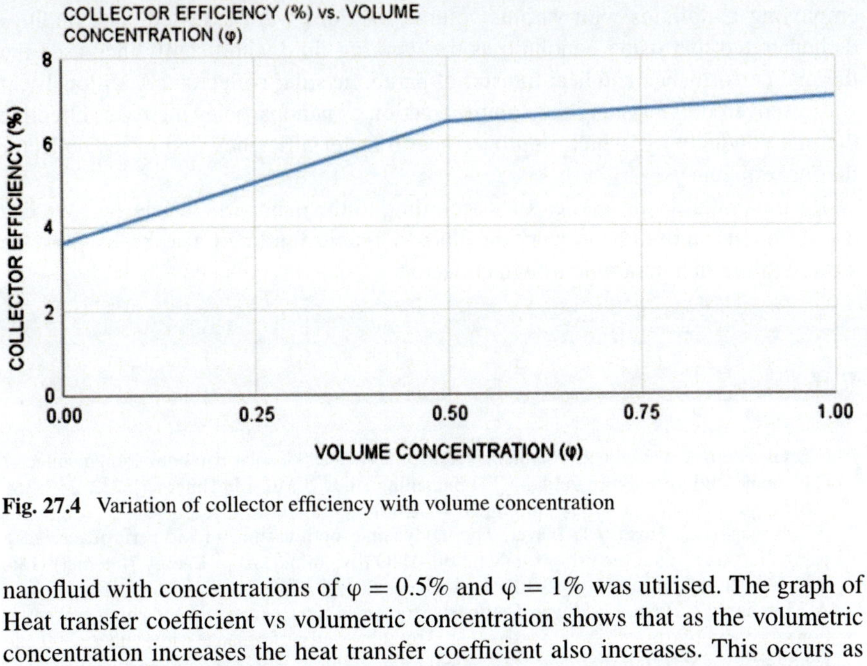

Fig. 27.4 Variation of collector efficiency with volume concentration

nanofluid with concentrations of $\varphi = 0.5\%$ and $\varphi = 1\%$ was utilised. The graph of Heat transfer coefficient vs volumetric concentration shows that as the volumetric concentration increases the heat transfer coefficient also increases. This occurs as a result of the nanoparticle's high thermal conductivity. As a result, the addition of these nanoparticles improved the thermal conductivity of the nanofluid, increasing the heat transfer coefficient. For $\varphi = 0.5\%$ there is 1.16% and for $\varphi = 1\%$ there is a 2.36% increase in the coefficient of heat transfer compared to water.

The relationship between Nusselt number and volumetric concentration may be seen in the graph, which demonstrates that the Nusselt number barely decreases as volumetric concentration rises. This is because the Nusselt number is the ratio between heat transfer due to convection and heat transfer due to conduction. As the volumetric concentration increases the amount of nano particle in the base fluid increases which result in more conduction heat transfer in the base fluid. So, there will be a decrease in Nusselt number.

The graph of collector thermal efficiency vs volumetric concentration shows that as the volumetric concentration increases, the efficiency also increases. This is due to the fact that as the volume concentration increases, the outlet temperature and the heat transfer coefficient also increase. So, the collector thermal efficiency increases. When φ changes from 0.5% to 1% there is 9.11% increase in collector efficiency.

27.6 Conclusion and Future Work

The project's main objective was the numerical analysis of nanofluid flow in a solar parabolic trough collector system and its impact on the collector's thermal performance when compared to water as the working fluid. Additionally, the impacts of employing nanofluids with various volume fractions were assessed. The findings demonstrated that using nanofluid as the working fluid significantly increased the thermal performance and heat transfer of parabolic solar collectors. Additionally, it was discovered that raising the volume fraction of nanoparticles increases effective thermal conductivity, which improves heat transfer efficiency and raises collector thermal efficiency.

Each nanofluid's properties vary according to the nanoparticle selected. We can use different nanofluids as working fluids. Organic fluid can also be used as the working fluid in a parabolic trough collector.

References

1. S.E. Ghasemi, A.A. Ranjbar, Thermal performance analysis of solar parabolic trough collector using nanofluid as working fluid: a CFD modelling study. J Mol Liq [Internet]. **222**, 159–166 (2016). https://doi.org/10.1016/j.molliq.2016.06.091
2. A. Mwesigye, Z. Huan, J.P. Meyer, Thermodynamic optimisation of the performance of a parabolic trough receiver using synthetic oil-Al2O3 nanofluid. Appl Energy [Internet]. **156**, 398–412 (2015). https://doi.org/10.1016/j.apenergy.2015.07.035
3. A. Mwesigye, Z. Huan, J.P. Meyer, Thermal performance of a receiver tube for a high concentration ratio parabolic trough system and potential for improved performance with syltherm800-cuo nanofluid, in *ASME International Mechanical Engineering Congress and Exposition* (2015)
4. H. Khakrah, A. Shamloo, S.K. Hannani, Determination of parabolic trough solar collector efficiency using nanofluid: a comprehensive numerical study. J. Sol. Energy Eng. Trans. ASME. **139**(5) (2017)
5. M.A. Rehan, M. Ali, N.A. Sheikh, M.S. Khalil, G.Q. Chaudhary, T. Rashid, et al., Experimental performance analysis of low concentration ratio solar parabolic trough collectors with nanofluids in winter conditions. Renew. Energy [Internet]. **118**, 742–751 (2018). https://doi.org/10.1016/j.renene.2017.11.062

6. A. Mwesigye, Z. Huan, J.P. Meyer, Thermal performance and entropy generation analysis of a high concentration ratio parabolic trough solar collector with Cu-Therminol®VP-1 nanofluid. Energy Convers Manag [Internet]. **120**, 449–465 (2016). https://doi.org/10.1016/j.enconman.2016.04.106
7. De los Rios, Mariana Soledad Bretado, R.-S. Carlos I, G.-C. Alejandro J. (2018). Thermal performance of a parabolic trough linear collector using Al2O3/H2O nanofluids. Renewable Energy. **122**, 665–673. https://doi.org/10.1016/j.renene.2018.01.094

Chapter 28
Thermal In-Orbit Performance and Evaluation of Multi-Staged Passive Radiant Cooler with Sun-Shield Configuration Variation

Sankalp Vishnoi, Subramanya, and P. Pradeep Kumar

Abstract Passive multi-staged Radiant coolers with appropriately configured sun-shield are an attractive choice to provide reliable cooling for electronic components onboard spacecrafts to cryogenic temperatures. If the tolerance on the temperature and allowable temperature fluctuations are not severe, these coolers and their variants like V-groove passive cooler offer a good choice, especially at higher orbits like Geosynchronous and Geostationary orbits. The current work focuses on the thermal performance evaluation of multi-staged IMAGER cooler with a rectangular sun-shield configuration, developed by ISRO for the IMAGER payload of INSAT-3D spacecraft placed in a Geostationary orbit around the Earth with a specified detector heat load of 18mW on patch. Evaluation of in-orbit thermal performance introduces complexity in view of handling distributed surface heat loads, diffuse-specular surface for radiant heat exchange, and non-homogeneous surface characteristics, which vary with the specific application requirements. In the current work, performance evaluation of a three-staged radiant cooler is carried out in commercial software SINDA/FLUINT 6.1 for thermal analysis with Thermal Desktop® as modeler and RadCAD® for surface radiation calculations. The geometry of the existing IMAGER cooler is modelled, and the in-orbit performance evaluation is benchmarked with reported data. The model boundary conditions are the solar heat load considering seasonal flipping of the cooler, planetary loads, and the spacecraft

S. Vishnoi (✉)
Vikram Sarabhai Space Centre, Trivandrum, India
e-mail: sankalpvishnoi99@gmail.com; sankalp_vishnoi@vssc.gov.in

Subramanya
U R Rao Satellite Centre, Bangalore, India
e-mail: sbhat@ursc.gov.in

P. Pradeep Kumar
Department of Aerospace Engineering, Indian Institute of Space Science and Technology, Trivandrum, India
e-mail: pradeepkumarp@iist.ac.in

A. Suryan et al. (eds.), *Proceedings of the 9th Asian Joint Workshop on Thermophysics and Fluid Science, 27–30 November 2022, Utsunomiya, Japan*, Springer Proceedings in Physics 396, https://doi.org/10.1007/978-981-99-9470-0_28

base temperature. We systematically present the influence of sun-shield angle variation on the thermal in-orbit performance, which helps the designer to choose an appropriate sun-shield opening angle based on the trade-off between patch temperature decrement and the associated mass penalties involved. The performance of a radiant cooler with an un-symmetrical sun-shield, which typically blocks all the planetary loads falling on the patch, is also presented. With curved sun-shield geometry posing implementational challenges in configuring a radiative surface with desired homogeneous characteristics, the current work presents an improved sun-shield design with a 12-sided sun-shield. In comparison with the original IMAGER cooler design, the dodecagon sun-shield is found to bring down the patch temperature and overall cooler mass by 7.51 and 9.17%, respectively. The projected surface area of the overall cooler is also reduced by 8.14%, making it more compact. The current work is an attempt toward performance assessment with a sun-shield configuration change, considering weight saving and improving the cooler's performance at cryogenic temperatures for the IR sensors as a part of satellites intended for high- altitude orbits.

28.1 Introduction

The temperature control of various components of a spacecraft in orbit is necessary in order to ensure their intended functioning and long life. Radiant coolers primarily provide thermal isolation of spacecraft electronic circuitry and crucial sensors from solar, planetary, and various other spacecraft heat loads and ensure that these operate at specified temperature ranges. Many such applications of components/sensors requiring temperature control are found in various Earth observation and meteorological satellites with a cryogenic operating temperature range of 1 to 100 K, depending on the heat load and specific requirement. Such cooling requirements has been successfully achieved using passive cryogenic radiant coolers up to 60 K [1]. Much lower temperature values are achieved introducing active components such as Stirling and Gifford–McMahon cryocoolers along with the passive radiant coolers.

The passive cryogenic radiant coolers use the cold space temperature as a sink to reject heat from the component to be cooled (patch) radiatively, while also protecting it from the solar and planetary heat loads using a sun-shield. The design of such passive coolers is highly dependent on specific requirements like the orbit and orientation of the spacecraft, the location of onboard sensors, and the patch heat load. Since the first radiant cooler flew in 1970 on NIMBUS-4 [1], there have been many design variations to improve thermal performance till date. One such design variation is the V-groove cooler developed by the Jet Propulsion Laboratory (JPL), capable of operating at a patch temperature near 60 K. The design of multi-staged passive radiant coolers involves using multiple reflective surfaces with desired non-homogenous radiative surface characteristics, specifically designed radiation insulation and coatings, instrument housings, intermediate radiator stages, and low-conductivity structural supports. A complete thermal performance evaluation of such a device involves solving a system of transient non-linear coupled equations taking a lumped system approach. The evaluation of specular exchange factors between different components for radiative interaction further complicates the formulation. Mantelli [2] in 2000 presented a semi-analytical model to predict the in-orbit performance of a three-staged radiant cooler, neglecting the effects of surface characteristics on the radiative exchange factors and making a few more assumptions on the variation of the environmental heat loads.

However, for an accurate thermal performance evaluation of such radiant coolers, it becomes necessary to capture the effects of non-homogenous surface characteristics and the varying environmental heat loads with solar declination. It becomes highly tedious to handle complexities using analytical models, leaving the designers with numerical techniques as the only choice in hand. Methods like FDM and FEM are used to evaluate the transient in-orbit thermal behaviour. Probabilistic approach such as the Monte Carlo ray trace is generally used to evaluate the specular exchange factors by designers. This approach helps the designers to analyse the behaviour of various designs for a particular requirement and, thus, optimize the geometry and properties of the radiant cooler, as the experimental setup for testing such devices is costly. Using the same approach, current work benchmarks the thermal performance of a three-staged radiant cooler for the IMAGER payload on board INSAT-3D developed by the Indian Space Research Organization (ISRO) [3]. In this paper, we present the geometrically improved design of the sun-shield for the same cooler patch load to meet the cooler thermal operational requirements and achieve some weight savings.

28.2 Thermal Performance Evaluation of Multi-Staged Passive Radiant Cooler

In the current work, the numerical modelling of a multi-staged radiant cooler, Thermal Desktop® [4], has been used as the modeler. RadCAD® module of the package solves for the specular exchange factors between all the nodes present in the model using the given input surface characteristics for radiative heat transfer. This is achieved using the Monte Carlo Ray trace algorithm built within the module. SINDA/FLUINT version 6.1 is then employed to solve the system for the nodal temperature and heat map. Other geometrical and mesh rectifications are performed in SpaceClaim and imported to Thermal Desktop®.

28.2.1 Numerical Model and Solution Methodology

The three-staged radiant cooler (Fig. 28.1) had been developed to maintain the patch at a temperature of 90 K, with a sensor heat dissipation of 18 mW as part of a meteorological satellite, INSAT-3D, developed by ISRO [3]. The cooler incorporates controlled heaters on the patch, radiator plate, top flanges, and vacuum housing to maintain a temperature of around 300 K during the initial degassing phase of the spacecraft. The design patch temperature of the cooler is 71 K, allowing a temperature margin of 19 K to be taken care of by the patch heater while operation [3].

The vacuum housing is directly mounted to the interface plate, which is bolted to the main spacecraft structure. Vacuum housing, along with the sun-shield and top flanges together, constitutes the first stage, whereas the radiator and patch are referred to as the second and third stages, respectively, as suggested in [5]. Additional vertical shields are provided outside the sun-shield to reduce the cooler's response to external load variation and minimize the sun-shield temperature cycling. These vertical shields are covered with MLI insulation on the external surface facing the environment.

Sun-shield internal surfaces are highly finished with an RMS surface roughness of less than 50 Å to achieve high specular reflectivity and coated with vapor-deposited aluminium (VDA), which provides low IR emittance to the surface. This is done to minimize the heat input to the patch. The top surface of the patch is a honeycomb structure painted in black to maximize the IR emittance using the concept of cavity effect, as mentioned in [3]. All the internal surfaces of the vacuum housing and the un-exposed radiator area are covered with vapor-deposited gold (VDG) to minimize the parasitic heat leakage to the coldest stage. The same VDG coating is also used on the sun-shield's external surface and the vertical shield's internal surface for similar reasons. Top flanges are used as an emitter area for the first stage to reject heat to the cold space radiatively. The top surface of these flanges is covered in optical solar reflector (OSR) tiles to reflect most of the solar fluxes falling onto them. The bottom surface of the same is covered with MLI insulation. Surface characteristics of all such insulations and coatings used are tabulated below [5] (Table 28.1).

The internal surface of the sun-shield panels has a specular reflectivity of 99%, giving it a mirror-like finish, and the effective conductance of the MLI insulation used is 0.04 W/mK. Aluminium material is used for all the cooler components except for the low-conductivity fibre-reinforced plastic (FRP) supports.

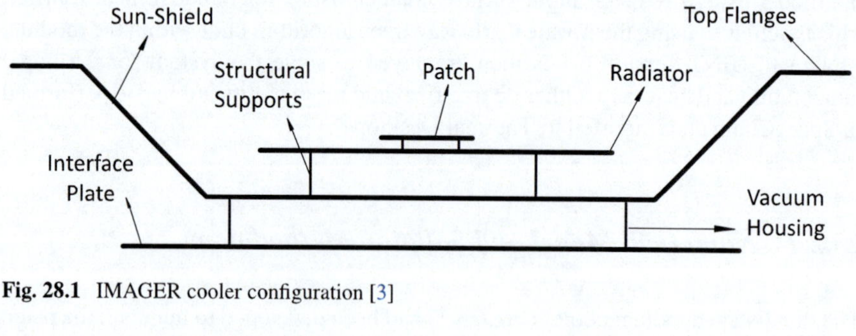

Fig. 28.1 IMAGER cooler configuration [3]

Table 28.1 Surface characteristics of IMAGER cooler [5]

Surfaces	Coating	α_{solar}	ϵ_{IR}
Patch top	Black honeycomb	0.9	0.94
Radiator top	White-paint	0.2	0.8
Patch bottom	VDG	0.3	0.03
Radiator bottom			
Vacuum housing			
Sun-shield outside			
Vertical shield inside			
Top flange (Top)	OSR	0.08	0.8
Vertical shield outside	MLI	0.2	0.8
Top flange (Bottom)			
Sun-shield inside	VDA	0.1	0.03

INSAT-3D is placed in a Geostationary orbit at roughly 36,000 km. The orbit is co-planar with the Earth's equator throughout the year; therefore, changing solar declination significantly affects the cooler performance. Figure 28.2 shows the variation of solar declination throughout the year.

For such meteorological payloads, the spacecraft's attitude is kept such that the cooler opening face is always tilted towards the Earth. This is achieved when the spacecraft undergoes a 180° yaw manoeuvre during the equinoxes. This results in similar environmental conditions for the cooler throughout the year.

In the current work, intricate minor components of ISRO's Imager cooler [3] present in the vacuum housing are not specifically modelled. However, the overall mass and radiative surface area, conductive contact, and radiative resistances have been appropriately accommodated for the thermal analysis. The design simplification does not compromise the requirement for thermal analysis. Moreover, the rectangular Imager cooler and the improved model preserve the component information. Finite elements are used to model all the cooler components except for simple rectangular planar objects like the cold patch and the vertical shields. Such simple components are discretized in space using finite difference nodes. Therefore, the thermal model

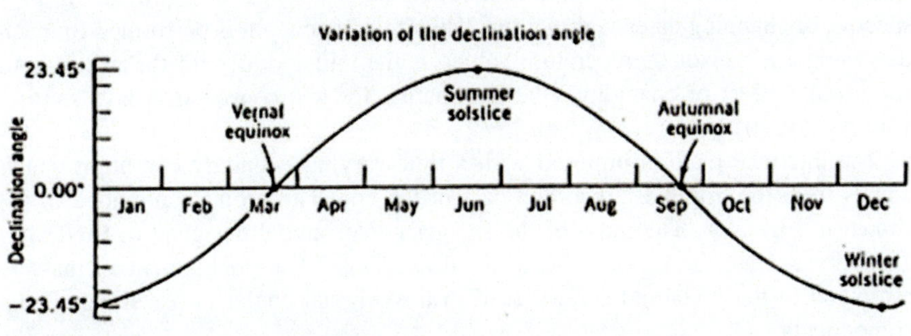

Fig. 28.2 Variation of solar declination throughout the year [6]

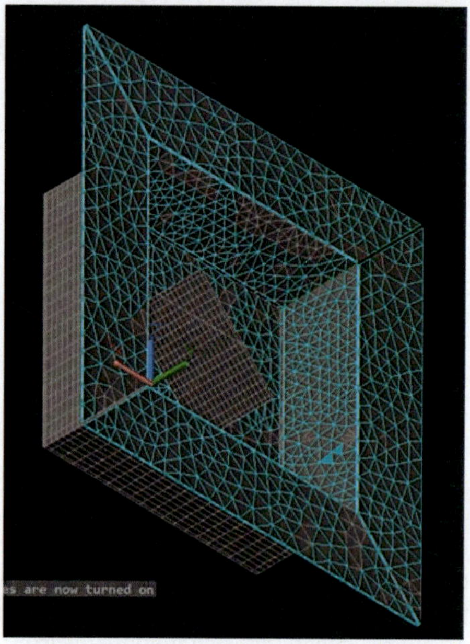

Fig. 28.3 IMAGER cooler SINDA/FLUINT model

is a combination of finite elements and finite difference surfaces and solids, as shown in Fig. 28.3.

The model handles the radiating surfaces so that all the FEM solids are surface coated with thin shell elements having the respective surface properties and material properties of Aluminium. The components in the model are bifurcated into three different radiation analysis groups for the sake of better accuracies, lesser computational time, and practical simulation database size. A solar flux of 1353 W/m^2 is used, along with an albedo factor of 0.35. For the IR-planet shine, the Earth's surface temperature of 250 K is assumed, both on the day and night sides. These values are typical for a Geostationary orbit around the Earth. Each orbit is discretized into 50 uniformly distributed points with a total duration of 24 h. The solar declination is automatically calculated based on the date the simulation is performed based on the J2000 epoch. However, this value remains constant throughout a simulation, irrespective of changing dates within a run. Therefore, a new run is performed for each day using the previous temperature values as the initial values for the next day to capture the effect of changing solar declination for five consecutive days starting 16th August 2013.

The interface plate is imposed with a time-varying boundary condition which comes from the originally recorded data of the cooler in orbit, as provided in [5] shown in Fig. 28.4. The mass of the Imager cooler model designed in SINDA is 18.10 kg against the original mass of 20 kg. This 1.9 kg decrement in mass is attributed to the bolts and screws used in the original cooler to assemble all the components.

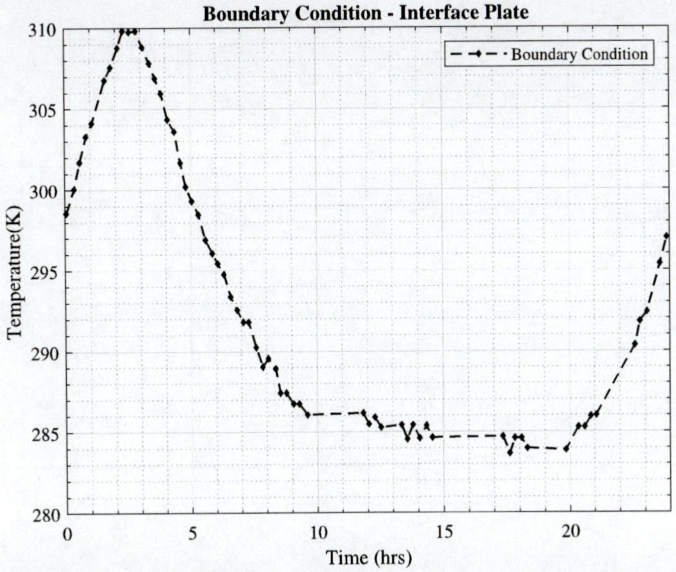

Fig. 28.4 Imposed time varying boundary condition on the interface plate

28.2.2 *In-Orbit Performance Benchmark*

Figure 28.5 shows the transient behaviour of the cold patch (3rd stage) for five consecutive orbits starting from 16th of August 2013, with an initial temperature of 300 K.

The steady state appears to be a straight line achieved in slightly less than two days. Although, a closer look at the later 3 days of the simulation shows a steady state as a fluctuating temperature behaviour between a minimum and a maximum temperature value which depends on the spacecraft's position in orbit, as in Fig. 28.6.

The minimum and maximum temperature values of the original cooler sent into orbit as of 19th August 2013 is also plotted in the same figure. Table 28.2 below summarizes the comparison between the simulation results and the originally recorded data for all the cooler components.

The above results correspond to the converged mesh after a mesh convergence study is performed. Table 28.3 presents the number of finite nodes/elements used for the study. The number of finite-difference nodes and finite elements is increased by 1.5 times on every surface, as shown.

The deviation in temperature values of all the components for improved mesh lies within 0.4% of the initial values. Therefore, the initial mesh is already a converged mesh that gives accurate enough results with a reasonable computational time.

The thermal behaviour of such coolers is governed by the sun-shield geometry surrounding its cold patch. A parametric study for the opening angle of the sun-shield in the existing design is one of the many approaches to arrive at a trade-off between

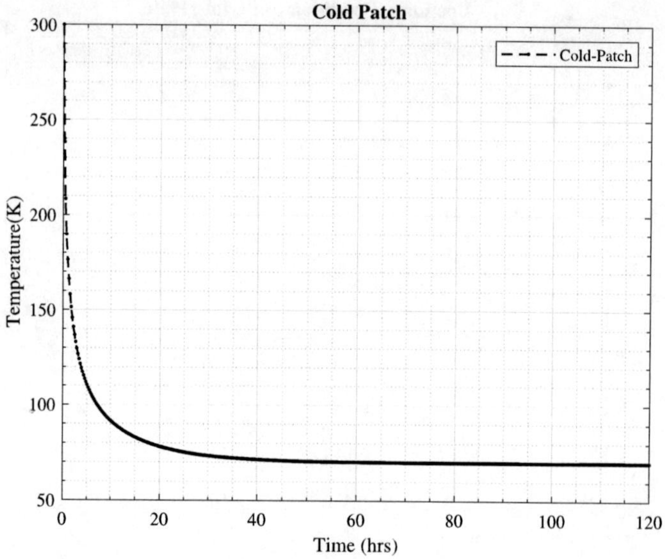

Fig. 28.5 Transient behaviour—patch

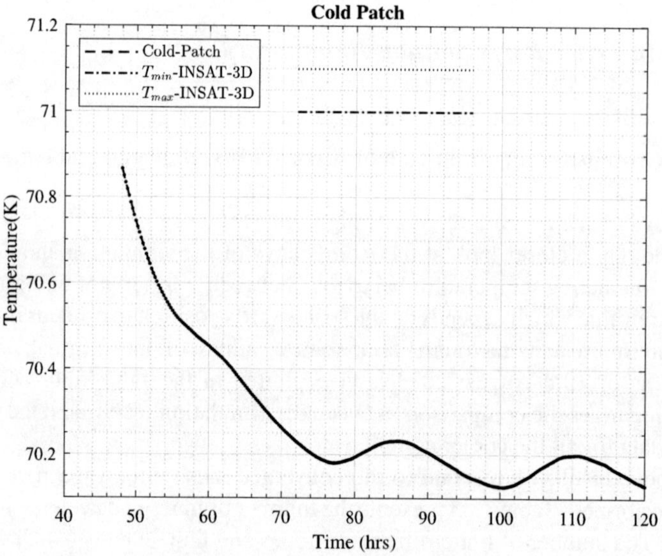

Fig. 28.6 Transient behaviour—patch for later 3 days

Table 28.2 Cooler elements temperature comparison

Cooler components	Calculated values (19th August 2013)		Observed values (19th August 2013) [3]	
	Min	Max	Min	Max
Patch	70.17	70.21	71	71.1
Radiator	111.46	111.81	108.2	108.8
Vacuum housing	178.9	181.8	178.5	183.7
Sun-shield	176.5	179.7	177.3	182.3

Table 28.3 Mesh convergence data

Component	Number of nodes/elements		Type
	Mesh-1	Mesh-2	
Patch	441	676	FDM
Radiator	3566 + 441	5349 + 676	FEM + FDM
Vacuum housing	19,744	29,616	FEM
Sun-shields	3032 × 4	4548 × 4	FEM
Top flanges	2592 × 4	3888 × 4	FEM
Vertical shields	625 × 4	841 × 4	FDM

the minimum cold plate temperature and the cooler mass. Furthermore, making the sun-shield more circular improves the patch view to cold space and improves the performance drastically. Making such a conical sun-shield is pragmatically infeasible owing to several polishing and coating requirements and therefore, sun-shields with multiple planar surfaces are used. Increasing the number of such planar surfaces in the sun-shield also improves the patch view to cold space but simultaneously also increases the mass of the connecting elements. Therefore, a limited number of sun-shield panels would result in an optimum cooler performance keeping total mass as per the requirement. The following sections present the thermal in-orbit performance of coolers. Finally, we present cooler performance with improved sun-shield designs (Fig. 28.7).

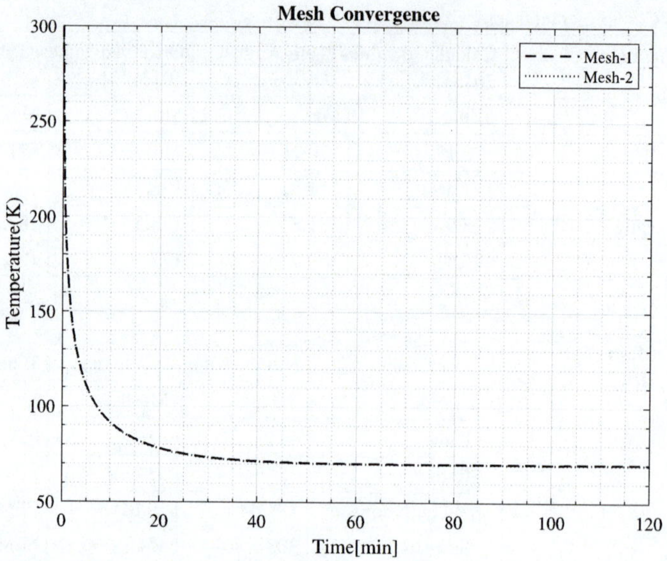

Fig. 28.7 Mesh convergence plot for cold plate

28.3 Influence of Sun-Shield Opening Angle and Improved Design

28.3.1 Radiant Cooler Operation with Sun-Shield Opening Angle Variation.

Changing the sun-shield opening angle directly impacts the incoming albedo, IR planet shine loads, view of the patch to the cold space and the overall mass of the cooler. Therefore, the designer must look for an optimum opening angle that leads to an acceptable minimum patch temperature with an overall cooler mass in the required range. A parametric study for the sun-shield opening angle is carried out for the radiant cooler developed by ISRO for the IMAGER payload [3] as discussed above.

While changing the sun-shield's opening angle, the first stage's emitter area, also referred to as the top flange area, is kept constant in all the designs. This constraint is required to radiate heat from the first stage to cold space. Keeping the height of the cooler same and adhering to the constraint mentioned above, the overall dimensions of the cooler vary. Table 28.4 above summarises the mass, top projected area, overall dimensions, and patch temperatures for each sun-shield opening angle case. The table also shows the % change in mass for each case compared to the benchmark case of $\phi = 25°$. Changes in the top projected area can be attributed to the varying dimensions of the sun-shield as the patch, radiator, and top flange areas are kept constant (Figs. 28.8 and 28.9).

Table 28.4 Sun-shield angle parametrization results

Case	Overall dimension (mm × mm)	Top projected area (m^2)	Mass (Kg)	% Mass change	Patch temperature (K)	
					Min	*Max*
$\phi = 0^o$	619.4 × 619.4	0.3837	17.26	−4.64	73	73.3
$\phi = 10^o$	629.5 × 629.5	0.3962	17.37	−4.03	70.3	70.4
$\phi = 25^o$	664 × 664	0.4408	18.10	0	70.17	70.21
$\phi = 40^o$	713 × 713	0.5083	19.30	+ 6.63	69.9	70
$\phi = 50^o$	762.4 × 762.4	0.5812	20.50	+ 13.26	70.31	70.4
$\phi = 60^o$	842 × 842	0.7091	22.67	+ 25.25	69.9	70.05

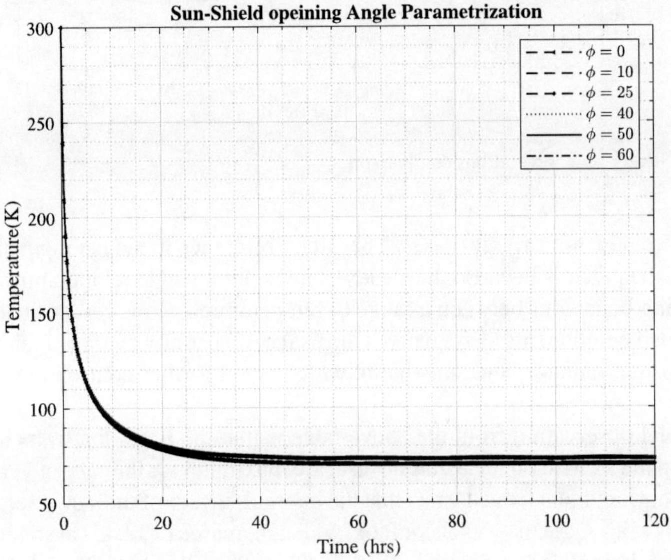

Fig. 28.8 Transient behaviour comparison

The temperature values for the case corresponding to $\phi = 0°$ are pretty high, with a maximum temperature of 73.3 K. This is because vertical sun-shields have a large view of the patch, thereby increasing the parasitic heat loads on it. A vertical sun-shield also reflects most of the incoming Earth loads on the patch. The minimum temperature is achieved for the sun-shield opening angle of $\phi = 40°$ and $\phi = 60°$, however there is a substantial increment in mass of the cooler for the later case owing to the increased dimensions.

A heat balance study is also carried out on the patch surface, similar to what is shown in [1] as a quick check for the simulation results. Energy contribution due to all interactions of the patch with surrounding surfaces and space is shown in

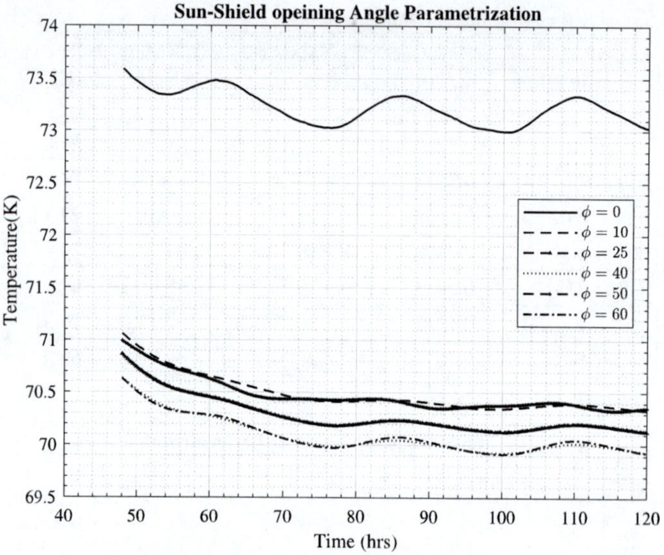

Fig. 28.9 Steady state fluctuations comparison

Table 28.5 below for different cases. The slight difference in the net input and output energy of the patch is because the steady-state is not a single temperature value but a fluctuating behaviour between a range of temperatures. However, evaluation of all the heat load components also involves the usage of spatially averaged temperatures over the patch, radiator, and sun-shield, which could further contribute to errors in such heat balance studies.

A crucial observation from the above table is that the Earth loads, including the albedo and the IR load on the patch surface, are maximum for the case of vertical sun-shields. This is evident from the fact that the internal surface of the sun-shield receives these loads at an angle of $\approx 9°$ at equinox, which increases to 32.45° at solstice. Thus, any sun-shield inclined at an angle less than $\phi = 9°$ will eventually reflect most of the incoming Earth loads onto the patch, thereby explaining high-temperature values of the same.

28.3.2 Un-Symmetrical Sun-Shield Design

A large opening angle for the sun-shield typically augments the Earth's heat loads to the patch. The cooler performance could be further improved if this load falling on the patch is reduced while still giving the patch a large view of cold space. This can be achieved by blocking some amount of Earth loads using the sun-shield panel facing the Earth.

Table 28.5 Heat Balance on Patch

Case	$\phi = 0^o$	$\phi = 10^o$	$\phi = 25^o$	$\phi = 40^o$	$\phi = 50^o$	$\phi = 60^o$
Earth loads (W)	0.00731	0.00055	0.00116	0.00117	0.00100	0.00111
Interstage radiative loads (W)	0.0093	0.0092	0.0092	0.0094	0.0106	0.0095
Sun-shield parasitic Loads (W)	3.7396×10^{-4}	2.4771×10^{-4}	3.729×10^{-5}	2.2747×10^{-5}	6.1066×10^{-6}	5.67×10^{-8}
Conductive loads (W)	0.0276	0.0289	0.0289	0.0293	0.0311	0.0295
Equipment load (W)	0.018	0.018	0.018	0.018	0.018	0.018
Total input load (W)	+0.0625	+0.0569	+0.05	+0.057	+0.0607	+0.0581
Output load (W)	−0.0597	−0.0525	−0.052	−0.0519	−0.0533	−0.0525
Net Difference (W)	***0.0028***	***0.0044***	***0.0052***	***0.0059***	***0.0074***	***0.0056***

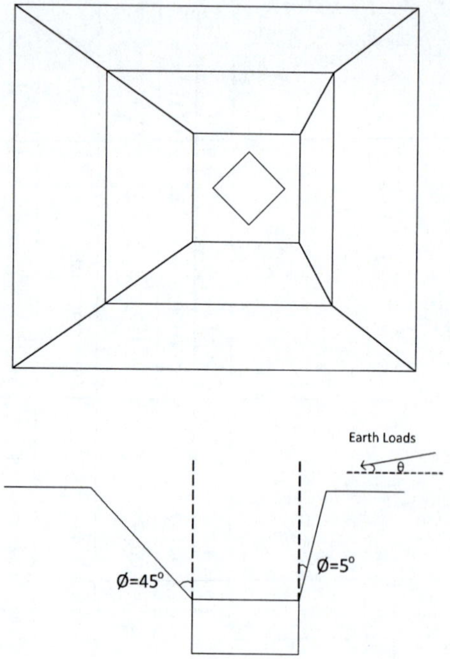

Fig. 28.10 Un-symmetrical sun-shield

Ideally, a perpendicular panel will block the maximum amount of such loads falling to the patch for a given sun-shield height. However, even a small amount of contaminants on the polished surface can scatter the loads falling onto the sun-shield at small angles, which, if reflected by a perpendicular sun-shield, could directly fall on the patch. Therefore, a small angle margin of 5° is given to the sun-shield panel facing the Earth. All the other three panels are kept at an angle of 45° so as to meet the second requirement for better performance, as shown in Fig. 28.10. The incoming Earth load is inclined at angle θ, which changes with the solar declination and has a value of around 9° at the equinox. Table 28.6 summarises the mass savings and the performance improvement of the un-symmetrical sun-shield case. However, the un-symmetrical sun-shield design improves the cooler performance marginally with a mass penalty of about + 5.63% as compared to the original design with all the panels inclined equally at $\phi = 25°$ (Fig. 28.11).

28.3.3 Do-Decagon Sun-Shield Design

See Fig. 28.12.

As discussed, a cooler configuration with a sun-shield design resembling the conical shape could further improve the cooler performance. Taking the IMAGER

Table 28.6 Summary for the improved sun-shield design

Case	Overall dimensions (mm × mm)	Mass (Kg)	Top-projected area (m^2)	% Mass change	Min temp (K)	Max temp (K)
$\phi = 25°$	664 × 664	18.10	0.4408	0	70.17	70.21
$\phi = 40°$	713 × 713	19.30	0.5098	+6.63	69.9	70
$\phi = 45°$, 45°, 45°, 5°	750 × 658	19.12	0.4935	+5.63	69.7	69.86

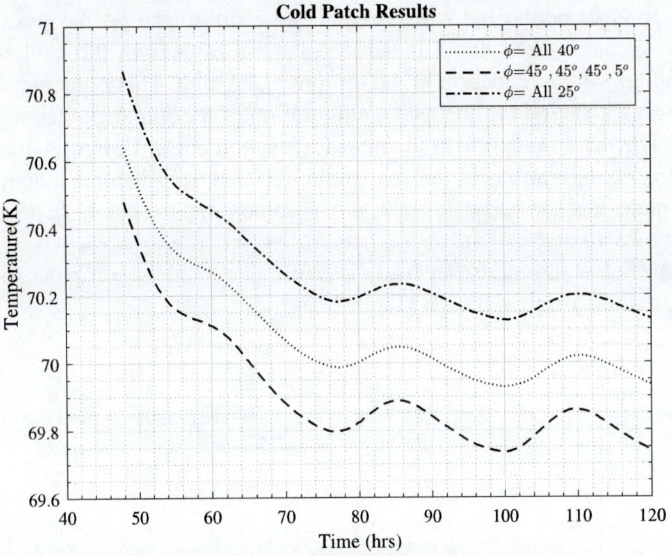

Fig. 28.11 Sun-shield performance

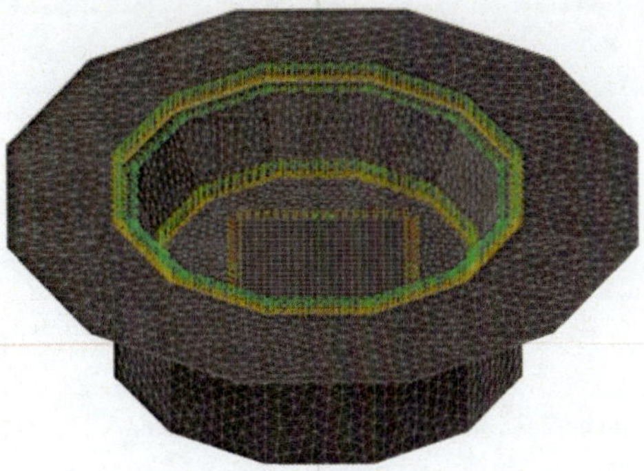

Fig. 28.12 Do-decagon cooler SINDA/FLUINT model

cooler [3] as the basis, a new cooler configuration is proposed with a 12-sided sun-shield and top flange for a considerable performance improvement. The following constraints are imposed while designing the 12-sided cooler:

a. The patch shape and dimensions are kept the same to fix the basis of comparison with the IMAGER cooler.
b. The Radiator and the first stage emitter areas (top flanges) also kept the same as the IMAGER cooler design, with their shapes changed to do-decagon.
c. Cooler height is kept the same.

The do-decagon cooler design with a sun-shield opening angle of $\phi = 25°$ outperforms the Imager cooler by $\approx$ 6 K temperature difference on the patch surface. Moreover, as anticipated, there is 9.17% of mass savings in the cooler's overall design and an 8.14% reduction in the top projected area. This reduction in the top projected area is due to the changed configuration of the sun-shield from 4-sided to 12-sided. The mass reduction in the new do-decagon cooler design is the net result of changing configurations of the sun-shield, top flange, radiator, vacuum housing, interface plate, and the vertical shield. At the same time, the patch shape is kept the same (Figs. 28.13 and 28.14).

The table below summarise the overall performance comparison of the do-decagon cooler against the original design (Table 28.7).

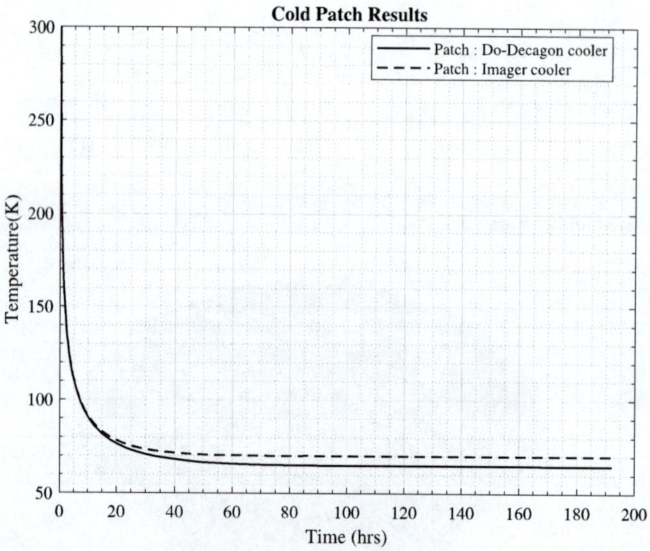

Fig. 28.13 Do-decagon cooler complete transient performance

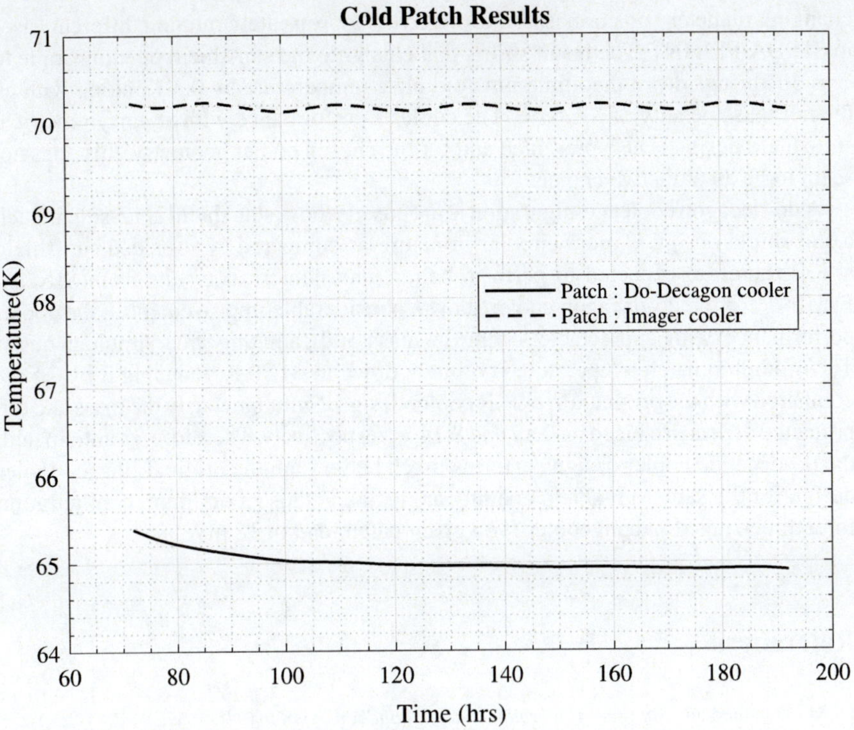

Fig. 28.14 Do-decagon cooler steady state performance

Table 28.7 Do-decagon cooler properties and performance

Cooler design	Overall dimensions (mm × mm)	Mass (Kg)	Top-projected area (m^2)	% Mass change	Min temperature (K)	Max temperature (K)
Imager Cooler	664 × 664	18.10	0.4408	0	70.17	70.21
Do-decagon cooler	734.8 × 734.8	16.44	0.4049	−9.17	64.90	64.92

28.4 Conclusion and Future Work

A three-staged passive radiant cooler developed by ISRO for the IMAGER payload on board INSAT-3D [5] is modelled in SINDA/FLUINT using Thermal Desktop® as the modeler. It is designed to operate the patch at a constant temperature of 90 K throughout the spacecraft's orbital life. The simulation results are benchmarked with the actual in-orbit temperature data as of 19th August 2013, presented in [3]. The variation in solar declination angle and, thus, the varying environmental load with changing planetary position is also captured by appropriately running different cases in SINDA/FLUINT. The result shows that changing the sun-shield opening angle to $\phi = 40°$ brings down the minimum cold plate temperature by 0.3 K, along with an overall mass penalty of +6.63%. The cooler's performance with an unsymmetrical sun-shield design is also presented and for the chosen cooler geometry, this does not seem to be an attractive choice.

A do-decagon cooler configuration is proposed whose sun-shield is 12-sided. Such a sun-shield design is more circular in nature as compared to a 4-sided sun-shield and is therefore expected to perform better. The numerical results from SINDA/FLUINT for the do-decagon cooler show a considerable improvement in the cooler performance with appreciable saving in mass and radiating area simultaneously. The minimum patch temperature is brought down to 64.90 K, reducing it by 7.51% compared with the original IMAGER cooler design. Simultaneously the mass and top projected area are reduced by 9.17 and 8.14%, respectively. Therefore, as anticipated, the do-decagon cooler design performs much better than the original Imager cooler design for the same sun-shield opening angle and is thus an optimum configuration for a three-staged radiant cooler for a given patch load of 18 mW.

References

1. M. Donabedian, *Spacecraft thermal control handbook, volume II: cryogenics.* (American Institute of Aeronautics and Astronautics, Inc., 2004)
2. P. Couto, M.B. Mantelli, Transient temperature behavior of multistage cryogenic radiators: model and experimental validation. J. Thermophys. Heat Transfer **14**(3), 313–321 (2000)
3. D.W. Tijare, N.K. Satya, A.K. Shrivastava, B. Akkimaradi, T. Selvan, V. Jaikumar, G. Tanneru et al., Performance of passive radiant cooler of imager payload through seven year designed mission life of a meteorological spacecraft, in *Proceedings of the 26th National and 4th International ISHMT-ASTFE Heat and Mass Transfer Conference*, December 17–20, 2021, IIT Madras, Chennai-600036, Tamil Nadu, India. (Begel House Inc., 2021)
4. D. Timothy, G. Steven, J. Mark, J. David, A. Brent, P. Douglas, Thermal desktop user's manual, 6th ver. (2017)
5. D.W. Tijare, Subrahmanya, Design, realization and performance of test set up with cryotarget for cooling of infrared detectors in meteorological payloads to 90k, in *Proceedings of the 22th National and 11th International ISHMT-ASME Heat and Mass Transfer Conference*, IIT Kharagpur, India (2013)
6. The renewable energy website. http://www.reuk.co.uk/wordpress/solar/solar-declination/. Accessed 10 April 2022

Chapter 29
Influence of Structural Parameters of Catalyst Support on Hydrogen Production Performance of Diesel Reforming

Sibo Wang and Jiang Qin

Abstract Low-carbon hydrogenation is the development trend of future energy. Hydrogen energy has received extensive attention, and the acquisition of hydrogen is particularly critical. Compared with hydrogen storage technology, diesel has a high theoretical mass hydrogen storage density, and its storage and transportation are more efficient and convenient. Moreover, diesel reforming hydrogen production technology has a high theoretical hydrogen production ratio, a wide range of applications, and relatively mature infrastructure. In this paper, honeycomb ceramic was used as the carrier of nickel based catalyst, and the diesel reforming reactor is numerically simulated and the computational model is experimentally verified to explore the influence of the structural parameters of the catalyst carrier on the reforming effect. The experimental results show that the numerical simulation using FLUENT has high accuracy, which is in good agreement with the experimental results, and the error does not exceed 20%. From the point of view of reducing the temperature gradient of the reactor, the reactor temperature is relatively uniform and the volume concentration of hydrogen in the product is relatively high when the reactor is arranged sparsely in the front and densely in the back. When the number of inlet mesh is 200 and the number of outlet mesh is 600, it can reach 70–72%. The numerical simulation and experimental research in this paper provide a theoretical basis for the optimization of reforming hydrogen production reaction, which has practical significance in engineering application.

S. Wang · J. Qin (✉)
School of Energy Science and Engineering, Harbin Institute of Technology, Harbin 150000, China
e-mail: qinjiang@hit.edu.cn

A. Suryan et al. (eds.), *Proceedings of the 9th Asian Joint Workshop on Thermophysics and Fluid Science, 27–30 November 2022, Utsunomiya, Japan*, Springer Proceedings in Physics 396, https://doi.org/10.1007/978-981-99-9470-0_29

29.1 Introduction

With the rapid development of the world today, it is inevitable that fossil energy will be consumed rapidly, which has created a huge burden on the environment. Accordingly, energy transformation has become the focus of global attention. Among them, hydrogen energy, as a green, clean, widely sourced and high energy density secondary energy, has attracted much attention [1–4]. The safer and more efficient way is to convert hydrogen energy into electric energy through fuel cells [5]. Fuel cell distributed generation technology has significant advantages such as high efficiency, clean, flexible, easy installation, etc. It has a very broad development prospect and can effectively solve the power supply problem in remote areas. It can not only meet regional power demand, but also reduce energy consumption and pollution caused by consumption [6–9]. In order to utilize hydrogen energy, the first thing to be concerned is the production of hydrogen. About 50% of hydrogen in the world is produced by natural gas steam reforming, about 30% from heavy oil reforming, about 18% from coal gasification, about 3.9% from water electrolysis, and 0.1% from other energy sources [10]. In recent years, with the vigorous development of fuel cells, the demand for hydrogen is growing. Fuel reforming is still the most common hydrogen production method [11]. Among hydrocarbons, diesel has a high energy density, storage and transportation are also very efficient and convenient, and the infrastructure is relatively mature, so it is necessary to study diesel reforming for hydrogen production [12–15]. In this paper, honeycomb ceramic was used as the carrier of nickel based catalyst, and the influence of structural parameters of honeycomb ceramic catalyst carrier on diesel steam reforming was investigated through experiments and numerical simulation. N-hexadecane, as a commonly used substitute for diesel oil, has a single composition, which can well simulate various properties of diesel oil. Therefore, in this paper, n-hexadecane is selected to replace diesel for simulation.

29.2 Establishment of Numerical Models

29.2.1 Physical and Mathematical Models

In order to reduce the calculation time and save the calculation resources, the physical model is simplified and a two-dimensional physical model is established. Based on relevant experimental data, the 200 mesh honeycomb ceramic reaction channel we adopted is 120 * 3 mm, 120 * 1.5 mm for 400 mesh reaction channel, 120 * 1 mm for 600 mesh reaction channel, the physical model is as shown in Fig. 29.1. The space scale of the selected reactor is very small, only the reaction on the catalyst coating is considered, and the reforming reaction only takes place in the channel on the honeycomb ceramic surface. Hexadecane and water are mixed evenly before entering the reaction channel, and enter the channel as gas. The reactants are calculated as incompressible fluid, ignoring the volume effect, the radiation heat transfer, the

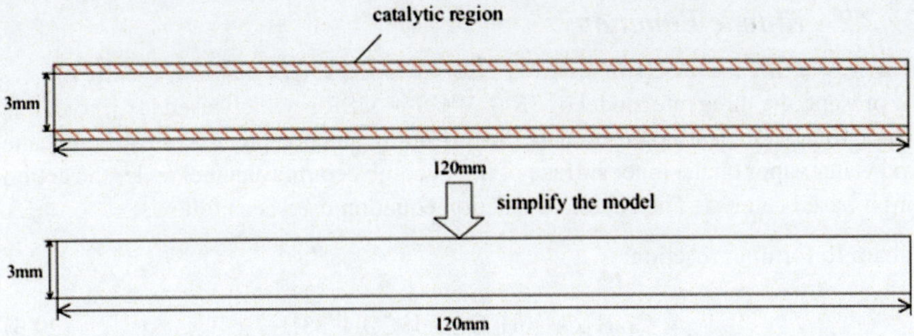

Fig. 29.1 Physical model of 200 mesh honeycomb ceramic reaction channel

volumetric force and the dissipation effect in the flow, and the influence of catalyst coating on the properties of honeycomb ceramic channel materials. Steady laminar flow is found in the reactor.

According to the above assumptions, there are continuity equations and N-S equations for the flow of incompressible fluid in honeycomb ceramic channels. The governing differential equations of component convection diffusion and chemical reaction are as follows:

Continuity equation:

$$\frac{\partial(\rho u_j)}{\partial x_j} = 0 \tag{29.1}$$

Component equation:

$$\rho u_j \frac{\partial Y_s}{\partial x_j} = \frac{\partial}{\partial x_j}\left(\rho D \frac{\partial Y_s}{\partial x_j}\right) + R_s \tag{29.2}$$

Momentum equation:

$$\frac{\partial(\rho u_j u_i)}{\partial x_j} = -\frac{\partial p}{\partial x_i} + \frac{\partial}{\partial x_j}\left[\mu\left(\frac{\partial u_i}{\partial x_j} + \frac{\partial u_j}{\partial x_i}\right)\right] \tag{29.3}$$

Energy equation:

$$\rho u_j \frac{\partial T}{\partial x_j} = \frac{\partial}{\partial x_j}\left(\frac{\lambda}{C_p} \frac{\partial T}{\partial x_j}\right) + q \tag{29.4}$$

29.2.2 Kinetic Equations

At present, the three rate model of SR + WGS + MSR is widely used [16], so in this paper, this model is used to simulate the reforming reaction process of n-hexadecane and water vapor on the inner surface of honeycomb ceramic channel under the action of Ni based catalyst. The chemical reaction equation used is as follows:

Steam Reforming reaction:

$$C_{16}H_{34} + 16H_2O \rightarrow 16CO + 33H_2 \tag{29.5}$$

Water Gas Shift reaction

$$CO + H_2O \leftrightarrow CO_2 + H_2 \tag{29.6}$$

Methanation reaction:

$$CH_4 + H_2O \leftrightarrow CO + 3H_2 \tag{29.7}$$

The reaction rate expression of each reaction is as follows:

$$r_{SR} = k_1 c_{C_{16}H_{34}} c_{H_2O} \tag{29.8}$$

$$r_{WGS} = \frac{k_2}{c_{H_2}}\left(c_{CO}c_{H_2O} - \frac{c_{CO_2}c_{H_2}}{\exp(\frac{4577.8}{T} - 4.33)}\right) \tag{29.9}$$

$$r_{MSR} = k_3\left(c_{CO}c_{H_2}^3 - \frac{c_{CH_4}c_{H_2O}}{\exp(\frac{26800}{T} - 29.8)}\right) \tag{29.10}$$

The definitions of k1, k2, and k3 are as follows:

$$k_i = k_i^0 \exp\left(-\frac{E_i}{RT}\right) \tag{29.11}$$

The parameter values in the above formula are as follows (Table 29.1).

29.2.3 Mesh-Independent Validation

Use ICEM to mesh the drawn model, as shown in Fig. 29.2:

Table 29.1 Parameter value

Parameter	Value
$k_1{}^0$	$5*10^{10}$
$k_2{}^0$	$3*10^8$
$k_3{}^0$	200
E_1	225 kJ/mol
E_2	125 kJ/mol
E_3	10 J/mol

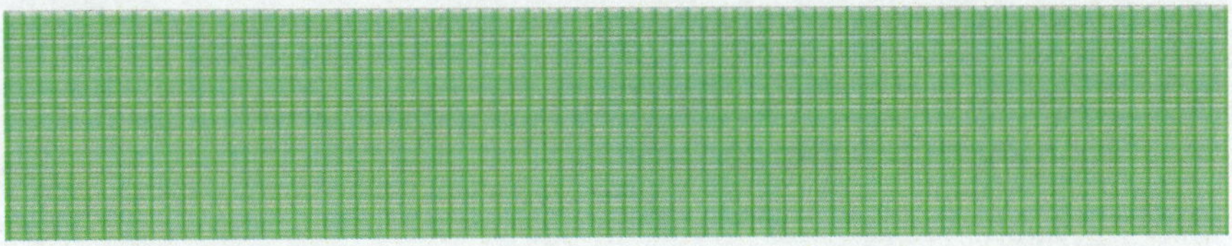

Fig. 29.2 Model meshing

The given mass flow inlet velocity is 0.0011g/s, the wall temperature is 1073K, and the molar water carbon ratio is 2.5. Calculate whether the volume fraction of hydrogen outlet changes by changing the number of grids (Table 29.2 and Fig. 29.3).

According to the above figure, when the number of grids reaches 54,000, the number of grids will no longer interfere with the results. In order to save computing

Table 29.2 Volume fraction of hydrogen outlet under different grid numbers

Number of grids	5760	9000	23,040	36,000	54,000	81,000
Hydrogen outlet volume fraction	0.355	0.378	0.385	0.394	0.411	0.412

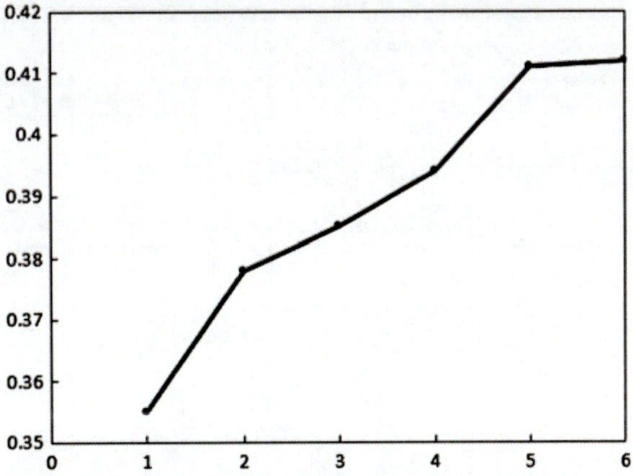

Fig. 29.3 Hydrogen outlet volume fraction under different grid numbers

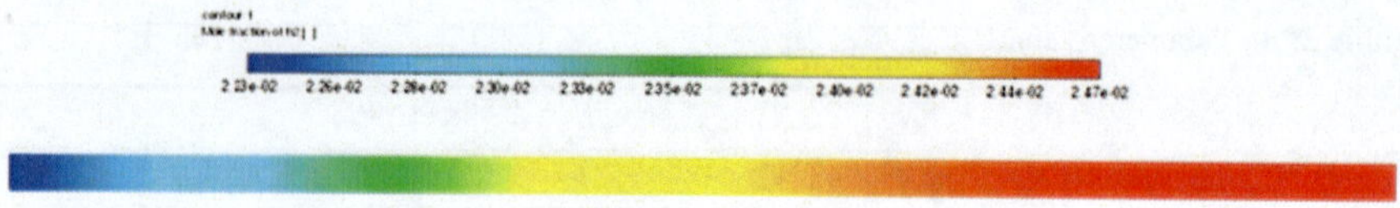

Fig. 29.4 Reaction nephogram of hydrogen

resources, this grid number is selected for subsequent calculation. The cloud diagram after calculation is shown in Fig. 29.4.

29.3 Experimental Study on Hydrogen Production by Steam Reforming

29.3.1 Design of Experimental Platform

Complete the design of diesel reforming hydrogen production experimental system, as shown in Fig. 29.5. It mainly includes two gas and liquid paths. The flow is controlled by using a high-precision constant flow pump. The preheating section is electrically heated. The temperature of the outer wall of the pipe at the end of the preheated pipe is measured by a thermocouple. The high-temperature needle valve between the preheating pipe, reactor and waste liquid tank is used to control the flow of reactants in the preheating stage. The gas generated from reforming reaction and unreacted reactants will be cooled through cooling pipes. The cooling pipes are water-cooled. Gas chromatograph is used to analyze the gas production components, and all sensor data are stored in the computer through the data acquisition box.

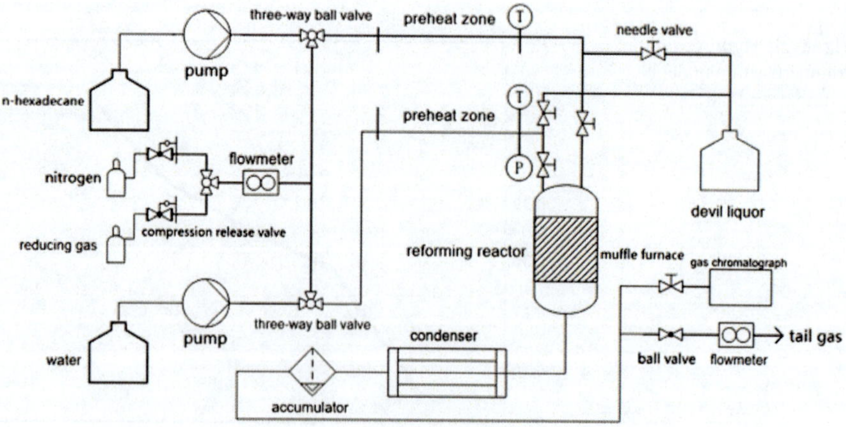

Fig. 29.5 Experimental system for hydrogen production from diesel reforming

29.3.2 Analysis of Hydrogen Production Experiment Results

Complete the construction of diesel reforming hydrogen production experimental system, as shown in Fig. 29.6. And the relevant experiments were carried out according to the finished reforming reactor. The flow rate of n-hexadecane is 2ml/min, the flow rate of water is 5ml/min, and the reaction temperature is 800°C. The selected catalyst supports are 200 mesh, 400 mesh and 600 mesh cordierite honeycomb ceramic supports.

Gas chromatograph is used to analyze the composition of produced gas, as shown in Fig. 29.7. It is the experimental data of several groups of reforming hydrogen production. It can be seen that the hydrogen production performance of 600 mesh honeycomb ceramic carrier is the best, the hydrogen production rate can reach 70%, and the carbon monoxide content is about 10%.

Fig. 29.6 DSR experimental station

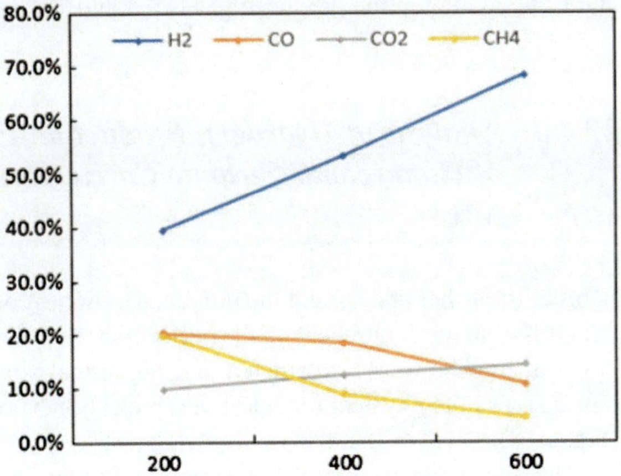

Fig. 29.7 Experimental results of composition of steam reforming products

Table 29.3 Comparison of experimental and simulation results

	H_2 (%)	CO (%)	CO_2 (%)	CH_4 (%)
Experiment	39.4	20.3	9.8	19.8
Simulation	41.1	19.7	10.8	16.1
Relative deviation	4.1	3.0	9.3	18.7

29.4 Analysis of Results Based on Numerical Simulations

29.4.1 Experimental Verification

Using the established numerical simulation model, the numerical calculation of reforming reaction was carried out under the appropriate grid density. Compared with the experimental results, the accuracy of the model is verified. Select reaction temperature T = 800 °C, S/C = 2.5 and honeycomb ceramic mesh number of 200 to compare the experimental results with the simulation results under the same working condition, so as to verify that the simulation is correct and reliable.

The comparison between the calculated results and the experimental results is shown in Table 29.3. The results show that the numerical model can simulate the process of n-hexadecane steam reforming more completely and correctly. In terms of CO and H2 content in the produced gas, the relative deviation between the experimental and simulation results is not more than 5%; The relative deviation of CO2 and CH4 content in the produced gas is 9.3 and 18.7% respectively.

29.4.2 Analysis of Hydrogen Production Performance of Honeycomb Ceramic Carriers with Different Mesh Size

In order to further understand the influence of honeycomb ceramic reaction channels with different mesh numbers on the performance of diesel steam reforming, different physical models were constructed and numerical simulation was carried out. The figure shows the physical model of 600 mesh honeycomb ceramic reaction channel (Fig. 29.8).

The simulation results of honeycomb ceramic reaction channels with different mesh numbers are shown in the following figure. It can be seen from the figure that with the increase of honeycomb ceramic carrier mesh, the volume fraction of

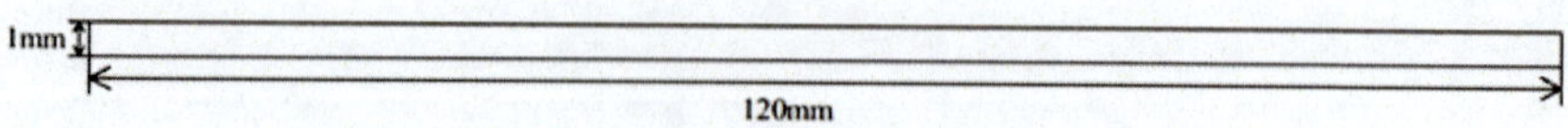

Fig. 29.8 Physical model of 600 mesh honeycomb ceramic reaction channel

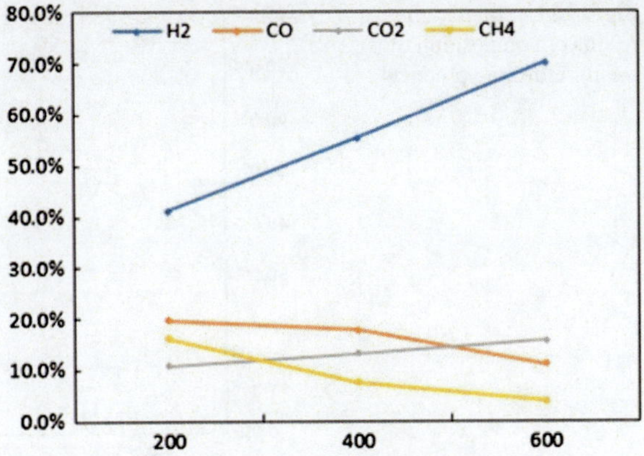

Fig. 29.9 Simulation results of composition of steam reforming products

hydrogen produced by steam reforming reaction is increasing, and the volume fraction of CO produced has a downward trend. The reason is that the larger the mesh number is, the larger the contact area between the reaction gas and the catalyst coated on the honeycomb ceramic support is, and the steam reforming reaction is more sufficient, which makes the hydrogen production effect of the reforming reaction better (Fig. 29.9).

At the same time, according to the above analysis, considering the different arrangement of honeycomb ceramic carriers with different mesh numbers, it may also affect the hydrogen production effect of steam reforming reaction. The reason is that honeycomb ceramic carriers with different mesh numbers will generate gas disturbance at the interface, and different mesh numbers will also affect the reaction adequacy. Therefore, the following calculation and analysis are made.

29.4.3 *Analysis of Hydrogen Production Performance of Honeycomb Ceramic Carriers with Different Structural Parameters*

In order to further understand the influence of different arrangement of honeycomb ceramic reaction channels on diesel steam reforming hydrogen production performance, different physical models were built and numerical simulation was carried out. The figure shows the physical model of 600 mesh and 400 mesh honeycomb ceramic reaction channel (Fig. 29.10).

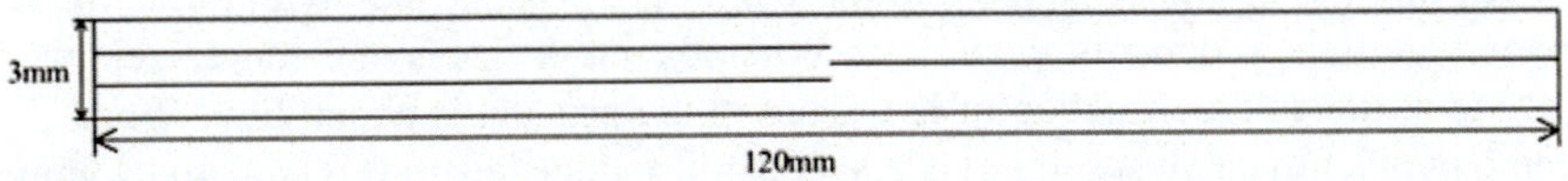

Fig. 29.10 Physical model of 600 mesh and 400 mesh honeycomb ceramic reaction channel

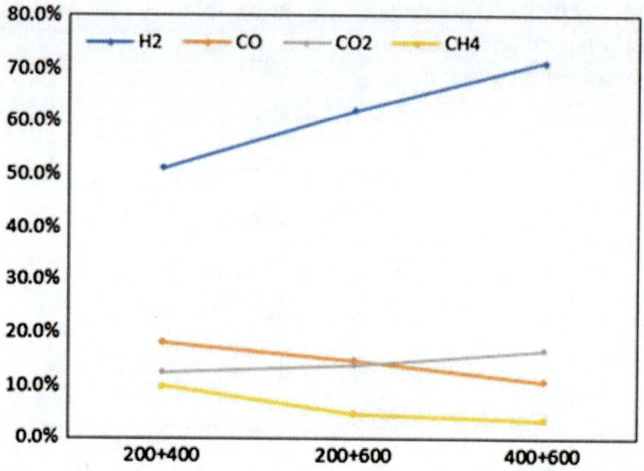

Fig. 29.11 Simulation results of composition of steam reforming products

The simulation results are shown in the figure below. It can be seen from the figure that when honeycomb ceramic carriers with mesh size of 400 and 600 are selected, the volume fraction of hydrogen in the generated gas reaches about 70%, which is roughly equal to the hydrogen fraction generated by reaction with honeycomb ceramic carriers with mesh size of 600. The reason is that the reaction channels with two different mesh numbers are arranged in front and behind, especially the arrangement of front sparse and back dense, which will allow the reactants with reduced concentration due to reaction to contact and react with the catalyst coated on the honeycomb ceramic carrier more fully in the future, thus generating more hydrogen. And since steam reforming reaction is endothermic reaction on the whole, more sufficient reaction later will further increase the heat absorption, thus making the overall temperature distribution of the reaction channel more uniform, which is conducive to the reaction and hydrogen production effect (Fig. 29.11).

29.5 Conclusion

In this paper, the numerical simulation of diesel reforming reactor was carried out using honeycomb ceramic as catalyst carrier, and the calculation model was verified by experiments to explore the influence of catalyst carrier structure parameters on the reforming effect. The results show that the FLUENT numerical simulation has a high accuracy, which is in good agreement with the experimental results, and the error is less than 20%. When the mesh number of catalyst support is 200 and the length is 120 mm, the volume fraction of hydrogen in the product is about 40%; When the mesh number is 400 and the length is 120 mm, the volume fraction of hydrogen in the product is about 55%; When the mesh size is 600 and the length is 120 mm, the volume fraction of hydrogen in the product is about 70%. When the mesh number of catalyst support is 200 and 400 and the length is 60mm, the volume fraction of hydrogen in the product is about 50%; When the mesh number is 200 and 600 and the length is 60mm, the volume fraction of hydrogen in the product is 60%; When the mesh size is 400 and 600 and the length is 60mm, the volume fraction of hydrogen in the product is about 70%. From the point of view of reducing the temperature gradient of the reactor, the reactor temperature is relatively uniform and the volume concentration of hydrogen in the product is relatively high when the reactor is arranged sparsely in the front and densely in the back. When the number of inlet mesh is 200 and the number of outlet mesh is 600, it can reach 70–72%. The numerical simulation and experimental research in this paper provide a theoretical basis for the optimization of reforming hydrogen production reaction, which has practical significance in engineering application.

References

1. Q. Wu, Analysis of several main hydrogen production technologies. IOP Conf. Ser. Earth Environ. Sci. **1011**(1) (2022)
2. L. Pingkuo, H. Xue, Comparative analysis on similarities and differences of hydrogen energy development in the World's top 4 largest economies: a novel framework. Int. J. Hydrogen Energy. **47**(16) (2022)
3. I.A. Makaryan, I.V. Sedov, The state and development prospects of the global hydrogen energy sector. Russian J. Gen. Chemi. **91**(9) (2021)
4. G. Xin, Application of hydrogen energy in low carbon cities. IOP Conf. Ser. Earth Environ. Sci. **766**(1) (2021)
5. G. Li, Y. Gou, J. Qiao, W. Sun, Z. Wang, K. Sun, Recent progress of tubular solid oxide fuel cell: From materials to applications. J. Power Sources **477** (2020)
6. Z. Pan, J. Shen, J. Wang, X. Xu, W.P. Chan, S. Liu, Y. Zhou, Y. Zilin, Z. Jiao, T.T. Lim, Z. Zhong, Thermodynamic analyses of a standalone diesel-fueled distributed power generation system based on solid oxide fuel cells. Appl. Energy **308** (2022)
7. S. Das, D. Das, A. Patra, Operation of solid oxide fuel cell based distributed generation. Energy Procedia **54** (2014)
8. J. Liu, H. Cui, L. Gao, J. Wang, M. Wang, F. Li, Research status of thermal management of hydrogen energy proton exchange membrane fuel cell. J. Phys. Conf. Ser. **2108**(1) (2021)
9. Z. Xu, Progress of proton exchange membrane fuel cell. J. Phys. Conf. Ser. **2009**(1) (2021)
10. R.M. Navarro Yerga, M.C. Álvarez-Galván, N. Mota, J.A. Villoria de la Mano, S.M. Al-Zahrani, J.L.G. Fierro, Catalysts for hydrogen production from heavy hydrocarbons. ChemCatChem **3**(3) (2011)
11. I. Dincer, C. Acar, Review and evaluation of hydrogen production methods for better sustainability. Int. J. Hydrogen Energy **40**(34) (2015)
12. W. Pan, T. Wan, Y. Han, S. Liu, J. Fu, Storage and transportation technology solutions selection for large-scale hydrogen energy utilization scenarios under the trend of carbon neutralization. IOP Conf. Ser. Earth Environ. Sci. **770**(1) (2021)
13. Q. Xue, Z. Li, Z. Jiang, M. Chen, Y. Wang, B. Yan, G. Luo, Diesel reforming to hydrogen over the mesoporous Ni–MgO catalyst synthesized in microfluidic platform. Int. J. Hydrogen Energy 46(74)
14. M. Bae, H. Cheon, J. Oh, D. Kim, J. Bae, S.P. Katikaneni, Rapid start-up strategy of 1 kWe diesel reformer by solid oxide fuel cell integration. Int. J. Hydrogen Energy **46**(52) (2021)
15. Steam diesel reforming for solid oxide fuel cell systems. ECS Meeting Abstracts (2005)
16. S.V. Zazhigalov, V.A. Shilov, V.N. Rogozhnikov, D.I. Potemkin, V.A. Sobyanin, A.N. Zagoruiko, P.V. Snytnikov, Modeling of hydrogen production by diesel reforming over Rh/Ce0.75Zr0.25O2-δ-η-Al2O3/FeCrAl wire mesh honeycomb catalytic module. Catalysis Today **378** (2020)

Chapter 30
Gasifier for Plastic Waste Using High-Temperature Superheated Steam

Junpei Maruo, M. M. A. Alam, and Osamu Yamada

Abstract In this study, gasification experiments of polyethylene (PE) samples were conducted under various conditions using superheated steam to produce gas fuel from plastic waste, and the amount of gas produced, gas composition, carbon conversion ratio, and cold gas efficiency were evaluated to clarify the trends. Moreover, a business model was proposed to further extend the research. From the experiments, it was clarified that in the gasification of PE samples under different superheated steam temperatures, the amount of gas produced decreases, the concentration of hydrocarbon components in the gasification gas increases, and the carbon conversion rate decreases with a decrease in temperature. The results suggest that the reason for this change is a decrease in the amount of steam reforming reaction due to the decrease in the temperature of the system. In addition, the effect of residence time in the system on the gasification reaction process under unsteady conditions was clarified by comparing the results of the mol ratio of carbon to water vapour (S/C) change in which the PE sample feed rate was continuously changed with the equilibrium calculations for CH_4 and C_2H_4. Finally, the equipment developed in this study has led to the possibility of developing a thermal energy recovery business for plastics waste that can reduce environmental impact, disposal costs, and energy consumption in islands and coastal areas that have problems with the disposal of drifted garbage.

30.1 Introduction

Nowadays, the modernization and expansion of the industrial economy market are advancing on a global scale, and therefore, the increase in the consumption of energy and fossil resources is remarkable. The use of fossil resources is regarded as a problem

J. Maruo · O. Yamada
OSU Company, 3-1-1 Nakagaito, Daito, Osaka, Japan

M. M. A. Alam (✉)
Department of Mechanical Engineering for Transportation, Faculty of Engineering, Osaka Sangyo University, 3-1-1 Nakagaito, Daito, Osaka, Japan
e-mail: alam@tm.osaka-sandai.ac.jp

A. Suryan et al. (eds.), *Proceedings of the 9th Asian Joint Workshop on Thermophysics and Fluid Science, 27–30 November 2022, Utsunomiya, Japan*, Springer Proceedings in Physics 396, https://doi.org/10.1007/978-981-99-9470-0_30

from the viewpoint of depletion of resources, air pollution caused by exhaust gas, and impact on global warming. Therefore, the reduction, reuse, and recycling of waste and the utilization of renewable energy (biomass, solar, wind, geothermal, etc.) are becoming increasingly active as efforts to create a sustainable, recycling-oriented society. However, the use of energy contained in the waste is limited due to some constraints such as energy conversion efficiency, equipment durability, and cost, although power generation and hot water supply using waste heat from incineration processes have been introduced.

In this research, a gasification of organic compounds by high-temperature superheated steam was carried out for the purpose of effective use of resources and development to a recycling society. Superheated steam refers to water vapor that has a temperature of over 100 °C at standard atmospheric pressure. It is a very active gas when compared to air and other gases at the same temperature, and the use of 100% superheated steam as a gasification agent, heat treatment can be performed in an almost oxygen-free state [1], makes it possible to obtain gasification gas to have superior H_2 content and calorific value compared to the partial oxidation gasification method using air and other gases. Polyethylene (PE) was used as a sample for the gasification, and the amount, composition, and carbon conversion of the gas produced by the high-temperature superheated steam gasification method were analyzed and evaluated under various experimental conditions to clarify changes in the behaviour of the gas produced. Moreover, the ways to improve system efficiency and optimize the processing conditions were investigated.

30.2 Experimental Method

A schematic of the experimental setup is shown in Fig. 30.1. Saturated water vapor at 100 °C is generated by the boiler and driven through a high-frequency induction heater. By contacting and passing through the ceramic heater pellets, which are heated to approximately 1200 °C by the high-frequency induction heating.

The heater pellets for heating the steam were made of Ni–Al porous material developed by the OSU Company, Ltd. The polyethylene (PE) samples were pyrolyzed and gasified by the superheated steam in a reaction vessel, and therefore, the generated gases were analyzed qualitatively and quantitatively for H_2, CO, CO_2, CH_4, C_2H_4, N_2 and O_2 using GAS Chromatograph GC-4000 and GC-323 manufactured by GL Sciences.

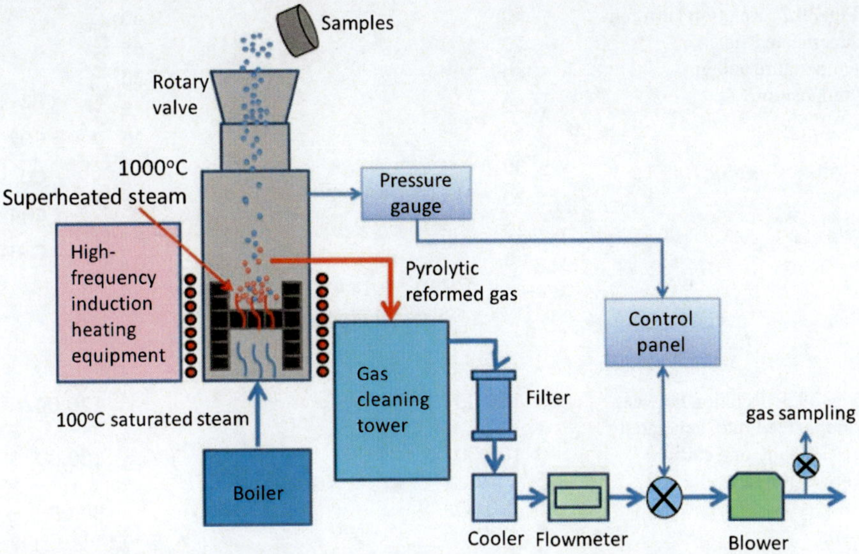

Fig. 30.1 Schematic diagram for experimental setup

30.3 Results and Discussion

30.3.1 Relationship Between Superheated Steam Temperature and Composition of Generated Gas

Figure 30.2 shows the relationship between superheated steam temperature and the composition of generated gases. Here, the measured gas compositions are H_2, CO_2, CO, methane, ethylene, oxygen and nitrogen. Oxygen and nitrogen were measured for the purpose of detecting equipment anomalies. As seen from the figure, as the superheated steam temperature increases, the volume percentages of generated H_2 and CO are increased, while the percentages of those of CO_2, CH_4, and C_2H_4 generated are decreased. Moreover, the carbon conversion rate, which is the weight ratio when the carbon in the raw material is converted to gas, and the amount of gas shows a tendency to decrease as the temperature decreased.

30.3.2 Relationship Between Sample Feed Rate, Generated Gas Amount, and Carbon Conversion Rate

Figure 30.3 shows the relationship between the amount of sample supplied, the amount of generated gas, and the carbon conversion rate. In doing this study, the sample feed rate was varied in a range of 18 g/min $\leq$ sample $\leq$ 36 g/min with an

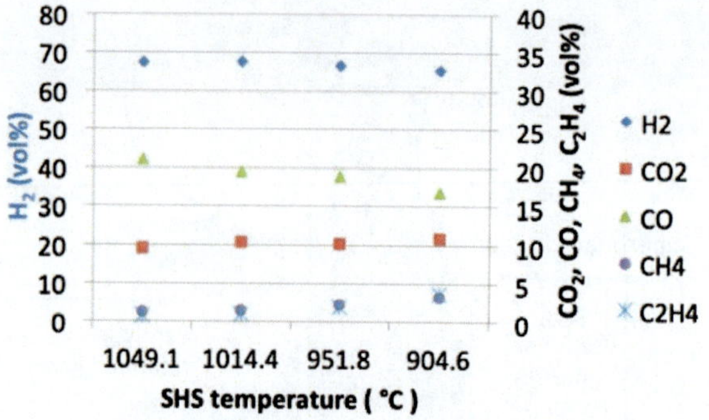

Fig. 30.2 Relation between superheated steam temperature and gas composition

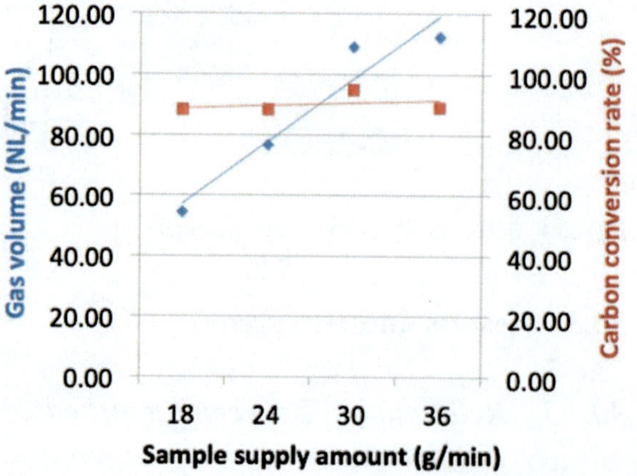

Fig. 30.3 Relation between sample feed rate, generated gas amount, and carbon conversion rate

increment of 6 g/min. Here, it should be mentioned that at sample supply amount of 18 g/min the superheated steam temperature was about 850 °C; the high-frequency induction heating output was not sufficiently maintained. While for sample feed rate of 30 g/min and 36 g/min, the heater was able to maintain the superheated steam temperatures around 1000 °C. As can be seen from the figure, the carbon conversion rate remains almost the same, however, the amount of generated gas is increased as the amount of sample supplied increases.

30.3.3 Relationship Between S/C and Composition of Generated Gas

The effect of residence time in the system on the gasification reaction process under unsteady conditions was clarified by comparing the results of varying mol ratio of water vapor to carbon (S/C), and the result for the relationship between S/C and generated gas composition is shown in Fig. 30.4. Here, it is mentioned that the sample

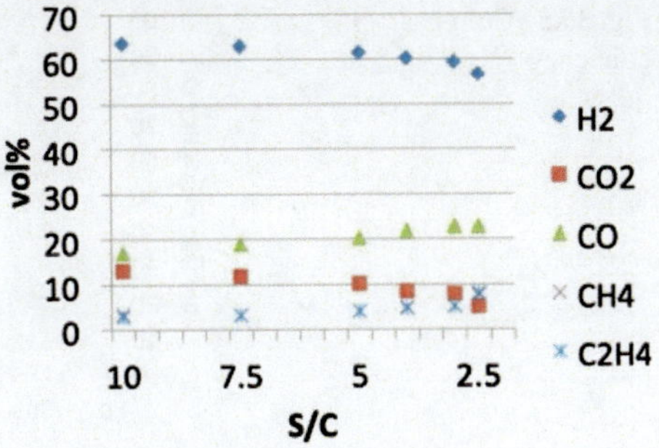

Fig. 30.4 Relationship between S/C and generated gas composition

supply amount was changed at 30 min intervals, and the S/C was varied like 10, 7.5, 5, 4, 3, 2.5 respectively. From the figure, as the S/C increases, the proportions of H_2 and CO_2 are increased, while the proportions of CO, CH_4 and C_2H_4 are decreased.

30.3.4 Generated Gas Energy

The generated gas energy in the long-time experiment (sample supply = 12 g/min, supply time = 180 min, S/C = 2.5, SHS temp = 1000 °C) is shown in Table 1. As a result of the steam reforming reaction, the gasification gas potential energy increased by 3.4% to 9.519 on average against the sample potential energy of 9.209 kW.

Figure 30.5 shows the cold gas efficiency in the the long-time experiment time which is calculated by this formula:

$$CDE = \frac{Generated\ Gas\ Energy}{Sample\ Material + Induction\ Heating\ Energy} \tag{30.1}$$

From the figure, it can be seen that the cold gas efficiency in long-term operation experiments is about 55% when the efficiency of the steam generator is taken into consideration, while the efficiency appears to be 60% under the condition that the steam generator is excluded assuming the use of waste heat. Here, it should be mentioned that in the experiments varying the for S/C (Steam/Carbon mol ratio) with a larger amount of sample supply than 12 g/min, the efficiency obtained is 68%, while the efficiency reaches to 77% if the steam generator is excluded.

Table 1 Generated gas energy at each elapsed time

Elapsed time (min)	30	45	75	105	135	165	195
Average gas volume (NL/min)	49.09	51.09	48.29	47.69	48.75	47.72	48.08
Gas calorific value (kJ/Nm^3)	11,852	11,643	11,590	11,705	11,589	11,817	11,946
Gas energy (kW)	9.697	9.915	9.328	9.303	9.417	9.399	9.573

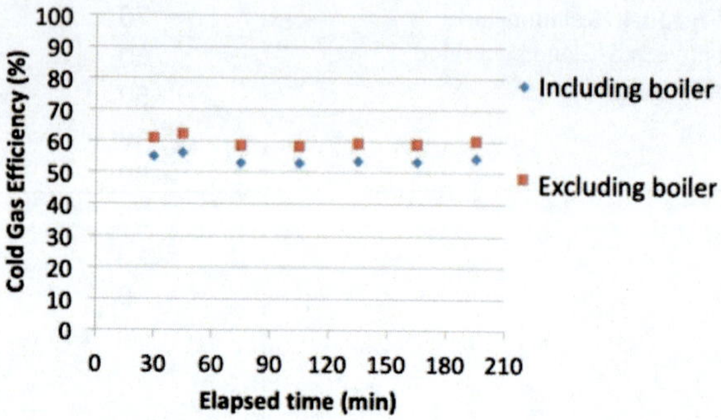

Fig. 30.5 Cold gas efficiency

30.4 Conclusions and Future Work

In the experimental work, it was clarified that, in the gasification of polyethylene (PE) samples under different superheated steam temperatures, the amount of gas produced decreases, while the concentration of hydrocarbon components in the gasified gas increases, and the carbon conversion rate decreases with a decrease in temperature. The results suggest that the reason for this change is a decrease in the amount of steam reforming reaction due to the decrease in the temperature of the system. In addition, the effect of residence time in the system on the gasification reaction process under unsteady conditions was clarified by comparing the results of the mol ratio of water vapour to carbon (S/C) change in which the PE sample feed rate was continuously changed with the equilibrium calculations for CH_4 and C_2H_4. In the future, data should be collected on other plastic samples (PP, PS, PET, etc.) as well, with the ultimate goal of developing a device capable of gasifying waste plastics mixed with various plastics, and utilizing the gas produced as renewable energy.

Reference

1. K. Suzuki, et al., *Integrated Super Heated Steam Technology*. (NTS Co., Ltd., 2005), p. 33 (in Japanese)

Chapter 31
Energy Efficient Drive for a Single-Phase Induction Motor Based on THD

A. Anuroopa, Sujatha Therese, and K. S. Biju

Abstract Due to high reliability and easy availability of single phase supply, single phase induction motors (SPIM) are very popular for commercial as well as industrial applications. Here a capacitor start capacitor run SPIM is considered. The efficiency of the motor can be improve by reducing the harmonic content present at the input supply. Variable speed operation of SPIM can be achieved by driving the motor with an adjustable speed drive. Among the different existing driving topologies, three important topologies based on pulse width modulation (PWM) methods are identified and Simulink model for the entire system is developed. For measuring the effect of harmonics on the performance of the motor and for selecting a suitable energy efficient drive for SPIM, total harmonic distortion (THD) of the different topologies are plotted and compared. The value of THD is minimum for an energy efficient drive. From the analysis, single phase PWM inverter with half bridge rectifier drives is having minimum THD of 38.11% and is identified as an energy efficient drive for the induction motors.

A. Anuroopa (✉)
Department of Electrical and Electronics Engineering, ACE College of Engineering, Thiruvananthapuram 695035, India
e-mail: anuroopa.a@acetvm.com

A. Anuroopa · S. Therese
Department of Electrical and Electronics Engineering, Noorul Islam Centre for Higher Education, Kanyakumari 629180, India

K. S. Biju (✉)
Department of Electronics and Communication Engineering, College of Engineering Trivandrum Thiruvananthapuram, Kerala 695016, India
e-mail: bijukarunnya@cet.ac.in

A. Suryan et al. (eds.), *Proceedings of the 9th Asian Joint Workshop on Thermophysics and Fluid Science, 27–30 November 2022, Utsunomiya, Japan*, Springer Proceedings in Physics 396, https://doi.org/10.1007/978-981-99-9470-0_31

31.1 Introduction

The induction motor is a commonly used in AC electrical machine. This type of machines have wide applications in commercial as well as industrial fields. Due to the easy availability of one phase supply one phase induction motors are most suitable for residential appliances [1], farming appliances etc. These motors are economical as compared to other AC machines. The construction and maintenance of these machines is simple and easy. The working and construction of a one phase induction motor is almost similar to a three phase induction motor. Capacitor one phase induction motors are most widely used for many of the applications [2]. SPIM should have one operating winding and a starting winding [3].

Speed of single phase induction machines can be changed by the use of an adjustable single phase motor drive. Among the different types of PWM strategies sinusoidal PWM technique is the most widely used technique. By using a proper adjustable speed drive, a considerable amount of energy can be saved and the performance of the induction motor can be improved [4]. One phase induction motors are more sensitive to the harmonics present in the input supply [5]. The effect of harmonics on the performance of the one phase induction motor will depend upon the amount of harmonics present at the input of the Induction motor [6]. In this paper, three different adjustable speed one phase induction motor drives along with SPIM have been simulated. The performance characteristics of SPIM, when these three drives are connected and are compared on the basis of total harmonic distortion.

31.2 Types of One Phase Induction Motor Drives

The efficiency of the SPIM is less when it is supplied with constant voltage and constant frequency inputs [7]. Variable speed drives for supplying input is a better alternative to improve the efficiency of the motor. Variable speed drives are very economical due to the reduction in the cost of power electronic devices [8].

31.2.1 One Phase AC-AC Chopper Drives

Figure 31.1 shows a one phase AC/AC chopper drive circuit. This circuit uses four distinct IGBTS connected as a bridge circuit to control a SIM and it is a PWM controlled AC converter [9]. Costly DC bus capacitors are not required in this circuit. To suppress the utility line spikes a small AC capacitor is included in the circuit. Four step switching strategy is used in this circuit.

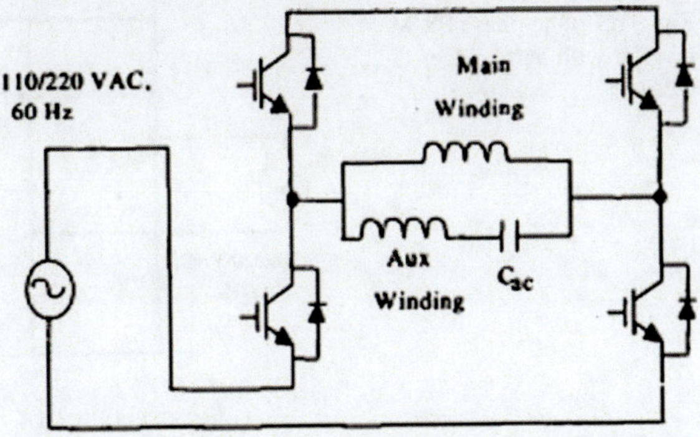

Fig. 31.1 One phase AC/AC chopper drive

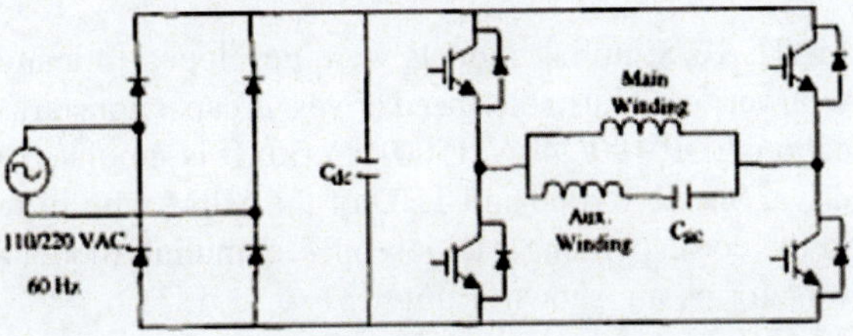

Fig. 31.2 Single phase PWM inverter with full bridge rectifier

31.2.2 One Phase PWM Inverter with Full Bridge Rectifier Drives

Single phase PWM inverter with full bridge rectifier drive circuit is shown in Fig. 31.2. Compared with other driving circuits this is an easy topology to control the motor voltage. Since the PWM inverters have constant amplitude pulses, pulse width can be changed for controlling inverters output voltage [10]. This is also helpful for reducing the distorting harmonics present in the signal.

31.2.3 Single Phase PWM Inverter with Half Bridge Rectifier Drives

Circuit of one phase PWM inverter with half bridge rectifier drive circuit is shown in Fig. 31.3. In this topology a half bridge controlled rectifier is used. In addition to the IGBTs a capacitor divisor is also connected.

Fig. 31.3 One phase PWM inverter with half bridge rectifier

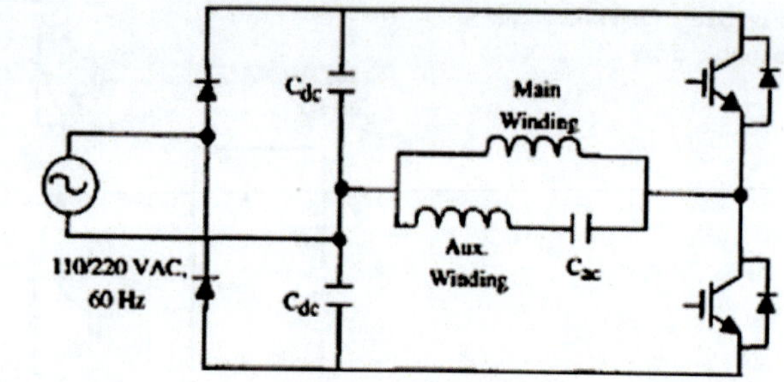

31.3 Results

MATLAB/Simulink models were developed to analyze the performance of SPIM with various adjustable speed drives. A capacitor start, capacitor run SPIM of specifications 2HP, 190/240 V, 1500 rpm, 50 Hz is simulated. Output of the adjustable speed drive, Speed, torque and THD of the SPIM with different topologies are displayed in the corresponding oscilloscopes. Simulink model and the respective Simulating waveforms are shown in Figs. 31.4, 31.5, 31.6, 31.7, 31.8, 31.9, 31.10, 31.11 and 31.12 Table 31.1.

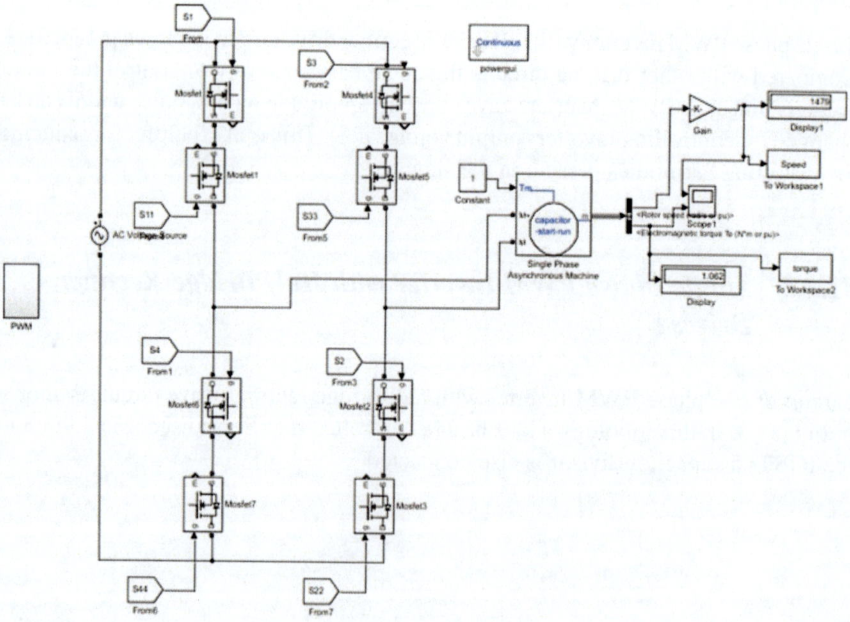

Fig. 31.4 Simulink model of one phase AC/AC chopper drive

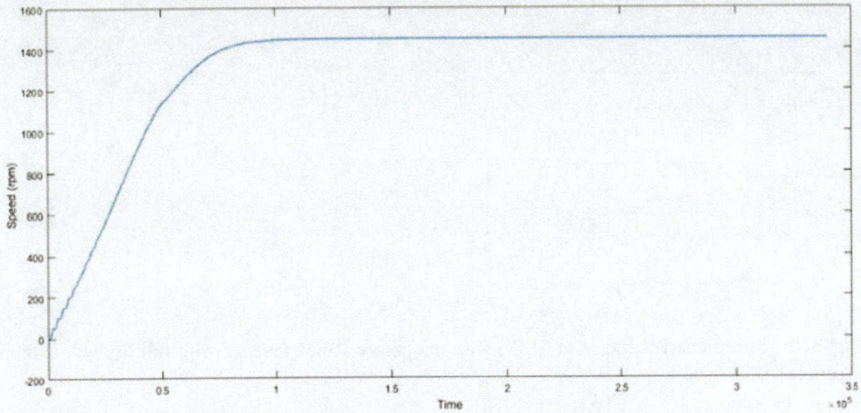

Fig. 31.5 Speed characteristics of SPIM with one phase AC/AC chopper drive

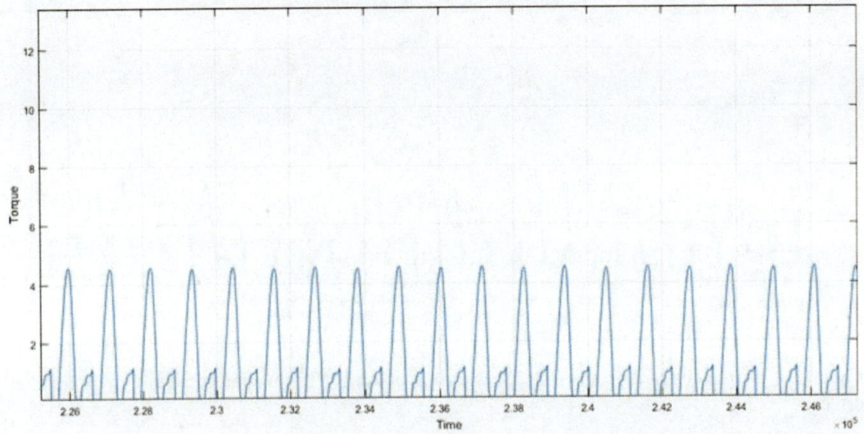

Fig. 31.6 Torque characteristics of SPIM with one phase AC/AC chopper drive

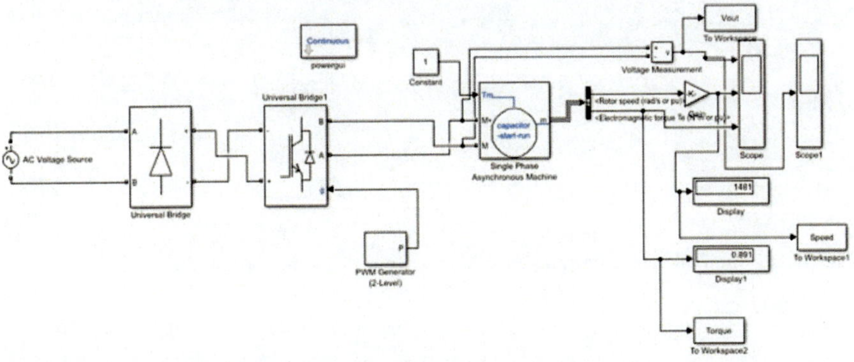

Fig. 31.7 Simulink model of single phase PWM inverter with full bridge rectifier

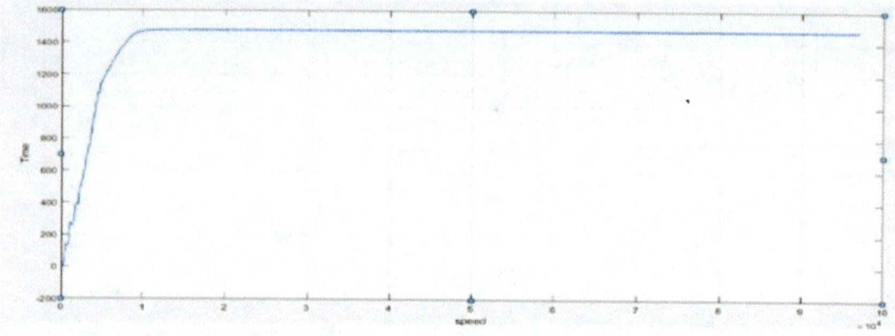

Fig. 31.8 Speed characteristics of SPIM with one phase PWM inverter with full bridge rectifier drive

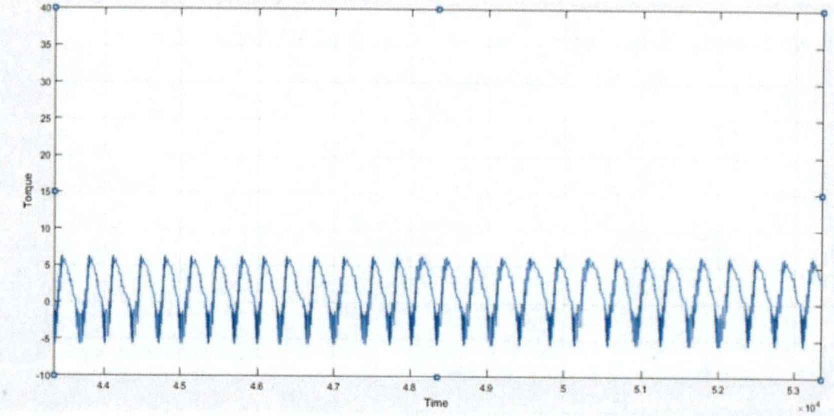

Fig. 31.9 Torque characteristics of SPIM with one phase PWM inverter with full bridge rectifier drive

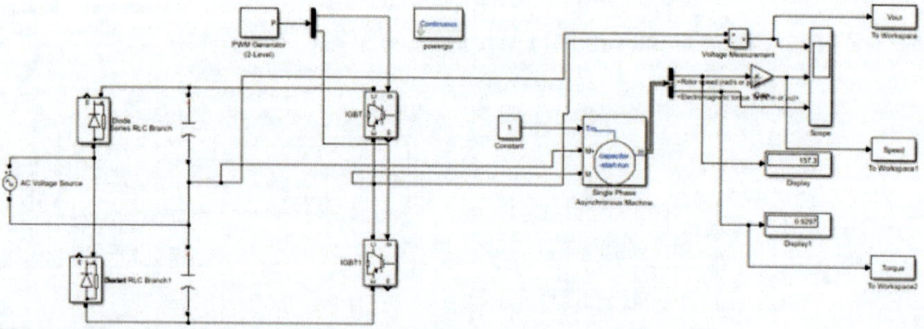

Fig. 31.10 Simulink model one phase PWM inverter with half bridge rectifier

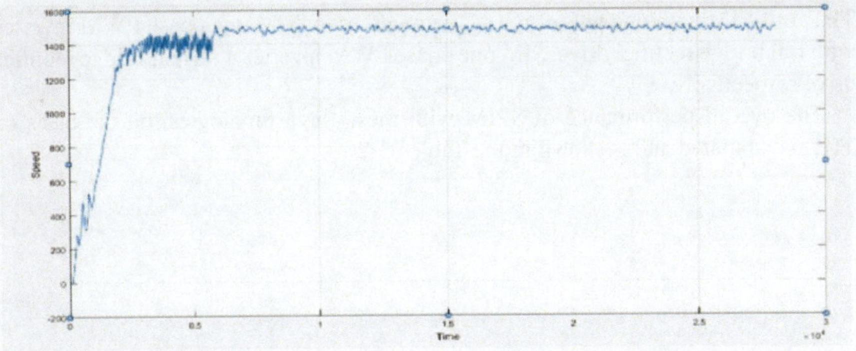

Fig. 31.11 Speed characteristics of SPIM with one phase PWM inverter with half bridge rectifier drives

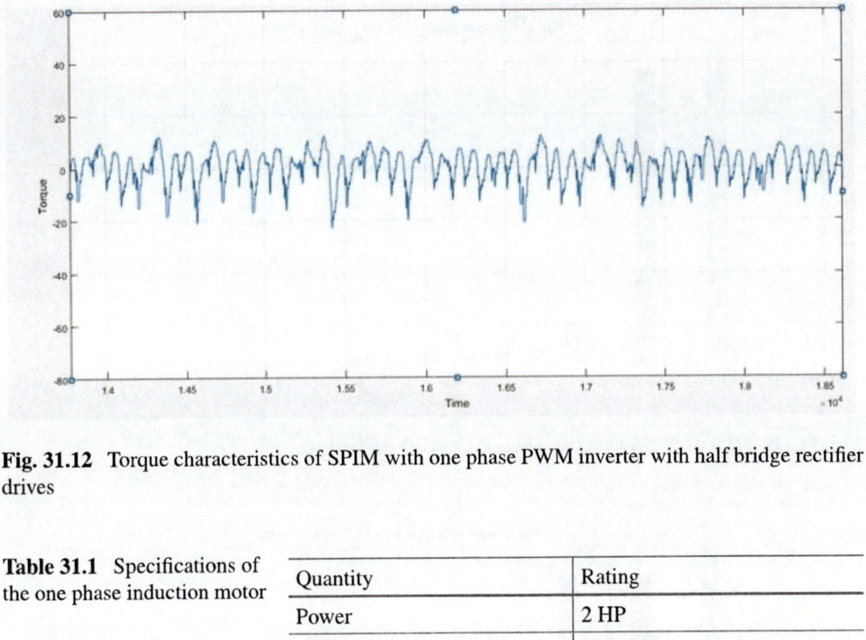

Fig. 31.12 Torque characteristics of SPIM with one phase PWM inverter with half bridge rectifier drives

Table 31.1 Specifications of the one phase induction motor

Quantity	Rating
Power	2 HP
Speed	1500 rpm
Current	10.5 A
Voltage	240 V

The amount of harmonics present in a signal can be represented in the form of THD. THD is the ratio of total powers of all harmonic elements to the power corresponding to the fundamental component. THD is expressed in percentage. Value of THD is small when the harmonic content is less. Simultaneously the amount of core loss and heating will be small. The Figs. 31.13, 31.14 and 31.15 are the different

THD of SPIM with one phase AC/AC chopper drive, SIM one phase PWM inverter with full bridge rectifier drive, SIM one phase PWM inverter with half bridge rectifier drive respectively.

The overall performance of SPIM with these three topologies, on the basis of THD is tabulated and is shown in the Table 31.2.

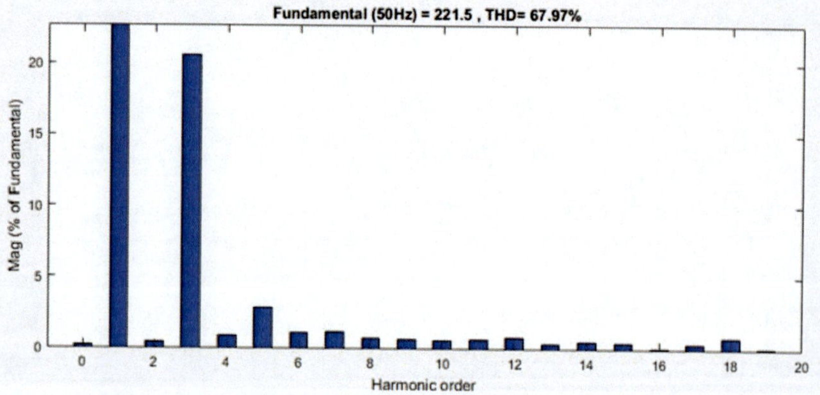

Fig. 31.13 THD of SPIM with one phase AC/AC chopper drive

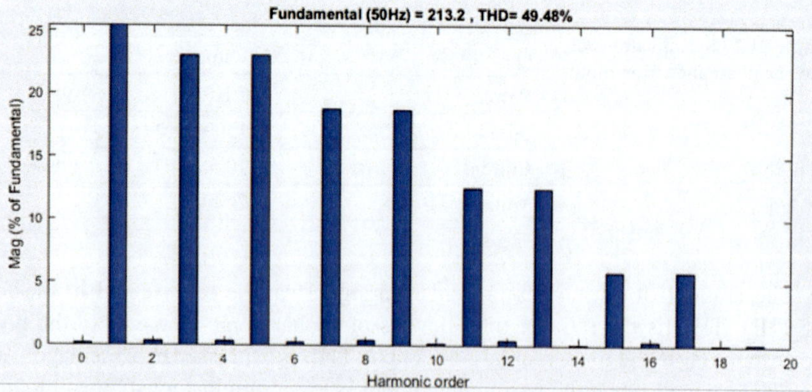

Fig. 31.14 THD of SPIM one phase PWM inverter with full bridge rectifier drive

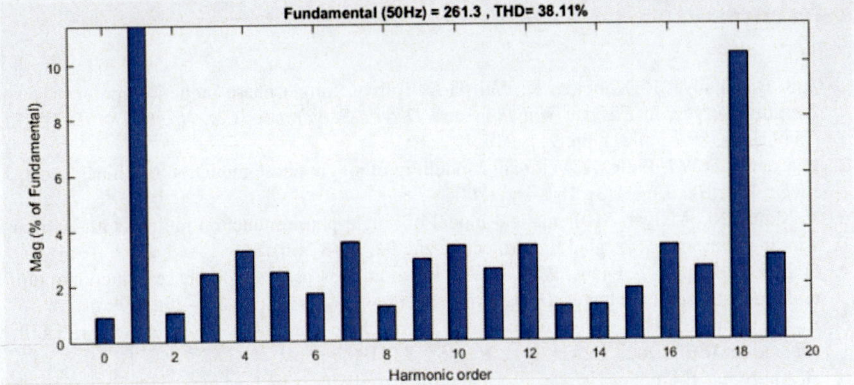

Fig. 31.15 THD of SPIM one phase PWM inverter with half bridge rectifier drive

Table 31.2 Comparison of performance characteristics of SPIM with different adjustable speed drives

Type of DRIVE	Speed (rpm)	Torque (Nm)	THD (%)	Efficiency of SIM (%)
One phase AC-AC chopper drives	1450	5	67.97	51.29
One phase PWM inverter with full bridge rectifier drives	1450	6	49.48	61.53
One phase PWM inverter with half bridge rectifier drives	1450	10	**38.11**	**98.95**

31.4 Conclusion

SPIM can be successfully operated for multispeed operations by a variable speed drive. The amount of harmonics present at the input supply can be reduced by selecting a suitable variable speed drive. In this paper three topologies for controlling the speed of SPIM are presented and the simulated results show the performance of SPIM with THD. The THD is the amount of harmonics present in a signal. It can be measured by the ratio of all harmonics components to the power corresponding to the fundamental component. The amount of harmonics present at the input supply can be reduced by selecting a suitable energy efficient variable speed drive. The value of THD is minimum for an energy efficient drive. From the analysis, one phase PWM inverter with half bridge rectifier drives has a minimum THD of 38.11% and efficiency of SIM is improved to 98.95%.

References

1. A.S. Ba-Thunya, R. Khopkar, K. Wei, H.A. Toliyat, Single phase induction motor drives-a literature survey, in *Electric Machines and Drives Conference (Cat. No. 01EX485)* (IEEE, Cambridge, USA, 2001), pp. 911–916
2. C. Coates, D.W.J. Pulle, A. Veltman, Modelling of single phase induction machines. Aust. J. Electr. Electron. Eng. **1**(3), 151–156 (2004)
3. V. Malyar, A. Malyar, Mathematical model of single-phase induction motors with auxiliary winding resistance. Przegląd Elektrotechniczny **94**, 33–37 (2018)
4. D. Lin, T. Batan, E.F. Fuchs, W.M. Grady, Harmonic losses of single-phase induction motors under nonsinusoidal voltages. IEEE Trans. Energy Convers. **11**(2), 273–286 (1996)
5. R. Billinton, E. Khan, A security based approach to composite power system reliability evaluation. IEEE Trans. Power Syst. **7**(1), 65–72 (1992)
6. A. Anuroopa, P. Sujatha Therese, K.S. Biju, Improved exergy in three phase induction motor by eliminating harmonics using wavelet packet transform, in *Energy and Exergy for Sustainable and Clean Environment*, vol. 1, ed. by V. Edwin Geo, F. Aloui. (Green Energy and Technology, Springer, Singapore, 2022), pp. 383–390
7. B. Zahedi, S. Vaez-Zadeh, Efficiency optimization control of single-phase induction motor drives. IEEE Trans. Power Electron. **24**(4), 1062–1070 (2009)
8. K. MakowskI, M.J. Wilk, Determination of dynamic characteristics of the single-phase capacitor induction motor. Prezglad Electrotechniczny **87**(5), 231–237 (2011)
9. L.J. Ganesan, D. Jeyadevi, D.E. Selvaraj, Harmonic analysis on single phase induction motor coated with Al_2O3 nano filler mixed enamel. Mater. Sci. Phys. (2014)
10. M. Popescu, C.B. Rasmussen, T.J.E. Miller, M. McGilp, Effect of MMF harmonics on single-phase induction motor performance-a unified approach, in *2007 IEEE Industry Applications Annual Meeting* (IEEE, New Orleans, USA, 2007), pp.1164–1170

Chapter 32
Aerospike Nozzles for Reusable Launch Vehicles: Prospects and Possibilities

Sandeep Soman, Antonius Luy, and Abhilash Suryan

Abstract When striving for frequent and cost-effective space transport, the paramount consideration is the reusability of various launch vehicle stages. In this context, aerospike engines offer a superior solution compared to conventional engines due to their innate capacity for continuous altitude adjustment. This paper delves into several aspects related to aerospike nozzles, including the comparative analysis of flow fields within linear and truncated plug configurations, the influence of base bleed mechanisms, and the utilization of aerospike engines in retro-propulsion scenarios. The examination encompasses both axi-symmetric and linear aerospike nozzle designs.

32.1 Introduction

Reusability is a pivotal research area in modern launch vehicles, profoundly influencing space transport economics and mission frequency. Ensuring both safe re-entry and landing, while achieving the desired orbit, is imperative for subsequent reuse with minimal modifications. Aerospike engines, known for their remarkable altitude compensation capability, stand out in this context. Unlike conventional bell nozzles, which experience around a 15% thrust loss due to over-expansion and under-expansion [1], Altitude Compensating Nozzles (ACNs), such as Aerospikes [1–3], Expansion Deflection nozzles [1, 4], Dual Bell nozzles [1, 5] and Double Divergent nozzles [1, 6–8] dynamically adjust their area ratio to adapt to changing altitude conditions. Among the various ACN concepts, Aerospike engines have garnered significant attention from both academia and industry. Notably, practical aerospike

S. Soman · A. Suryan (✉)
College of Engineering Trivandrum, Thiruvananthapuram, Kerala 695016, India
e-mail: suryan@cet.ac.in

A. Luy
Intern, College of Engineering Trivandrum, Thiruvananthapuram, Kerala 695016, India

Karlsruhe Institute of Technology, Karlsruhe, Germany

A. Suryan et al. (eds.), *Proceedings of the 9th Asian Joint Workshop on Thermophysics and Fluid Science, 27–30 November 2022, Utsunomiya, Japan*, Springer Proceedings in Physics 396, https://doi.org/10.1007/978-981-99-9470-0_32

engines are currently in development by pioneering companies such as Stoke Space [9] and Pangea Aerospace [10].

The concept of an aerospike was originally patented in 1954 by Arnold Griffith [11]. It swiftly attracted interest for aircraft applications and was studied by NACA [12]. Subsequently, its application to rocket propulsion was proposed, and contour design methods were introduced by Rao [13] and Angelino [14]. In the 1990s, three major space agencies—NASA [15], ESA [16–18], and NAL Japan [19–21] intensively explored aerospike technology, revealing critical insights into aspects like flow-field behavior, shock dynamics, and the benefits of base bleed in enhancing thrust. Neither of these studies resulted in a practical engine. However, academic interest in aerospike engines continued to rise, and with advancements in computational and experimental methods, more specialized studies emerged from different parts of the world. These studies notably concentrated on areas such as the flow-field associated with clustering [22, 23], side wall placement [24], acoustics [25–27], free stream effects [28, 29], unsteadiness [23, 30] and base bleed effects [28, 31, 32].

This study investigates the potential and opportunities for implementing aerospike engines in reusable launch vehicles, analyzing various flow-field dynamics and performance variables.

32.2 Flow-Field in Full Length and Truncated Plugs

Aerospikes can be broadly categorized into two types: axi-symmetric and linear aerospikes. These nozzles can be further designed in either full-length or truncated plug configurations. Full-length contours are primarily optimized for thrust, while a truncated contour results when a portion of the full-length contour is removed for various practical considerations.

In the case of both full-length and truncated plugs, whether they are axi-symmetric or linear in design, a high degree of over-expansion (low pressure ratio) tends to favor the occurrence of a Mach reflection of over-expansion shocks (Fig. 1a). As the pressure ratio increases, this phenomenon transitions into a regular reflection (Fig. 1b), marking the presence of two distinct flow regimes within aerospike nozzles. Downstream of the over-expansion shock, reattachment occurs after the recirculation region. However, once the bubble has advanced significantly downstream, there is no longer reattachment along the plug surface (Fig. 1c), giving rise to what can be described as a third flow regime in aerospike nozzles [28–32].

As depicted in Fig. 32.2, in the case of a truncated plug, the location where the over-expansion shock from primary nozzle's tip impinges plays a pivotal role in shaping the flow characteristics within the truncated plug. When the point of impingement is within the subsonic region behind the base, this flow type is referred to as the 'open wake mode' (Fig. 2a). Conversely, if the point of impingement occurs downstream of the subsonic region, the flow field is categorized as the 'closed wake mode' (Fig. 2b). In their study, Sandeep et al. documented flow-field unsteadiness in a full-length plug nozzle within a narrow range of pressure ratios. This unsteadiness

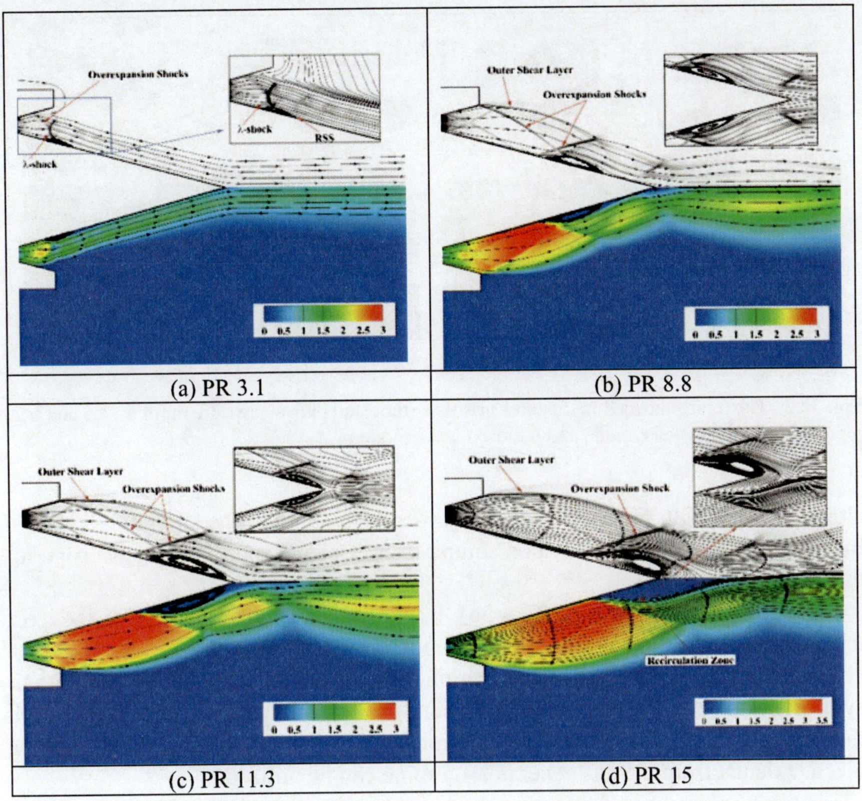

Fig. 32.1 The figure shows the evolution of flow-field with increase in Pressure Ratios (PR) **a** 3.1, **b** 8.8, **c** 11.3 and **d** 15 [30]

stemmed from the interaction between recirculation bubbles on both sides of the plug surfaces, resulting in fluctuations in the separation shock. Additionally, they found that by strategically truncating the nozzle to an optimal length, this unsteadiness could be mitigated, ultimately achieving stable operation.

32.3 Effects of Base Bleed on Improving the Thrust

As previously mentioned truncation is a necessary design choice driven by factors such as mitigating unsteadiness and addressing considerations like weight and friction. While there is a minor loss in thrust associated with truncation, this lost thrust can be effectively recovered through the implementation of base bleed at the truncated base. Figure 32.3 illustrates a comparison of thrust coefficients for both axi-symmetric and linear plug nozzles when base bleed is employed. In the case of the axi-symmetric aerospike Fig. 3a, base bleed is implemented at two different locations

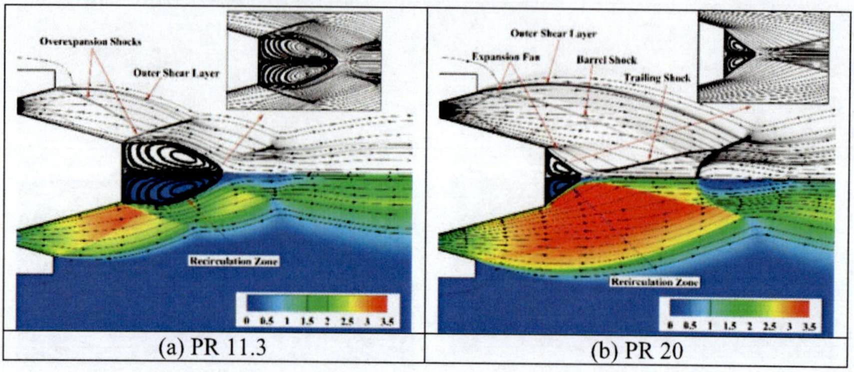

Fig. 32.2 The figure shows flow-features in a 50% truncated plug at pressure ratios **a** 11.3 and **b** 20 representing **a** open wake mode and **b** closed wake mode [30]

along a 20% length plug: one near the axis of the plug ('Case 1') and the other at the tip of the plug ('Case 2'). Upon comparison, 'Case 2' appears to yield a superior thrust coefficient.

For the linear plug design (Fig. 3b), base bleed is applied away from the axis, towards the plug's tip, covering a 40% length of the plug, resulting in a higher thrust coefficient compared to the full-length plug configuration. Therefore, in both axi-symmetric and linear plug scenarios, the configurations employing base bleed exhibit thrust coefficients even greater than their full-length counterparts. To implement base bleed, exhaust from the gas generator's turbine can be utilized.

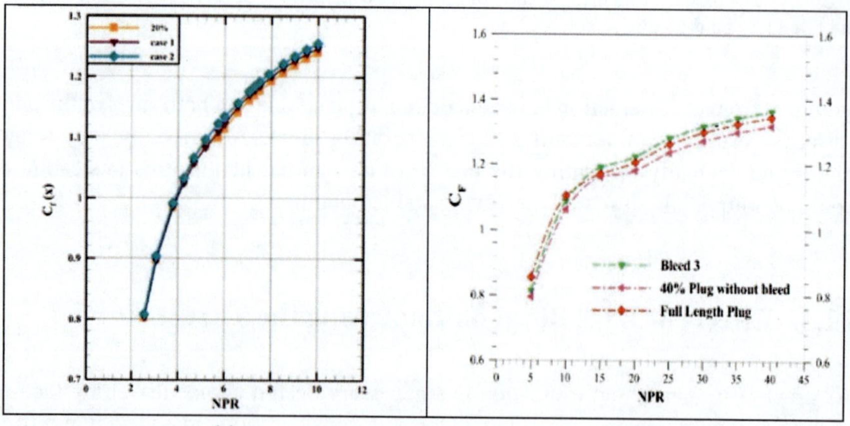

(a) (b)

Fig. 32.3 The figure shows a comparison of thrust coefficient (C_F) with change in Nozzle Pressure Ratio (NPR) for an **a** Axi-symmetric [28] and **b** linear plug nozzle with base bleed [31]

32.4 Aerospike Engines During Retro-Propulsion

In the context of reusable launch vehicle applications, the study of aerospike engines extends to their performance during the descent phase. This scenario essentially involves the interaction of supersonic exhaust expelled from the nozzle with a counterflowing jet. Early insights into this challenge were provided by Finley et al. [33]. Scarlatella et al. [34] conducted a comparative analysis of various altitude compensation nozzles (ACNs) within a retro propulsion scenario, presenting preliminary findings. The total drag experienced by the vehicle results from two key contributors: the effective thrust coefficient arising from this exhaust-jet interaction and the aerodynamic drag. They also underscore the significance of the drag coefficient relative to the (d/D) ratio, where 'd' represents the distance between the free stagnation point and the vehicle's base plate to which it is attached, and 'D' denotes the overall diameter of the base plate (or the vehicle).

32.5 Conclusion

This paper explores multiple facets of aerospike nozzles in the context of their utility in reusable launch vehicle applications. Full-length aerospike nozzles exhibit inherent instability within a specific range of pressure ratios, a problem effectively mitigated through truncation. The slight reduction in thrust resulting from truncation can be offset through the implementation of base bleed. This observation holds true for both axi-symmetric and linear aerospike configurations, each finding distinct applications based on their design characteristics. Additionally, the paper delves into fundamental aspects of aerospike engine performance during retro-propulsion scenarios.

References

1. G. Hagemann, H. Immich, T. Van Nguyen, G.E. Dumnov, Advanced rocket nozzles. J. Propul. Power **14**(5), 620–634 (1998)
2. J. Ruf, P. McConaughey, J. Ruf, P. McConaughey, The plume physics behind aerospike nozzle altitude compensation and slipstream effect, in *33rd Joint Propulsion Conference and Exhibit* (1997), p. 3218
3. P.P. Nair, A. Suryan, H.D. Kim, Computational study of performance characteristics for truncated conical aerospike nozzles. J. Thermal Sci. **26**, 483–489 (2017)
4. J. Paul P, P.P. Nair, A. Suryan, M.J.P. Martin, H. Dong Kim, Numerical simulation on optimization of pintle base shape in planar expansion-deflection nozzles. J. Spacecraft Rock. **57**(3), 539–548 (2020)
5. S.P. Choudhury, A. Suryan, J.C. Pisharady, A. Jayashree, K. Rashid, Parametric study of supersonic film cooling in dual bell nozzle for an experimental air–kerosene engine. Aerosp. Sci. Technol. **78**, 364–376 (2018)

6. P.P. Nair, A. Suryan, H.D. Kim, Computational study on reducing flow asymmetry in over-expanded planar nozzle by incorporating double divergence. Aerosp. Sci. Technol. **100**, 105790 (2020)
7. J. George, P.P. Nair, S. Soman, A. Suryan, H.D. Kim, Visualization of flow through planar double divergent nozzles by computational method. J. Visual. **24**, 711–732 (2021)
8. S. Soman, J. George, P.P. Nair, A. Suryan, Numerical study of flow through planar double divergent nozzles, in *AIP Conference Proceedings*, vol. 2134(1), (AIP Publishing, 2019)
9. Stoke Space Homepage, https://www.stokespace.com. Accessed 09 Aug 2023
10. Pangea Aerospace Homepage, https://pangeaaerospace.com. Accessed 09 Aug 2023
11. G.A. Arnold, P.L.C. Rolls Royce, Jet-propulsion nozzle for use at supersonic jet velocities. U.S. Patent 2,683,962 (1954)
12. H.G. Krull, W.T. Beale, Effect of plug design on performance characteristics of convergent-plug exhaust nozzles (No. NACA-RM-E54H05) (1954)
13. G.V.R. Rao, Spike nozzle contour for optimum thrust. Planet. Space Sci. **4**, 92–101 (1961)
14. G. Angelino, Approximate method for plug nozzle design. AIAA J. **2**(10), 1834–1835 (1964)
15. R.A. Thompson, Review of X-33 hypersonic aerodynamic and aerothermodynamic development, in International Congress of Aeronautical Science, Harrogate, United Kingdom (2000)
16. H. Immich, M. Caporicci, H. Immich, M. Caporicci, Status of the FESTIP rocket propulsion technology programme. In: *33rd joint propulsion conference and exhibit* (1977), p. 3311
17. H. Immich, M. Caporicci, FESTIP technology developments in liquid rocket propulsion for reusable launch vehicles, in *32nd joint propulsion conference and exhibit* (1997), p. 3113
18. M. Onofri, M. Calabro, G. Hagemann, H. Immich, P. Sacher, F. Nasuti, P. Reijasse, Plug nozzles: summary of flow features and engine performance-overview of RTO/AVT WG 10 subgroup 1, in *40th AIAA Aerospace Sciences Meeting & Exhibit* (2002), p. 584
19. T. Tomita, H. Tamura, M. Takahashi, An experimental evaluation of plug nozzle flow field, in *32nd Joint Propulsion Conference and Exhibit* (1996), p. 2632
20. T. Tomita, M. Takahashi, H. Tamura, T. Tomita, M. Takahashi, H. Tamura, Flow field of clustered plug nozzles, in *33rd Joint Propulsion Conference and Exhibit* (1997), p. 3219
21. T. Tomita, M. Takahashi, T. Onodera, H. Tamura, Visualization of shock wave interaction on the surface of aerospike nozzles, in *34th AIAA/ASME/SAE/ASEE Joint Propulsion Conference and Exhibit* (1998), p. 3523
22. K. Chutkey, B. Vasudevan, N. Balakrishnan, Flow investigation of linear cluster plug nozzle. J. Spacecr. Rocket. **53**(1), 39–45 (2016)
23. K. Chutkey, M. Viji, S.B. Verma, Interaction of external flow with linear cluster plug nozzle jet. Shock Waves **28**, 1207–1221 (2018)
24. M. Chaudhary, T.V. Krishna, S.R. Nanda, S.K. Karthick, A. Khan, A. De, I.M. Sugarno, On the fluidic behavior of an over-expanded planar plug nozzle under lateral confinement. Phys. Fluids **32**(8), 086106 (2020)
25. N. Karthikeyan, S.B. Verma, L. Venkatakrishnan, Experimental investigation of the acoustics of an annular aerospike nozzle flow, in *15th AIAA/CEAS Aeroacoustics Conference (30th AIAA Aeroacoustics Conference)* (2009), p. 3375
26. N. Karthikeyan, A. Kumar, S.B. Verma, L. Venkatakrishnan, Effect of spike truncation on the acoustic behavior of annular aerospike nozzles. AIAA J. **51**(9), 2168–2182 (2013)
27. A. Khan, R. Kumar, Experimental study and passive control of overexpanded plug nozzle jet. J. Spacecr. Rocket. **55**(3), 778–782 (2018)
28. P.P. Nair, A. Suryan, H.D. Kim, Study of conical aerospike nozzles with base-bleed and freestream effects. J. Spacecr. Rocket. **56**(4), 990–1005 (2019)
29. S.B. Verma, M. Viji, Linear-plug flowfield and base pressure development in freestream flow. J. Propul. Power **27**(6), 1247–1258 (2011)
30. S. Soman, A. Suryan, P.P. Nair, M. Ritter, H.D. Kim, Study on control of unsteadiness and flow asymmetry in linear aerospike nozzles. J. Spacecr. Rocket. **60**(1), 326–338 (2023)
31. S. Soman, A. Suryan, P.P. Nair, H.D. Kim, Numerical analysis of flowfield in linear plug nozzle with base bleed. J. Spacecr. Rocket. **58**(6), 1786–1798 (2021)

32. P.P. Nair, A. Suryan, H.D. Kim, Computational study on flow through truncated conical plug nozzle with base bleed. Propulsion Power Res. **8**(2), 108–120 (2019)
33. P.J. Finley, The flow of a jet from a body opposing a supersonic free stream. J. Fluid Mech. **26**(2), 337–368 (1966)
34. G. Scarlatella, T. Heuling, J. Sieder-Katzmann, F. Weber, M. Portolani, M. Garutti, D. Bianchi, A. Ferrero, D.G. Pastrone, M. Tajmar, C. Bach, Cold-gas experiments on advanced nozzles in subsonic counter-flows, in *Aerospace Europe Conference* (2023)